Systems of Nonlinear Partial Differential Equations

Systems of Nonlinear Partial Differential Equations

Applications to Biology and Engineering

by

Anthony W. Leung
Department of Mathematical Sciences,
University of Cincinnati, Cincinnati, Ohio, U.S.A.

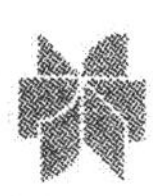

KLUWER ACADEMIC PUBLISHERS
DORDRECHT / BOSTON / LONDON

Library of Congress Cataloging in Publication Data

Leung, Anthony W., 1946-
Systems of nonlinear partial differential equations
applications to biology and engineering.

(Mathematics and its applications)
Bibliography: p.
Includes index.
1. Differential equations, Partial. 2. Differential
equations, Nonlinear. I. Title. II. Series: Mathematics
and its applications (Kluwer Academic Publishers)
QA377.L4155 1989 515'.353 89-2623

ISBN 0-7923-0138-2

Published by Kluwer Academic Publishers,
P.O. Box 17, 3300 AA Dordrecht, The Netherlands.

Kluwer Academic Publishers incorporates
the publishing programmes of
D. Reidel, Martinus Nijhoff, Dr W. Junk and MTP Press.

Sold and distributed in the U.S.A. and Canada
by Kluwer Academic Publishers,
101 Philip Drive, Norwell, MA 02061, U.S.A.

In all other countries, sold and distributed
by Kluwer Academic Publishers Group,
P.O. Box 322, 3300 AH Dordrecht, The Netherlands.

printed on acid free paper

Printed in The Netherlands

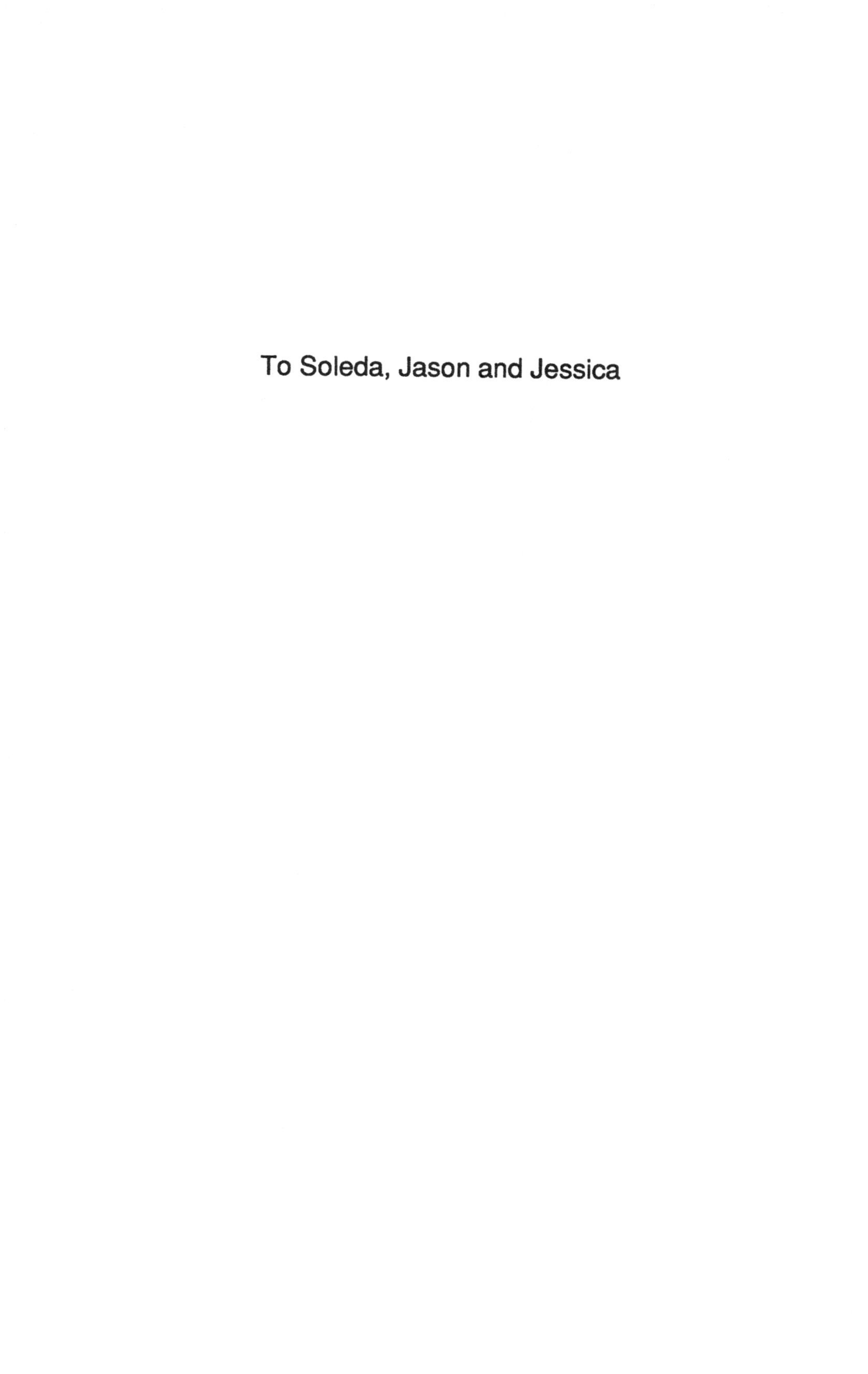

To Soleda, Jason and Jessica

SERIES EDITOR'S PREFACE

'Et moi, ..., si j'avait su comment en revenir, je n'y serais point allé.'

Jules Verne

The series is divergent; therefore we may be able to do something with it.

O. Heaviside

One service mathematics has rendered the human race. It has put common sense back where it belongs, on the topmost shelf next to the dusty canister labelled 'discarded non-sense'.

Eric T. Bell

Mathematics is a tool for thought. A highly necessary tool in a world where both feedback and non-linearities abound. Similarly, all kinds of parts of mathematics serve as tools for other parts and for other sciences.

Applying a simple rewriting rule to the quote on the right above one finds such statements as: 'One service topology has rendered mathematical physics ...'; 'One service logic has rendered computer science ...'; 'One service category theory has rendered mathematics ...'. All arguably true. And all statements obtainable this way form part of the raison d'être of this series.

This series, *Mathematics and Its Applications*, started in 1977. Now that over one hundred volumes have appeared it seems opportune to reexamine its scope. At the time I wrote

> "Growing specialization and diversification have brought a host of monographs and textbooks on increasingly specialized topics. However, the 'tree' of knowledge of mathematics and related fields does not grow only by putting forth new branches. It also happens, quite often in fact, that branches which were thought to be completely disparate are suddenly seen to be related. Further, the kind and level of sophistication of mathematics applied in various sciences has changed drastically in recent years: measure theory is used (non-trivially) in regional and theoretical economics; algebraic geometry interacts with physics; the Minkowsky lemma, coding theory and the structure of water meet one another in packing and covering theory; quantum fields, crystal defects and mathematical programming profit from homotopy theory; Lie algebras are relevant to filtering; and prediction and electrical engineering can use Stein spaces. And in addition to this there are such new emerging subdisciplines as 'experimental mathematics', 'CFD', 'completely integrable systems', 'chaos, synergetics and large-scale order', which are almost impossible to fit into the existing classification schemes. They draw upon widely different sections of mathematics."

By and large, all this still applies today. It is still true that at first sight mathematics seems rather fragmented and that to find, see, and exploit the deeper underlying interrelations more effort is needed and so are books that can help mathematicians and scientists do so. Accordingly MIA will continue to try to make such books available.

If anything, the description I gave in 1977 is now an understatement. To the examples of interaction areas one should add string theory where Riemann surfaces, algebraic geometry, modular functions, knots, quantum field theory, Kac-Moody algebras, monstrous moonshine (and more) all come together. And to the examples of things which can be usefully applied let me add the topic 'finite geometry'; a combination of words which sounds like it might not even exist, let alone be applicable. And yet it is being applied: to statistics via designs, to radar/sonar detection arrays (via finite projective planes), and to bus connections of VLSI chips (via difference sets). There seems to be no part of (so-called pure) mathematics that is not in immediate danger of being applied. And, accordingly, the applied mathematician needs to be aware of much more. Besides analysis and numerics, the traditional workhorses, he may need all kinds of combinatorics, algebra, probability, and so on.

In addition, the applied scientist needs to cope increasingly with the nonlinear world and the extra mathematical sophistication that this requires. For that is where the rewards are. Linear models are honest and a bit sad and depressing: proportional efforts and results. It is in the non-linear world that infinitesimal inputs may result in macroscopic outputs (or vice versa). To appreciate what I am hinting at: if electronics were linear we would have no fun with transistors and computers; we would have no TV; in fact you would not be reading these lines.

There is also no safety in ignoring such outlandish things as nonstandard analysis, superspace and anticommuting integration, p-adic and ultrametric space. All three have applications in both electrical engineering and physics. Once, complex numbers were equally outlandish, but they frequently proved the shortest path between 'real' results. Similarly, the first two topics named have already provided a number of 'wormhole' paths. There is no telling where all this is leading - fortunately.

Thus the original scope of the series, which for various (sound) reasons now comprises five subseries: white (Japan), yellow (China), red (USSR), blue (Eastern Europe), and green (everything else), still applies. It has been enlarged a bit to include books treating of the tools from one subdiscipline which are used in others. Thus the series still aims at books dealing with:

- a central concept which plays an important role in several different mathematical and/or scientific specialization areas;
- new applications of the results and ideas from one area of scientific endeavour into another;
- influences which the results, problems and concepts of one field of enquiry have, and have had, on the development of another.

Reaction-diffusion equations, as the name indicates, came from the mathematical modelling of chemical reactions. As a recognized specialism the subject took form and shape in the 1970's and in spite of its relative youth it is now a well developed substantial research field with a host of applications for instance in population dynamics and ecology, where the individual particles tend to be a bit larger, and in reactor engineering. These two application areas are the main ones considered in this volume.

It is also a field in which applications and 'pure' mathematical structures and notions like order, graphs, topological degree interact nontrivially and beautifully. Finally, despite its importance, it is a field that does not yet have many books devoted to it especially at a level of nonsuperspecialists. These are two excellent reasons for welcoming this volume in this series.

Perusing the present volume is not guaranteed to turn you into an instant expert, but it will help, though perhaps only in the sense of the last quote on the right below.

The shortest path between two truths in the real domain passes through the complex domain.

J. Hadamard

La physique ne nous donne pas seulement l'occasion de résoudre des problèmes ... elle nous fait pressentir la solution.

H. Poincaré

Never lend books, for no one ever returns them; the only books I have in my library are books that other folk have lent me.

Anatole France

The function of an expert is not to be more right than other people, but to be wrong for more sophisticated reasons.

David Butler

Bussum, February 1989

Michiel Hazewinkel

CONTENTS

PREFACE

In the last twelve years, much progress was made in the use of systems of reaction-diffusion equations in the study of a variety of applied topics: ecological systems, fission reactors, chemical reactions and many others. Although several excellent books related to such systems are available, yet numerous useful results in the last twelve years are not readily accessible in a book form for convenient study and reference. In the mean time the need for applications encourages us to enhance the understanding and improve the skill in analyzing such nonlinear systems of parabolic and elliptic partial differential equations.

Several methods had been extremely fruitful in the analysis and are extensively used in this book: (a) Intermediate-value type existence theorem for elliptic system (cf. section 1.4) is valuable for showing the occurence of steady states together with estimates of their sizes; such theorem actually includes the use of Leray-Schauder topological degree. (b) Differential inequalities for parabolic systems are suitable for considering the time stability of steady state solutions (cf. sections 1.3 and 2.3). (c) Upper and lower solutions combined with suitable monotone schemes provide a constructive approach to obtain the existence of solutions for systems (cf. chapter 5); moreover this method is adaptable to numerical approximations (cf. chapter 6). (d) Bifurcation techniques in functional analysis combined with estimations by means of maximum principles provides understanding of structural changes of positive solutions as various parameters varies globally (cf. section 7.4). (e) For large ecological systems with Neumann boundary conditions, the use of Lyapunov functions together with tehniques in graph theory gives extremely keen insight into the interactions between the various components (cf. sections 7.2 and 7.3). (f) Recent results in strongly order-preserving dynamical systems provide a powerful method to analyze the global behavior of solutions of parabolic systems, as time tends to infinity (cf. section 4.6).

All the above methods are carefully explained in the book. One clearly sees how they successfully lead to applicable results in nonlinear elliptic and parabolic partial differential systems related to many ecological interactions and reactor engineering problems. Chapters 2 and 3 contain many recent theorems in the study of prey-predator and competing species under diffusion. Various types of results are considered, involving a variety of assumptions on the population models. Research in such systems are progressing in such a fast pace that it is impossible to include all interesting discoveries in a few short chapters. Scattered throughout chapters 2, 3, 5 and 7 are some recent results which I believe should be useful to future research in this field. These problems had aroused my attention from the viewpoint of ecology as well as mathematics. Chapter 5 futher gives a systematic study of the method of upper and lower solutions, coupled with general monotone schemes recently developed for large elliptic systems. The technique blends beautifully with the analysis of interacting population models. Moreover, there is a section on time-periodic solutions on parabolic systems. Chapter 4 considers reaction difussion systems for reactor engineering. It studies the multigroup neutron fission models, collecting some results obtained by the methods described above. There is also a section on transport systems. The last section presents some recent elegent results in strongly order-preserving dynamical systems. Such systems are applicable to the study of reactor models, genectics, and coupling cooperating and competing species. Chapter 6 is concerned with computational and numerical analysis. It adapts the monotone scheme method to study finite difference systems of equations. Practical procedures as well as convergence theorems are presented. The first part of chapter 7 combines the use of Lyapunov function and graph theory to give very elegant and general results for large Volterra-Lotka type diffusive systems under Neumann boundary conditions. It summarizes the efforts related to many researchers in the last decade. The second part of chapter 7 employs some results in the earlier chapters and some bifurcation techniques in functional analysis to obtain interesting bifurcating solutions in elliptic prey-predator systems. It analyzes the changes in positive solution structures under Dirichlet boundary conditions as the parameters vary globally.

One of the the aims of the book is to gather many useful materials for researchers in reaction-diffusion systems. It is also hoped that by studying applications and pure mathematical methods simultaneously, it is easier to motivate and teach a variety of students the many difficult relevant subjects. I have tried to present the subject in a way which is accessible to advanced undergraduates in mathematics and beginning graduate students. The students are supposed to have a background in advanced calculus together with only some elementary knowledge of differential equations. Hence, the book begins with basic maximum principles in partial differential equations, differential inequalities, introduction of Hölder spaces, Schauder's estimates for solutions of linear scalar equations etc. Although not all the proofs of these preliminary topics are included, those not commonly accessible in standard text books or are involved with techniques which we will employ extensively in later chapters are all presented in detail. The applications should also be understandable to practical researchers who are not fundamentally concerned with the pure mathematics.

The book primarily uses classical solutons in Hölder spaces, and the methods of generalized solutions are not emphasized. Although $W^{k,p}$ estimates are used a couple of times in the proof of convergence of approximate solutions, the thourough understanding of $W^{k,p}$ strong solutions is not absolutely necessary, if one accepts the validity of such estimates. For completeness, such results together with Sobolev's embedding theorem, and some functional analytic bifurcation theorems are included in the appendix. Moreover, with the use of generalized solutions, much of the theorems in this book can be stated in more general terms. Consequently, they can be extended to be applicable to more general practical situations. However, such a task is not our present emphasis.

Some of the materials had been used in several classes in partial differential equations and seminar courses. They had stimulate students to proceed to further work in various directions of their own interests. The range of application of the methods should not be limited only to ecology and reactors, although they are the prime concerns here. The book should be suitable for a two quarters or one semester course in applied mathematics, or for a nonlinear theory part of a sequence of partial differential equations courses. I had regretably omitted many interesting topics in reaction-diffusion systems, for example travelling waves, combustion, free boundary, etc. A treatment of these and other topics is too lengthy, and is beyond the scope of this present manuscript.

I am grateful to many colleagues, students and friends who had visited me at Cincinnati. They include in chronolgical order: Dr. A. Lazer, Dr. D. Clark, Dr. D. Murio, Dr. G. S. Chen, Dr. B. Benjilali, Dr. P. Korman, and Dr. Z. M. Zhou. Their stimulations are valuable in the development of the subject matter of this book. I also wish to thank Miss June Anderson for her typing of most of the manuscript.

Cincinnati, October 1988 Anthony W. Leung

CHAPTER I

Background And Fundamental Methods

1.1 Maximum Principles

We begin with an introduction to background methods and techniques which will be widely used in this book. Among the most fundamental and important tools are the maximum principles. In the calculus of one variable we know that a function which is concave up in (a,b) and continuous in $[a,b]$ must attain its maximum at $x = a$ or b at the boundary. In more than one independent variables, similar situations occur. We will generalize and clarify such principles for twice continuously differentiable functions in this section.

(I) Elliptic Equations

Let Ω be a domain in R^n (not necessarily bounded), $n \geq 2$, $x = (x_1, \ldots, x_n)$. Suppose that $a_{ij}(x)$, $b_i(x)$, $1 \leq i, j \leq n$, are functions defined on Ω; and let L be an operator on functions which have all partial derivatives up to second order in Ω, defined by:

$$L \equiv \sum_{i,j=1}^{n} a_{ij}(x) \frac{\partial^2}{\partial x_i \partial x_j} + \sum_{i=1}^{n} b_i(x) \frac{\partial}{\partial x_i} . \tag{1.1-1}$$

<u>Definition</u>. L <u>is elliptic at a point</u> $\bar{x}$ <u>if and only if there is a positive number</u> $\mu(\bar{x})$ <u>such that</u>

$$\sum_{i,j=1}^{n} a_{ij}(\bar{x})\, \xi_i \xi_j \geq \mu(\bar{x}) \sum_{i=1}^{n} \xi_i^2 \tag{1.1-2}$$

<u>for all</u> n-<u>tuples of real numbers</u> $(\xi_1, \ldots, \xi_n)$. L <u>is elliptic in a domain</u> Ω

<u>if it is elliptic at each point of Ω.</u> L <u>is uniformly elliptic in Ω if</u> (1.1-2) <u>holds for each $\bar{x} \in \Omega$, and $\mu(\bar{x}) \geq \mu_0$ for all $\bar{x} \in \Omega$, where $\mu_0 > 0$ is some constant.</u>

Suppose $u \in C^2(\Omega)$, the set of functions which have all derivatives up to 2nd order continous in Ω, then we may assume that the coefficients a_{ij} in $L[u]$ satisfy $a_{ij} = a_{ji}$. Consequently at each point where L is elliptic, there is an orthogonal matrix (d_{ij}) such that under the linear transformation $z_k = \sum_{j=1}^{n} d_{kj}x_j$, the expression $\sum_{i,j=1}^{n} a_{ij} \frac{\partial^2 u}{\partial x_i \partial x_j}$ takes the form $\sum_{k=1}^{n} \sigma_k \frac{\partial^2 u}{\partial z_k^2}$, where $\sigma_k > 0$, $k = 1, \ldots, n$. Let u satisfies the strict inequality $L[u] > 0$ in Ω, where L is elliptic in Ω, and suppose that u attains a relative maximum at a point $p \in \Omega$. At p, we perform the linear transformation described above to obtain coordinates $(z_1, \ldots, z_n)$. From calculus, we have $\frac{\partial u}{\partial z_k} = 0$ and $\frac{\partial^2 u}{\partial z_k^2} \leq 0$ for $k = 1, \ldots, n$ at the point p; thus contradicting $\sum_{k=1}^{n} \sigma_k \frac{\partial^2 u}{\partial z_k^2} = L[u] > 0$ at p. In other words, if $u \in C^2(\Omega)$, $L[u] > 0$ in Ω, where L is elliptic in Ω, then u cannot attain an interior maximum in Ω. We now extend the above principle to the case $L[u] \geq 0$ in Ω. We will always assume $u \in C^2(\Omega)$ in this entire subsection.

<u>Theorem</u> 1.1-1. <u>Let</u> L <u>be uniformly elliptic in</u> Ω, <u>and</u> $a_{ij}(x)$, $b_i(x)$ <u>are bounded functions in bounded subsets of</u> Ω, <u>for</u> $1 \leq i, j \leq n$. <u>Suppose that</u>

$$L[u] \equiv \sum_{i,j=1}^{n} a_{ij}(x) \frac{\partial^2 u}{\partial x_i \partial x_j} + \sum_{i=1}^{n} b_i(x) \frac{\partial u}{\partial x_i} \geq 0 \text{ in } \Omega, \tag{1.1-3}$$

<u>and</u> u <u>attains a maximum</u> M <u>at a point in</u> Ω. <u>Then</u> $u(x) \equiv M$ <u>for all</u> $x \in \Omega$.

<u>Proof</u>. Suppose that the conclusion is false. Let $\hat{x}_0$, x_0 be points in Ω where $u(\hat{x}_0) < M$ and $u(x_0) = M$. Let $\gamma(t)$, $0 \leq t \leq 1$ be a path in Ω joining $\hat{x}_0$ to x_0 (with $\gamma(0) = \hat{x}_0$, $\gamma(1) = x_0$). There exists t_1 which is the first positive t where $u(\gamma(t)) = M$. For convenience, let $B_r(\bar{x})$ denotes the open ball $\{x: |x-\bar{x}|$

$< r\}$ and $\bar{B}_r(\bar{x})$ denotes its closure. There is a small enough $\delta > 0$ so that the set $\bigcup_{0 \le t \le 1} B_\delta(\gamma(t))$ is contained in Ω. Choose a point $Q \in B_{\delta/2}(\gamma(t_1)) \cap \{\gamma(t): 0 \le t < t_1\}$, and let $\rho_1 > 0$ be the first positive number so that $\bar{B}_{\rho_1}(Q)$ contains a point P where $u(P) = M$. Clearly $P \in \delta B_{\rho_1}(Q)$ and $\rho_1 < \delta/2$. We now modify u to w near P so that $L[w] > 0$ in a ball centered at P, and we can apply the principle for the case of strict inequality described before this theorem (cf. Fig. 1.1-1).

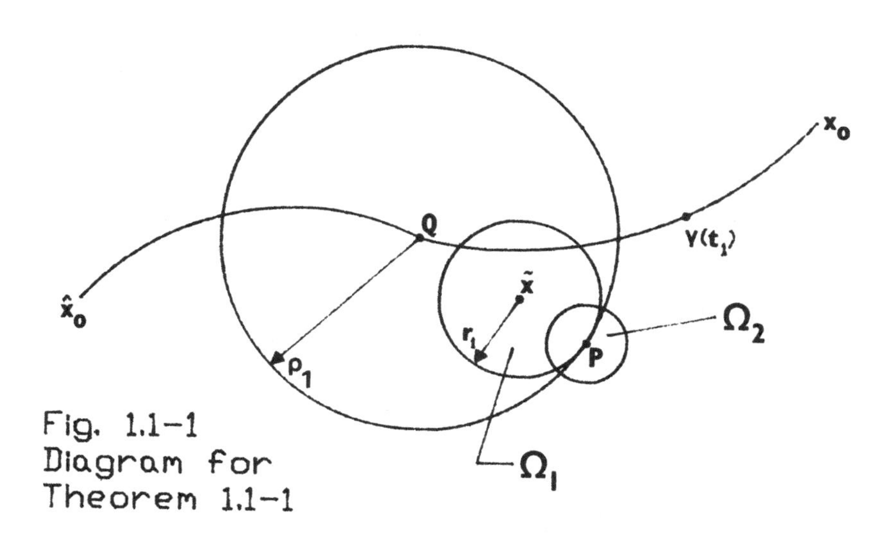

Fig. 1.1-1 Diagram for Theorem 1.1-1

Let $\Omega_1 \overset{(def)}{\equiv} B_{r_1}(\tilde{x}) \subset B_{\rho_1}(Q)$ with $r_1 < \rho_1$ so that Ω_1 and $B_{\rho_1}(Q)$ are tangent to each other at P; and finally draw the ball $\Omega_2 \overset{(def)}{\equiv} B_{r_1/2}(P)$.

Define $z(x) = \exp\{-\alpha \sum_{i=1}^{n} (x_i - \tilde{x}_i)^2\} - e^{-\alpha r_1^2}$, where $\tilde{x} = (\tilde{x}_1, \ldots, \tilde{x}_n)$ and α is a positive constant to be determined. Clearly, we have $z(x) > 0$ for $x \in \Omega_1$, $z = 0$ on $\delta\Omega_1$, $z < 0$ outside Ω_1, and

$$L[z] = \exp\{-\alpha \sum_{i=1}^{n} (x_i-\tilde{x}_i)^2\}\{4\alpha^2 \sum_{i,j=1}^{n} a_{ij}(x_i-\tilde{x}_i)(x_j-\tilde{x}_j) - 2\alpha \sum_{i=1}^{n} [a_{ii} + b_i(x_i-\tilde{x}_i)]\}.$$

By the uniform ellipticity of L, we thus have the inequality

$$L[z] \geq \alpha \exp\{-\alpha \sum_{i=1}^{n} (x_i-\tilde{x}_i)^2\} \{\alpha\mu_0 r_1^2 - 2 \sum_{i=1}^{n} [a_{ii}+b_i(x_i-\tilde{x}_i)]\}$$

for all $x \in \Omega_2 \equiv B_{r_1/2}(P)$ for a positive constant μ_0, and therefore $L[z] > 0$ in Ω_2 for $\alpha > 0$ chosen sufficiently large.

By construction, we have $u(x) \leq M-\hat{\delta}$ for all $x \in \delta\Omega_2 \cap \bar{\Omega}_1$, where $\hat{\delta} > 0$ is a small constant. Finally, we define $w(x) = u(x) + \varepsilon z(x)$ for a small $\varepsilon > 0$ to be determined. Choose ε small enough so that $w(x) < M$ for all $x \in \delta\Omega_2 \cap \bar{\Omega}_1$. Since $z < 0$ outside Ω_1, we have $w(x) < M$ on the part of $\delta\Omega_2$ outside Ω_1, we have $w(x) < M$ on the part of $\delta\Omega_2$ outside Ω_1. We therefore have $w(P) = M$ and $w < M$ on $\delta B_{r_1/2}(P) \equiv \delta\Omega_2$, implying w must have a maximum in Ω_2. However, this contradicts the fact that $L[w] = L[u] + \varepsilon L[z] > 0$ in Ω_2. This proves the theorem.

Remarks. For the validity of the above theorem, note: (i) No assumption had been made on the continuity of $a_{ij}(x)$ and $b_i(x)$, although we assumed $u \in C^2(\Omega)$. (ii) The uniform ellipticity of L and the boundedness assumption on a_{ij} and b_i can be weakened to: $\sum_{i=1}^{n} |a_{ii}(x)|/\mu(x)$ and $\sum_{i=1}^{n} |b_i(x)|/\mu(x)$ are bounded on every closed ball contained in Ω.

The following is a direct consequence of Theorem 1.1-1.

Theorem 1.1-2. Let L and a_{ij}, b_i be as described in Theorem 1.1-1, and h(x) is a bounded function with $h(x) \leq 0$ in Ω. Suppose that (1.1-3) is modified to

(1.1-4) $\quad (L + h)[u] \geq 0$ *in* Ω ,

and u *attains a nonnegative maximum* M *at a point in* Ω. *Then* $u(x) \equiv M$ *for* $x \in \Omega$.

In the next two theorems we assume that the domain Ω has smooth boundary $\delta\Omega$ (i.e., at every point on $\delta\Omega$, for some k, $\delta\Omega$ is locally representable as $x_k = \phi(x_1, \ldots, x_{k-1}, x_{k+1}, \ldots, x_n)$ where ϕ has all partial derivatives continuous). Let $P \in \delta\Omega$ and η be the unit outward normal at P. We say that the vector ν points outward from Ω at the point P if $\nu\cdot\eta > 0$, and define the outward directional derivative $\frac{\partial u}{\partial \nu} = \lim_{x\to P}[\nu \text{ grad } u(x)]$ if it exists at P (where the limit is taken for $x \in \Omega$).

Theorem 1.1-3. *Let* L *be elliptic in* Ω, *and* $a_{ij}(x)/\mu(x)$, $b_i(x)/\mu(x)$ *are bounded in a neighborhood* in Ω *of a point* $P \in \delta\Omega$. *Assume that* P *lies on the boundary of a ball* Ω_1 *in* Ω. *Suppose that* u(x) *is continuous* in $\Omega \cup P$, $u(x) \leq M = u(P)$ *for all* $x \in \Omega$, *and* $L[u] \geq 0$ *in* Ω. *Then, if the outward directional derivative* $\partial u/\partial\nu$ *exists at* P, *we must have*

(1.1-5) $\quad \frac{\partial u}{\partial \nu} > 0$ *at* P,

unless $u(x) \equiv M$ *in* Ω. (*Here* $\mu(x)$ *is defined in* (1.1-2)).

The proof is similar to that of Theorem 1.1. Suppose $u \not\equiv M$ (hence $u < M$ in Ω, by Theorem 1.1. By shrinking Ω_1 if necessary, we may assume that $\bar{\Omega}_1 \subset \Omega \cup P$. Let r_1 and $\tilde{x}$ be the radius and center of Ω_1 respectively. Define $\Omega_2 \equiv B_{r_1/2}(P)$, and z(x) in exactly the same formula as in Theorem 1.1. Choose $\varepsilon > 0$ sufficiently small so that $w = u + \varepsilon z$ satisfies $w \leq M$ on $\delta(\Omega_1 \cap \Omega_2)$. Deduce from the definition of z that $\frac{\partial z}{\partial \nu} < 0$ at P, and $\frac{\partial u}{\partial \nu} + \varepsilon\frac{\partial z}{\partial \nu} = \frac{\partial w}{\partial \nu} \geq 0$ at P (using $L[w] > 0$ in $\Omega_1 \cap \Omega_2$). Consequently, we have $\frac{\partial u}{\partial \nu} > 0$ at P (cf. Fig. 1.1-2).

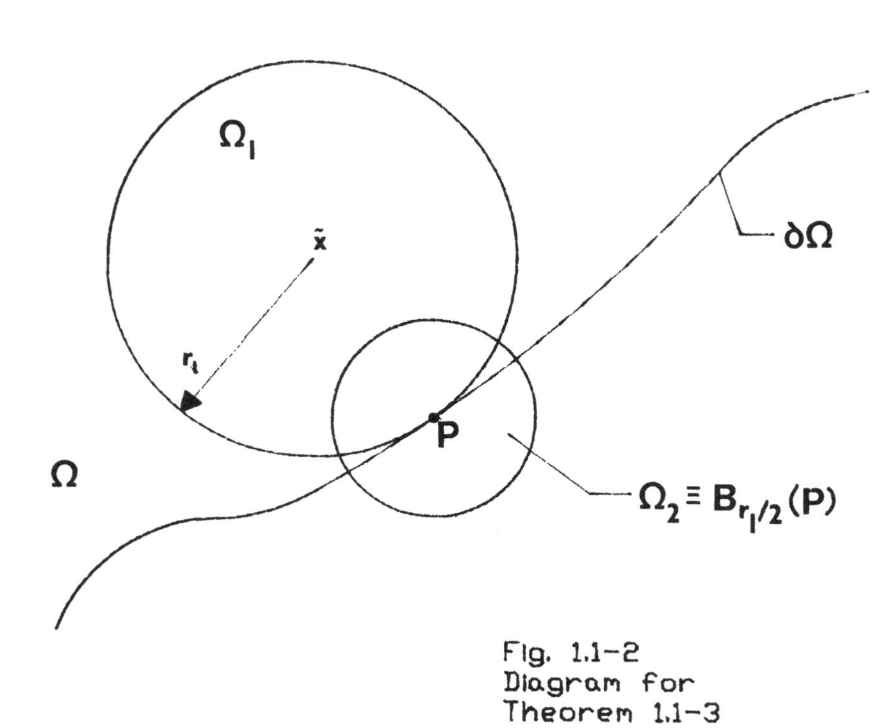

Fig. 1.1-2
Diagram for
Theorem 1.1-3

The following is a slight extension of the last theorem.

Theorem 1.1-4. *Let* L *and its coefficients satisfy the condtions in Theorem 1.1-3, and* $P \in \delta\Omega$ *lies on the boundary of a ball* Ω_1 *in* Ω. *Let* $h(x) \le 0$ *in* Ω *and* $h(x)/\mu(x)$ *is bounded in a neighborhood in* Ω *of* P. *Suppose that* $u(x)$ *is continuous in* $\Omega \cup P$, $u(x) \le M = u(P)$ *for all* $x \in \Omega$, $(L+h)[u] \ge 0$ *in* Ω, *and* $M \ge 0$. *Then the outward normal derivative* $\frac{\partial u}{\partial \nu} > 0$ *at* P *if it exists, unless* $u \equiv M$ *in* Ω.

(II) *Parabolic Equations*

We first consider the simple case of one (space) variable x and a (time) variable t. Let $I = (\alpha,\beta)$, $T > 0$, and $a(x,t)$, $b(x,t)$ be defined on $I\times(0,T]$. Let L be an operator on functions which have second and first derivatives with respect to x and t respectively on $I \times (0,T]$ defined by

(1.1-6) $$L \equiv a(x,t)\frac{\partial^2}{\partial x^2} + b(x,t)\frac{\partial}{\partial x} - \frac{\partial}{\partial t}.$$

We say L is parabolic at $(x_0, t_0) \in I\times[0,T)$ if $a(x_0,t_0) > 0$. If $a(x,t) \geq \mu_0 > 0$ for some constant μ_0 and all $(x,t) \in I\times(0,T]$ we say L is uniformly parabolic in $I\times(0,T]$.

Suppose that L is parabolic at each point in $I\times(0,T]$ and $u(x,t)$, defined on $I\times(0,T]$, satisfies

$$L[u] > 0 \quad \text{in } I\times(0,T].$$

It is clear from calculus that u cannot have a local maximum at a point in $I\times(0,T]$. (Since $\frac{\partial^2 u}{\partial x^2} \leq 0$, $\frac{\partial u}{\partial x} = 0$ and $\frac{\partial u}{\partial x} \geq 0$ at that point). We now extend this principle to the case $L[u] \geq 0$, through the following lemmas.

Lemma 1.1-1. *Let L given by (1.1-6) be uniformly parabolic and $a(x,t)$, $b(x,t)$ be bounded in $I\times(0,T]$, and B is an open ball with $\bar{B} \subset I\times(0,T]$. Suppose that*

(1.1-7) $$L[u] \geq 0 \quad \text{in } I\times(0,T],$$

$u(x,t) < M \overset{(\text{def})}{=} \max\{u(x,t) \mid (x,t) \in \Omega\times(0,T]\}$ for $(x,t) \in B$; and $u = M$ at some point $P \in \delta B$. Then the tangent to B at P is parallel to the x-axis (i.e., P is directly above or below the center of B).

Proof. Suppose that P is not directly above or below the center of B as described. We will arrive at a contradiction. We next assume temporarily that P is the only point on δB where $u = M$. Let $P = (\bar{x}, \bar{t})$, the radius of B be r, and the center of B has coordinates $(\hat{x}, \hat{t})$. Choose $r_1 > 0$ sufficiently small so that the ball $B_1 = \{(x,t)|\sqrt{[(x-\bar{x})^2 + (t-\bar{t})^2]} < r_1\}$ has its closure contained in $I\times(0,T]$, and $0 < r_1 < |\bar{x}-\hat{x}|$. Observe that $u \le M-\delta$ on $\delta B_1 \cap \bar{B}$ for a small $\delta > 0$, and $u \le M$ on δB_1 outside $\bar{B}$ (cf. Fig. 1.1-3).

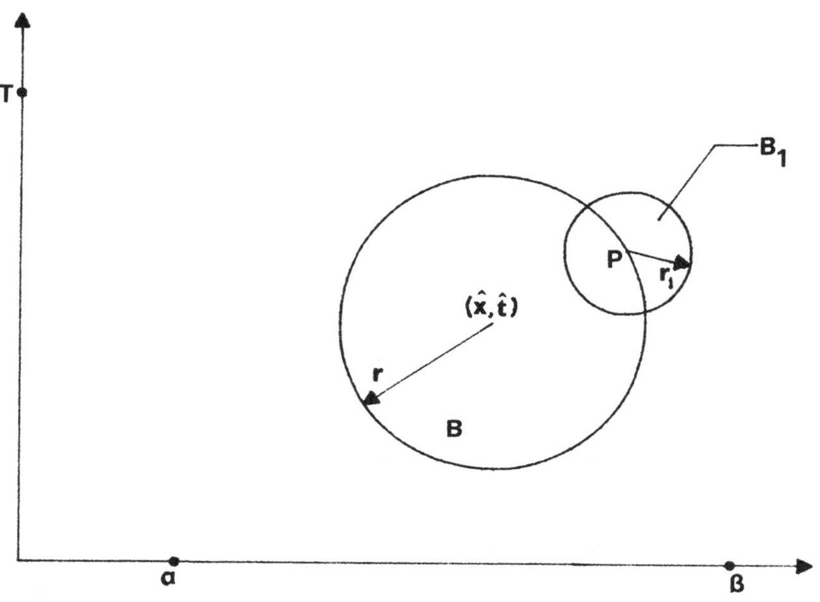

Fig 1.1-3 Diagram for Lemma 1.1-1

Define $v(x,t) = e^{-\alpha[(x-\hat{x})^2+(t-\hat{t})^2]} - e^{-\alpha r^2}$, which satisfies $v > 0$ in B, $v = 0$ on δB, $v < 0$ outside B (here $\alpha > 0$ is to be determined later). Direct computation gives

$$L[v] = 2\alpha e^{-\alpha[(x-\hat{x})^2+(t-\hat{t})^2]}\{2\alpha a(x-\hat{x})^2 - a - b(x-\hat{x}) + (t-\hat{t})\}.$$

For (x,t) on $\bar{B}_1$, we have $|x-\hat{x}| \ge |\bar{x}-\hat{x}| - r_1 > 0$, enabling us to choose $\alpha > 0$ so that $L[v] > 0$ on $\bar{B}_1$. We now define $w = u + \varepsilon v$ with $\varepsilon > 0$ chosen sufficiently small, so that $w < M$ on δB_1. However, $w(P) = u(P) = M$, because $P \in \delta B$; hence the maximum of w on $\bar{B}_1$ must be attained inside B_1. This contradicts the fact that $L[w] > 0$ in B_1, consequently we must have $P = (\bar{x},\bar{t})$

satisfying $|\bar{x}-\hat{x}| = 0$.

Finally, suppose that P is not the only point on δB where $u = M$. We draw a ball $\tilde{B}$ with its closure inside $B \cup P$, and touching δB at P. The ball $\tilde{B}$ has the property that P is the only point at its boundary where $u = M$. Repeat the above arguments with the role of B replaced by $\tilde{B}$ to again arrive at a contradiction.

Lemma 1.1-2. Let L, $a(x,t)$, $b(x,t)$ satisfy the conditions in Lemma 1.1-1 and $L[u] \geq 0$ in $I\times(0,T]$. Suppose that $u(x,t) \leq M$ in $I\times(0,T]$; and $u(x_1,t_1) = M$ with $x_1 \in I$, $0 < t_1 < T$. Then $u(x,t_1) = M$ for all $x \in I$.

Proof. Suppose the conclusion is false, let $u(x_2,t_1) < M$, $x_2 \in I$. For convenience, suppose $x_1 < x_2$. There is a $\bar{x}_1$ with $x_1 < \bar{x}_1 < x_2$ so that $u(\bar{x}_1,t_1) = M$ and $u(x,t_1) < M$ for $\bar{x}_1 < x \leq x_2$. Let $d < \min\{x_2-\bar{x}_1, \bar{x}_1-\alpha, \beta-x_2, T-t_1, t_1\}$ and $d > 0$. For $\bar{x}_1 < x < \bar{x}_1 + d$, let $\rho(x)$ be the distance from (x,t_1) to the nearest point in $I\times(0,T]$ where $u = M$. We must have $\rho(x) < x - \bar{x}_1 < d$ for such x; and the nearest such point should be $(x,t_1+\rho(x))$ or $(x,t_1-\rho(x))$, by Lemma 1.1. Hence, for any $\bar{x}_1 < x < x+\delta < \bar{x}_1 + d$, the triangular inequality gives

$$\rho(x+\delta) \leq \sqrt{[\rho(x)^2 + \delta^2]} < \rho(x) + \frac{\delta^2}{2\rho(x)} . \tag{1.1-8}$$

Interchanging the role of x, $x+\delta$ we have $\rho(x+\delta) \geq \sqrt{[\rho(x)^2-\delta^2]}$. Choosing $0 < \delta < \rho(x)$, we can further improve inequality (1.1-8) by subdividing $(x,x+\delta)$ into n equal parts. Apply (1.1-8) with δ replaced by δ/n, we obtain for each $j = 0, \ldots, n-1$:

$$\rho(x + \frac{j+1}{n}\delta) - \rho(x + \frac{j\delta}{n}) \leq \frac{\delta^2}{2n^2\rho(x+\frac{j\delta}{n})} \leq \frac{\delta^2}{2n^2\sqrt{[\rho(x)^2-\delta^2]}}$$

Summing j from 0 to n-1, we conclude that

$$\rho(x+\delta) - \rho(x) \le \frac{\delta^2}{2n\sqrt{[\rho(x)^2-\delta^2]}} .$$

Letting $n \to \infty$, this implies $\rho(x+\delta) \le \rho(x)$. We thus must have $\rho(x)$ nonincreasing in $\bar{x}_1 < x < \bar{x}_1 + d$. However, from definition $\rho(x) \le x - \bar{x}_1$ which tends to zero as $x \to \bar{x}_1$, implying $\rho(x) \equiv 0$ in $\bar{x}_1 < x < \bar{x}_1 + d$. This means $u(x) \equiv M$ for $\bar{x}_1 < x < \bar{x}_1 + d$, which is a contradiction.

Theorem 1.1-5. *Let L given by (1.1-6) be uniformly parabolic, with $a(x,t)$ and $b(x,t)$ being bounded in $Ix(0,T]$. Suppose that*

$$L[u] \ge 0 \quad \text{in } Ix(0,T];$$

and the maximum M of u in $Ix(0,T]$ is attained at (x_0, t_0) with $\alpha < x_0 < \beta$, $0 < t_0 \le T$. Then $u(x,t) \equiv M$ for all $(x,t) \in Ix(0,t_0]$.

Proof. Suppose that the conclusion is false. There must be a $\hat{t} < t_0$ where $u(x_0,\hat{t}) < M$, by Lemma 1.1-2. Let $S = \{\tilde{t}: \tilde{t} > \hat{t}$ and $u(x,t) < M$ for all $t \in [\hat{t},\tilde{t}]\}$, and let the least upper bound of S be denoted by τ. We must have $u(x_0,\tau) = M$ and $u(x_0,t) < M$ for $\hat{t} < t < \tau$. Moreover, Lemma 1.1-2 implies that $u(x,t) < M$ for $(x,t) \in Ix(\hat{t},\tau)$; and we will show that this contradicts $u(x_0,\tau) = M$.

Draw a ball Ω_1 centered at (x_0,τ) so small so that the lower half of Ω_1 is entirely contained in $Ix(\hat{t},\tau]$. Define $v(x,t) = e^{-[(x-x_0)^2+\alpha(t-\tau)]} - 1$, with $\alpha > 0$ chosen so that

$$L[v] = e^{-[(x-x_0)^2+\alpha(t-\tau)]}[4a(x-x_0)^2 - 2a - 2b(x-x_0) + \alpha] > 0$$

for $(x,t) \in \Omega_1$, $t \le \tau$. We have $v = 0$ on the parabola $(x-x_0)^2 + \alpha(t-\tau) = 0$, $v > 0$ below the parabola, and $v < 0$ above it. For (x,t) on the part of $\delta\Omega_1$ which lies on and below the parabola, we have $u(x,t) \le M-\delta$ for some $\delta > 0$. Consequently for sufficiently small $\varepsilon > 0$, the function

$$w(x,t) = u(x,t) + \varepsilon v(x,t)$$

must have $w \le M$ on the boundary of $\tilde{\Omega} \overset{\text{def}}{=} \Omega_1 \cap \{(x,t) \mid (x-x_0)^2 + \alpha(t-\tau) < 0\}$ (cf. Fig. 1.1-4). However $L[w] = L[u] + \varepsilon L[v] > 0$ on $\tilde{\Omega}$ implies that w cannot attain interior maximum in $\tilde{\Omega}$. Therefore the maximum of w in the closure of $\tilde{\Omega}$ is attained at (x_0,τ) where $w = M$, and we must have $\frac{\partial w}{\partial t} \ge 0$ at (x_0,τ). This means $\frac{\partial u}{\partial t} \ge -\varepsilon \frac{\partial v}{\partial t} = \varepsilon\alpha > 0$ at the point (x_0,τ). Moreover $\frac{\partial u}{\partial x} = 0$, $\frac{\partial^2 u}{\partial x^2} \le 0$ at (x_0,τ) where $u = M$, implying $L[u] < 0$ and contradicting $L[u] \ge 0$. This shows that $u(x_0,\tau)$ cannot be M, and consequently the conclusion of the theorem must be true.

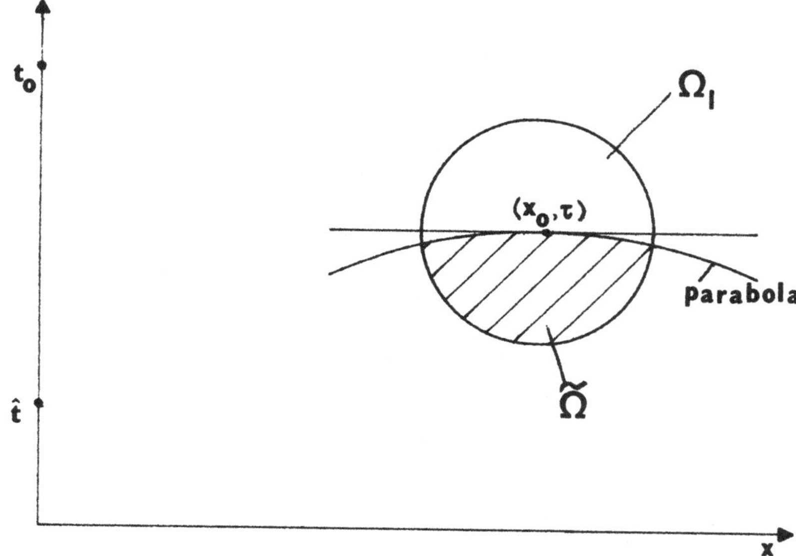

Fig. 1.1-4 Diagram for Theorem 1.1-5

Remark. An immediate consequence of Theorem 1.1-5 is that: under the conditions for L and a(x,t), b(x,t) as stated in Theorem 1.1-5, if $L[u] \geq 0$ in $I \times (0,T]$ and u is continuous in $\bar{I} \times [0,T]$, then the maximum of u in $\bar{I} \times [0,T]$ must be attained at $t = 0$ or at $x = \alpha$ or β, $t \in [0,T]$.

The preceding principle can be extended to the case of n space variables. Let Ω be a domain in R^n (not necessarily bounded), $x = (x_1, \ldots, x_n)$, $T > 0$. Let L be the operator

$$L \equiv \sum_{i,j=1}^{n} a_{ij}(x,t) \frac{\partial^2}{\partial x_i \partial x_j} + \sum_{i=1}^{n} b_i(x,t) \frac{\partial}{\partial x_i} - \frac{\partial}{\partial t} \tag{1.1-9}$$

for (x,t) in $\Omega \times (0,T]$.

Definition. We say L is parabolic at $(\bar{x},\bar{t})$ if there is a positive number $\mu(\bar{x},\bar{t})$ so that

$$\sum_{i,j=1}^{n} a_{ij}(\bar{x},\bar{t})\, \xi_i \xi_j \geq \mu(\bar{x},\bar{t}) \sum_{i=1}^{n} \xi_i^2 \tag{1.1-10}$$

for all n-tuples of real numbers $(\xi_1, \xi_2, \ldots, \xi_n)$. L is uniformly parabolic in $\Omega \times (0,T]$ if (1.1-10) holds for each $(\bar{x},\bar{t}) \in \Omega \times (0,T]$, and $\mu(\bar{x},\bar{t}) \geq \mu_0$ for all $(\bar{x},\bar{t}) \in \Omega \times (0,T]$, where $\mu_0 > 0$ is some constant.

As in the part for elliptic equations, we will assume that all the second partial derivatives of u with respect to the space variables $x_1, \ldots, x_n$, $\frac{\partial^2 u}{\partial x_i \partial x_j}(x,t)$, are continuous in $\Omega \times (0,T]$, in Theorem 1.1-6 to 1.1-8. On the other hand, we only assume that $\frac{\partial u}{\partial t}(x,t)$ exists in $\Omega \times (0,T]$. The proofs of Theorems 1.1-6 - 1.1-8 are either similar to that of Theorem 1.1-5 or are direct generalizations and consequences. Their detailed proofs can be found in e.g. [188], and will be omitted. They are stated in forms which are most commonly used.

Theorem 1.1-6. *Let* L *given by* (1.1-9) *be uniformly parabolic in* $\Omega\times(0,T]$, *with its coefficients* $a_{ij}(x,t)$, $b_i(x,t)$, $1 \le i, j \le n$ *being bounded in* $\Omega\times(0,T]$. *Suppose that*

$$L[u] \ge 0 \text{ in } \Omega\times(0,T],$$

and the maximum M *of* u *in* $\Omega\times(0,T]$ *is attained at* (x_0,t_0), *with* $x_0 \in \Omega$, $0 < t_0 \le T$. *Then* $u(x,t) \equiv M$ *for all* $(x,t) \in \Omega\times(0,t_0]$.

Theorem 1.1-7. *Let* L *and its coefficients satisfy the conditions in Theorem* 1.1-6 *in* $\Omega\times(0,T]$, *and* $L[u] \ge 0$ *in* $\Omega\times(0,T]$. *Suppose that* $P \in \partial\Omega\times(0,T)$ *and* u *is continuous in* $P \cup \{\Omega\times(0,T]\}$; $u(P) = M$, *and* $u(x,t) \le M$ *for all* (x,t) *in* $\Omega\times(0,T]$. *If there is a sphere through* P *with its interior contained in* $\Omega\times(0,T)$, *and inside which* $u < M$. *Then any directional derivative* $\frac{\partial}{\partial\nu}$ *in an outward direction satisfies*

$$\frac{\partial u}{\partial\nu} > 0 \text{ at } P, \text{ if it exist .}$$

(Here we mean the projection of ν *on the* x-plane *points outward from* Ω).

Theorem 1.1-8. *Let* $h(x,t) \le 0$ *in* $\Omega\times(0,T]$ *be a bounded function. Suppose that the assumptions in Theorems* 1.1-6 *and* 1.1-7 *are unchanged except that* $L[u] \ge 0$ *in* $\Omega\times(0,T]$ *is changed to*

$$(L+h)[u] \ge 0 \quad \text{in } \Omega\times(0,T], \tag{1.1-11}$$

and moreover $M \ge 0$. *Then the corresponding conclusions of Theorem* 1.1-6 *and* 1.1-7 *are valid.*

Corollary 1.1-9. Let $h(x,t)$ be a bounded function in $\Omega\times(0,T]$. Suppose that the assumptions in Theorem 1.1-6 and 1.1-7 are unchanged except that $L[u] \geq 0$ in $\Omega\times(0,T]$ is changed to (1.1-11), and moreover $M = 0$. Then the corresponding conclusions of Theorem 1.1-6 and 1.1-7 are valid.

To prove the corollary, we let $v = ue^{-\lambda t}$ where λ is a large positive constant so that $h(x,t) - \lambda \leq 0$ in $\Omega\times(0,T]$. The assumption $(L+h)[u] \geq 0$ in $\Omega\times(0,T]$ implies that $(L+h-\lambda)[v] \geq 0$ then. Apply Theorem 1.1-8 with the role of u and h replaced respectively by v and $h-\lambda$ respectively.

1.2. Differential Inequalities for Parabolic Equations and Systems

We next consider methods which compare functions which satisfy parabolic differential equations with those which satisfy differential inequalities of the same type. These are useful for proving existence of solutions within prescribed bounds, and for studying the time stability properties of steady-state solutions. In this section, Ω is a bounded domain in R^n, $x = (x_1, \ldots, x_n)$, $T > 0$. L denotes a parabolic operator defined for $(x,t) \in \Omega\times(0,T]$:

$$L \equiv \sum_{i,j=1}^{n} a_{ij}(x,t) \frac{\partial^2}{\partial x_i \partial x_j} + \sum_{i=1}^{n} b_i(x,t) \frac{\partial}{\partial x_i} - \frac{\partial}{\partial t}, \tag{1.2-1}$$

where $a_{ij}(x,t)$ and $b_i(x,t)$, $1 \leq i, j \leq n$, are bounded functions in $\Omega\times(0,T]$. ∇ denotes the gradient operator, $\nabla = (\frac{\partial}{\partial x_1}, \ldots, \frac{\partial}{\partial x_n})$.

(I) Scalar Comparisons

Theorem 1.2-1. Let $v(x,t)$, $w(x,t)$ be continuous functions defined in $\bar{\Omega}\times[0,T]$, with their first derivatives with respect to t exist in $\Omega x(0,T]$, their first and second derivatives with respect to each x_i exist and continuous in $\Omega x(0,T]$. Suppose that

$$(1.2\text{-}2)\qquad \begin{aligned} &Lw + f(x,t,\nabla w,w) < Lv + f(x,t,\nabla v,v) \ \text{in}\ \Omega x(0,T]\ \text{and} \\ &v < w \quad \text{on}\ (\bar{\Omega}x\{0\}) \cup (\delta\Omega x[0,T]), \end{aligned}$$

where f is a function defined on $\Omega x(0,T] \times R^n \times R$. Then

$$v(x,t) < w(x,t) \text{ in } \bar{\Omega}x[0,T].$$

Proof. Let $z(x,t) = v(x,t) - w(x,t)$ on $\bar{\Omega}x[0,T]$. Suppose that the conclusion is false. Let $t_0 > 0$ be the first t such that there is a point $(x_0,t_0) \in \Omega x(0,T]$ where $z(x_0,t_0) = 0$. At (x_0,t_0), we have $\sum_{i,j=1}^{n} a_{ij}\frac{\partial^2 z}{\partial x_i \partial x_j} \le 0$ (since the second partial derivatives here are continuous, we may assume $a_{ij} = a_{ji}$ and perform an appropriate orthogonal linear transformation of independent variables at (x_0,t_0)), and $\frac{\partial z}{\partial x_i} = 0$; thus a simple computation shows

$$\begin{aligned} \frac{\partial z}{\partial t}(x_0,t_0) < &\sum_{i,j=1}^{n} a_{ij}(x_0,t_0)\frac{\partial^2 z}{\partial x_i \partial x_j}(x_0,t_0) + \sum_{i=1}^{n} b_i(x_0,t_0)\frac{\partial z}{\partial x_i} + \\ &f(x_0,t_0,\nabla v(x_0,t_0), v(x_0,t_0)) - f(x_0,t_0,\nabla w(x_0,t_0), w(x_0,t_0)) \\ \le\ & 0. \end{aligned}$$

This contradicts the definition of t_0. Consequently $v < w$ in $\bar{\Omega}x[0,T]$.

For convenience, we say the boundary $\delta\Omega$ (if it exists) is C^2 smooth if it can be locally represented as $x_k = \phi(x_1, \ldots, x_{k-1}, x_{k+1}, \ldots, x_n)$ where ϕ has continuous second derivatives. Let η denotes the outward unit normal on $\partial\Omega$.

<u>Theorem</u> 1.2-2. <u>Let $\delta\Omega$ be C^2 smooth and</u> v, w <u>be continuous functions in</u> $\bar{\Omega}\times[0,T]$, <u>with their first derivatives with respect to</u> t <u>exist in</u> $\Omega\times(0,T]$, <u>their first and second derivatives with respect to each</u> x_i <u>exist and continuous in</u> $\Omega\times(0,T]$. <u>Suppose that</u>

$$(1.2\text{-}3)\quad \begin{cases} Lw + f(x,t,\nabla w,w) < Lv + f(x,t,\nabla v,v) & \text{in } \Omega\times(0,T], \\ \alpha(x,t)\dfrac{\partial v}{\partial \eta} + \beta(x,t)v < \alpha(x,t)\dfrac{\partial w}{\partial \eta} + \beta(x,t)w & \text{on } \delta\Omega\times(0,T], \\ v(x,0) < w(x,0) \quad \text{for } x \in \bar{\Omega}. & \end{cases}$$

<u>where</u> f <u>is defined on</u> $\Omega\times(0,T]\times R^n\times R$ <u>and</u> α, β <u>are defined on</u> $\delta\Omega\times(0,T]$ <u>with</u> $\alpha \ge 0$, $\beta \ge 0$, $\alpha^2 + \beta^2 \ne 0$. <u>Then</u>

$$v(x,t) < w(x,t) \quad \underline{\text{in}} \quad \bar{\Omega}\times[0,t].$$

(<u>Here</u>, <u>we assume</u> $\frac{\partial u}{\partial \eta}$ <u>and</u> $\frac{\partial w}{\partial \eta}$ <u>exist on</u> $\delta\Omega\times(0,t]$, if $\alpha \not\equiv 0$.)

<u>Proof</u>. The proof is the same as the last theorem. It only remains to show that at the first $t = t_0$ where there is a point (x_0,t_0) such that $v(x_0,t_0) = w(x_0,t_0)$, one cannot have $x_0 \in \delta\Omega$. Suppose that $(x_0,t_0) \in \delta\Omega\times(0,T]$. If $\beta(x_0,t_0) > 0$, then the second inequality in (1.2-3) implies that $\beta(x_0,t_0)\cdot[v(x_0,t_0) - w(x_0,t_0)] < \alpha \frac{\partial}{\partial \eta}(w-v)(x_0,t_0) \le 0$ (because $w - v \ge 0$ on $\bar{\Omega}\times[0,t_0]$). This contracdicts the definition of (x_0,t_0). If $\beta(x_0,t_0) = 0$, (1.2-3) implies that $\alpha(x_0,t_0)\frac{\partial}{\partial \eta}(v-w)(x_0,t_0) < 0$, and thus $\frac{\partial}{\partial \eta}(v-w)(x_0,t_0) < 0$. However, $v-w \le 0$ on $\bar{\Omega}\times\{t_0\}$ implies that $\frac{\partial}{\partial \eta}(v-w)(x_0,t_0) \ge 0$; and again leading to contradiction. Consequently (x_0,t_0) cannot be on $\delta\Omega\times(0,T]$.

Theorem 1.2-3. *Let* v, w *and* f *be defined with the same properties concerning their continuity and smoothness as in Theorem 1.2-1. In addition, let* f *satisfies* $|f(x,t,p,u_1) - f(x,t,p,u_2)| \le K|u_1-u_2|$ *for all* $(x,t,p) \in \Omega \times (0,T] \times R^n$, u_1, u_2 *arbitrary, where* K *is a positive constant. Suppose that*

$$(1.2\text{-}4) \qquad \begin{aligned} &Lw + f(x,t,\nabla w,w) \le Lv + f(x,t,\nabla v,v) \text{ in } \Omega \times (0,T], \text{ and} \\ &v \le w \text{ on } (\bar{\Omega} \times \{0\}) \cup (\delta\Omega \times [0,T]). \end{aligned}$$

Then $v \le w$ *on* $\bar{\Omega} \times [0,T]$. *(Note that (1.2-4) is the same as (1.2-2) except that the inequalities become nonstrict).*

Proof. Let $w_\varepsilon^+(x,t) = w(x,t) + \varepsilon[1+3Kt]$ for $(x,t) \in \bar{\Omega} \times [0,T]$, $\varepsilon > 0$. By hypothesis, $v < w_\varepsilon^+$ on $(\bar{\Omega} \times \{0\}) \cup (\delta\Omega \times [0,T])$. Suppose that $v = w_\varepsilon^+$ at some point in $\Omega \times (0,\tau_1)$ where $\tau_1 = \min\{T, \frac{1}{3K}\}$; and (x_1,t_1) is such a point in $\Omega \times (0,\tau_1)$ with minimal t_1 where $v(x_1,t_1) = w_\varepsilon^+(x_1,t_1)$. We evaluate at (x_1,t_1) to find:

$$(1.2\text{-}5) \qquad \begin{aligned} \frac{\partial}{\partial t}(v-w_\varepsilon^+)|_{(x_1,t_1)} &= \frac{\partial v}{\partial t} - \frac{\partial w}{\partial t} - 3K\varepsilon \le \sum_{i,j=1}^{n} a_{ij} \frac{\partial^2}{\partial x_i \partial x_j}(v-w) + \sum_{i=1}^{n} b_i \frac{\partial}{\partial x_i}(v-w) \\ &+ f(x_1,t_1,\nabla v(x_1,t_1),v(x_1,t_1)) - f(x_1,t_1,\nabla w(x_1,t_1),w(x_1,t_1)) - 3K\varepsilon \\ &\le K|v(x_1,t_1) - w(x_1,t_1)| - 3K\varepsilon = K\varepsilon[1+3Kt_1] - 3K\varepsilon < 2K\varepsilon - 3K\varepsilon < 0, \end{aligned}$$

since $\sum_{i,j=1}^{n} a_{ij} \frac{\partial^2}{\partial x_i \partial x_j}(v-w)|_{(x_1,t_1)} = \sum_{i,j=1}^{n} a_{ij} \frac{\partial^2}{\partial x_i \partial x_j}(v-w_\varepsilon^+)|_{(x_1,t_1)} \le 0$, and $\nabla(v(x_1,t_1) - w(x_1,t_1)) = \nabla(v(x_1,t_1) - w_\varepsilon^+(x_1,t_1)) = 0$. Inequality (1.2-5) contradicts the definition of (x_1,t_1). Passing to the limit as $\varepsilon \to 0^+$, we obtain $v \le w$ on $\bar{\Omega} \times [0,\tau_1]$.

If $\tau_1 < T$, we define $w_\varepsilon^+(x,t) = w(x,t) + \varepsilon[1+3K(t-\tau_1)]$ for $(x,t) \in \bar{\Omega} \times [\tau,T]$, $\varepsilon > 0$. From above, we have $v < w_\varepsilon^+$ on $(\bar{\Omega} \times \{\tau_1\}) \cup (\delta\Omega \times [\tau_1,T])$. Suppose $v = w_\varepsilon^+$ at some point in $\Omega \times (\tau_1, 2\tau_1)$ and $2\tau_1 < T$, we repeat the above arguments to deduce that $v \le w$ on $\bar{\Omega}\times[\tau_1, 2\tau_1]$ through an inequality analogous to (1.2-5). Eventually, we obtain $v \le w$ on $\bar{\Omega}\times[0,T]$.

The following is a variant of Theorem 1.2-3 with different boundary conditions.

Theorem 1.2-4. Let δD be C^2 smooth. Suppose that v, w and f satisfy all the conditions as described in Theorem 1.2-3, and inequalities (1.2-4) are replaced by

$$
(1.2\text{-}6)\qquad \begin{cases} Lw + f(x,t,\nabla w, w) \le Lv + f(x,t,\nabla v, v) \text{ in } \Omega\times(0,T] \text{ and} \\ \alpha(x,t)\dfrac{\partial v}{\partial \eta} + \beta(x,t)v \le \alpha(x,t)\dfrac{\partial w}{\partial \eta} + \beta(x,t)w \text{ on } \delta\Omega\times(0,T], \\ v(x,0) \le w(x,0) \qquad \text{for } x \in \bar{\Omega}, \end{cases}
$$

where α, β are defined on $\delta\Omega\ \times(0,T]$ with $\alpha \ge 0$, $\beta > 0$. Then

$$
(1.2\text{-}7)\qquad v(x,t) \le w(x,t) \text{ on } \bar{\Omega}\times[0,T].
$$

(Here, we assume $\frac{\partial v}{\partial \eta}$ and $\frac{\partial w}{\partial \eta}$ exist on $\delta\Omega\ \times(0,T]$, if $\alpha \not\equiv 0$).

The proof uses the arguments as presented in Theorem 1.2-2 and 1.2-3, and will be omitted

Remark 1.2-1. Suppose that in Theorem 1.2-4, f is further assumed to be independent of p, i.e., $f = f(x,t,u)$ defined on $\Omega\times(0,T]\times R$, and $\alpha(x,t) > 0$, $\beta \equiv 0$. If all the other hypothese remain unchanged, then (1.2-7) will be valid. We follow the proof in Theorem 1.2-3. At t_1 in $(0,\tau_1)$, we consider the situation $v(x_1,t_1) = w_\varepsilon^+(x_1,t_1)$ with $x_1 \in \delta\Omega$. In a small neighborhood N of (x_1,t_1) in $\Omega\times(0,T]$, we have $v > w$ in $N \cap \{\Omega\times(0,t_1]\}$. In this set, one has $w \le v \le w_\varepsilon^+$. Thus $|v-w| \le w_\varepsilon^+ - w = \varepsilon(1+3Kt) \le 2\varepsilon$. Consequently in $N \cap \{\Omega\times(0,t_1]\}$, one has $L(v-w_\varepsilon^+) = L(v-w) + 3K\varepsilon \ge f(x,t,w) - f(x,t,v) + 3K\varepsilon \ge 3K\varepsilon - K|w-v| \ge$

$3K\varepsilon - 2K\varepsilon = \varepsilon > 0$. Apply a variant of Theorem 1.1-7 at the point (x_1, t_1), we deduce that $\frac{\partial}{\partial\eta}(v-w_\varepsilon^+)|_{(x_1,t_1)} > 0$. However, from (1.2-6) with $\beta \equiv 0$, we have $\frac{\partial}{\partial\eta}(v-w_\varepsilon^+) = \frac{\partial}{\partial\eta}(v-w) \le 0$ at (x_1, t_1). This gives rise to a contradiction. Thus, we obtain $v \le w$ in $\bar{\Omega}\times[0, \tau_1]$. Then we complete the proof as in Theorem 1.2-3.

<u>Remark</u> 1.2-2. For an outline of the proof of the assertion that $\frac{\partial}{\partial\eta}(v-w_\varepsilon^+)|_{(x_1,t_1)} > 0$ in the last remark, we first construct a ball B_1 centered at $(\hat{x}, t_1)$, $\hat{x} \in \Omega$, touching (x_1, t_1) with radius R so that $B_1 \cap \{\Omega\times(0, t_1]\} \subset N \cap \{\Omega\times(0, t_1]\}$. Then construct another ball B_2 centered at (x_1, t_1) with radius less than R, and define $D = B_1 \cap B_2 \cap \{\Omega\times(0, t_1]\}$. Introduce an auxilary function $z(x,t) = e^{-\alpha[\sum_{i=1}^{n}(x_i-\hat{x}_i)^2+(t-t_1)^2]} - e^{-\alpha R^2}$, where $\hat{x} = (\hat{x}_1, \ldots, \hat{x}_n)$, for large enough positive α so that $L[z] > 0$ for (x,t) on $D \cup \delta D$. Note that $z = 0$ on δB_1. Let $u = [v-w_\varepsilon^+] + \delta z$, for a positive constant δ, we have $L[u] > 0$ in D; and $u < 0$ on $\delta D \cap (\delta B_1 \cup \delta B_2)$ for sufficiently small $\delta > 0$, except at (x_1, t_1) where equality holds. On the top boundary of D, use a variant of Theorem 1.1-6 to deduct that $u < 0$ (otherwise $u \equiv 0$ in $\bar{D}$). Thus $\frac{\partial u}{\partial\eta}|_{(x_1,t_1)} \ge 0$. However, $\frac{\partial z}{\partial\eta}|_{(x_1,t_1)} < 0$ and $\frac{\partial u}{\partial\eta} = \frac{\partial}{\partial\eta}[v-w_\varepsilon^+] + \delta\frac{\partial z}{\partial\eta}$.
Consequently, we must have $\frac{\partial}{\partial\eta}[v-w_\varepsilon^+]|_{(x_1,t_1)} > 0$.

<u>Remark</u> 1.2-3. For an initial-boundary value problem:

$$
\begin{aligned}
&Lu + f(x,t,\nabla u,u) = 0 && \text{in } \Omega\times(0,T],\\
&\alpha(x,t)\frac{\partial u}{\partial\eta} + \beta(x,t)u = h(x,t) && \text{on } \delta\Omega\times(0,T],\\
&u(x,0) = g(x) && \text{for } x \in \bar{\Omega},
\end{aligned}
$$

the functions $\psi(x,t)$ and $\phi(x,t)$ on $\bar{\Omega}\times[0,T]$ are respectively called upper and lower solutions for the above problem if they satisfy

$$L\psi + f(x,t,\nabla\psi,\psi) \le 0 \le L\phi + f(x,t,\nabla\phi,\phi) \qquad \text{in } \Omega\times(0,T],$$

$$\alpha\frac{\partial\phi}{\partial\eta} + \beta\phi \le h(x,t) \le \alpha\frac{\partial\psi}{\partial\eta} + \beta\psi \qquad \text{on } \delta\Omega\times(0,T],$$

$$\phi(x,0) \le g(x) \le \psi(x,0) \qquad \text{for } x \in \bar{\Omega}.$$

Note that ψ and ϕ are defined independently here.

(II) Quasimonotone Systems

A function g: $\Omega\times(0,T] \times R^n \times R^m \to R^m$ where $g = (g_1, \ldots, g_m)$ is said to be quasimonotone increasing in a set $S \subset \Omega\times(0,T] \times R^m$ if for an arbitrary $p \in R^n$, we have

$$g_i(x,t,p,\tilde{u}) \ge g_i(x,t,p,u)$$

for each $i = 1, \ldots, m$, any pair (x,t,u), $(x,t,\tilde{u}) \in S$ satisfying $u_i = \tilde{u}_i$ and $\tilde{u}_j \ge u_j$, $j \ne i$ (in the inequality for g_i).

Let f: $\Omega\times(0,T] \times R^n \times R^m \to R^m$ be defined, where $f = (f_1, \ldots, f_m)$. Assume $v(x,t) = (v_1, \ldots, v_m)$, $w(x,t) = (w_1, \ldots, w_m)$ have all their components continuously defined in $\bar{\Omega}\times[0,T]$, first derivatives with respect to t exist in $\Omega\times(0,T]$, first and second derivatives with respect to x_i continuous in $\Omega\times(0,T]$. For each i, let

$$(1.2\text{-}8) \qquad Lv_i + f_i(x,t,\nabla v_i,v) \ge 0 \qquad \text{for all } (x,t) \in \Omega\times(0,T],$$

$$(1.2\text{-}9) \qquad Lw_i + f_i(x,t,\nabla w_i,w) \le 0 \qquad \text{for all } (x,t) \in \Omega\times(0,T],$$

(1.2-10) $v_i \le w_i$ for all $(x,t) \in \bar{\Omega}\times[0,T]$.

Let $Q = \{(x,t,u)\colon v_i(x,t) \le u_i \le w_i(x,t)$ each $i = 1, \ldots, m$, $(x,t) \in \Omega\times(0,T]\}$, and for any $\delta > 0$, let $Q^\delta = \{(x,t,u)\colon v_i(x,t) - \delta \le u_i \le w_i(x,t) + \delta$ each $i = 1, \ldots, m$, $(x,t) \in \Omega\times(0,T]\}$

Theorem 1.2-5. *Let $\delta\Omega$ be C^2 smooth, η denotes unit outward normal at $\delta\Omega$, and v, w, f be as described above satisfying (1.2-8) to (1.2-10). Moreover, we assume that f is quasimonotone increasing in Q; and for each i, $(x,t) \in \Omega\times(0,T]$, $p \in R^n$, $(x,t,\tilde{u})$ and $(x,t,\hat{u})$ in Q^δ (where $\delta > 0$ is a small constant), it satisfies*

(1.2-11) $$|f_i(x,t,p,\tilde{u}) - f_i(x,t,p,\hat{u})| \le K\sqrt{\sum_{i=1}^{m} (\tilde{u}_i-\hat{u}_i)^2}$$

for a positive constant K. Suppose that $u(x,t)$ is defined on $\bar{\Omega}\times[0,T]$, has the same continuity and smoothness properties as v and w; and for each $i = 1, \ldots, m$, it satisfies:

(1.2-12) $$Lu_i + f_i(x,t,\nabla u_i,u) = 0 \quad \text{in } \Omega\times(0,T],$$

(1.2-13) $$\alpha_i(x,t)\frac{\partial v_i}{\partial\eta} + \beta_i(x,t)v_i \le \alpha_i\frac{\partial u_i}{\partial\eta} + \beta_i u_i \le \alpha_i\frac{\partial w_i}{\partial\eta} + \beta_i w_i \text{ on } \partial\Omega\times(0,T],$$

where $\alpha_i(x,t)$, $\beta_i(x,t)$ are defined on $\delta\Omega\times(0,T]$ with $\alpha_i \ge 0$, $\beta_i > 0$, and

(1.2-14) $$v_i(x,0) \le u_i(x,0) \le w_i(x,0) \quad \text{for } x \in \bar{\Omega}.$$

Then we have the inequalities

$$v_i(x,t) \le u_i(x,t) \le w_i(x,t) \quad \text{on } \bar{\Omega}\times[0,T].$$

(Here, we assume $\frac{\partial v_i}{\partial \eta}$, $\frac{\partial u_i}{\partial \eta}$ and $\frac{\partial w_i}{\partial \eta}$ exist on $\delta\Omega x(0,T]$, if $\alpha_i \neq 0$).

Remark: Functions v and w satisfying (1.2-8), (1.2-9), (1.2-10), (1.2-13) are said to be lower and upper solutions respectively for the boundary value problem (1.2-12) together with $\alpha_i \frac{\partial u_i}{\partial \eta} + \beta_i u_i = h_i$. Here, we assume that f is quasimonotone increasing in Q.

Proof. For $0 < \varepsilon < \delta/2$, $i = 1, \ldots, m$, let

$$u_i^{\pm\varepsilon}(x,t) = u_i(x,t) \pm [1 + 3K\sqrt{n}\ t]\varepsilon \qquad \text{for } (x,t) \in \bar{\Omega}x[0,T].$$

By hypothesis, we have

$$v_i(x,t) < u_i^{+\varepsilon}(x,t) \quad \text{and} \quad u_i^{-\varepsilon}(x,t) < w_i(x,t) \tag{1.2-15}$$

for $i = 1, \ldots, m$, $t = 0$. Suppose that one of these inequalities fails at some point in $\bar{\Omega}x(0,\tau_1)$, where $\tau_1 = \min\{T, \frac{1}{3K\sqrt{n}}\}$; and (x_1,t_1) is a point in $\bar{\Omega}x(0,\tau_1)$, with minimal t_1 where (1.2-15) fails. At (x_1,t_1), $v_i = u_i^{+\varepsilon}$ or $u_i^{-\varepsilon} = w_i$ for some i. Assume that former is the case; a similar proof holds for the latter case.

First, let $v_j(x_1,t_1) = u_j^{+\varepsilon}(x_1,t_1)$ and $x_1 \in \Omega$. At (x_1,t_1), we evaluate

$$\frac{\partial}{\partial t}(v_j - u_j^{+\varepsilon})|_{(x_1,t_1)} = \frac{\partial v_j}{\partial t} - \frac{\partial u_j}{\partial t} - 3\varepsilon K\sqrt{n} \le \sum_{i,k=1}^{n} a_{ik} \frac{\partial^2}{\partial x_i \partial x_k}(v_j - u_j) \tag{1.2-16}$$

$$+ \sum_{i=1}^{n} b_i \frac{\partial}{\partial x_i}(v_j - u_j) + f_j(x_1,t_1,\nabla v_j,v) - f_j(x_1,t_1,\nabla u_j,u) - 3\varepsilon K\sqrt{n}$$

$$\le f_j(x_1,t_1,\nabla u_j,\tilde{v}) - f_j(x_1,t_1,\nabla u_j,u) - 3\varepsilon K\sqrt{n}$$

where $\tilde{v} = (\tilde{u}_1^+(x_1,t_1), \ldots, \tilde{u}_{j-1}^+(x_1,t_1), v_j(x_1,t_1), \tilde{u}_{j+1}^+(x_1,t_1), \ldots,$ $\tilde{u}_m^+(x_1,t_1))$, $\tilde{u}_k^+(x_1,t_1) = \min\{u_k^{+\varepsilon}(x_1,t_1), w_k(x_1,t_1)\}$. (Note that at (x_1,t_1), $\nabla u_j = \nabla v_j$). This is due to the quasimonotonicity of f. Now, $(x_1,t_1,\tilde{v})$ and (x_1,t_1,u) are in Q^δ, since $0 < v_j(x_1,t_1) - u_j(x_1,t_1) = u_j^{+\varepsilon}(x_1,t_1) - u_j(x_1,t_1)$ $< 2\varepsilon < \delta$, and we can apply (1.2-11). We thus deduce from (1.2-16) and (1.2-11) that $\frac{\partial}{\partial t}(v_j-u_j^{+\varepsilon})|_{(x_1,t_1)} \le K\sqrt{n}\ \varepsilon\ (1+3K\sqrt{n}\ t_1) - 3\varepsilon K\sqrt{n} < K\sqrt{n}\varepsilon 2 - 3\varepsilon K\sqrt{n}$ < 0, contradicting the definition of (x_1,t_1). Hence, we must have $x_1 \in \delta\Omega$. Inequality (1.2-13) implies that

$$(1.2\text{-}17) \qquad \beta_j(x_1,t_1)[v_j(x_1,t_1) - u_j^{+\varepsilon}(x_1,t_1)] < \alpha_j(x_1,t_1)\frac{\partial}{\partial\eta}(u_j^{+\varepsilon}-v_j)(x_1,t_1).$$

The right side in (1.2-17) is ≤ 0, because $u_j^{+\varepsilon} - v_j > 0$ for $(x,t) \in \Omega\times\{t_1\}$ and $u_j^{+\varepsilon}(x_1,t_1) - v_j(x_1,t_1) = 0$. This contradicts the fact that the left hand side of (1.2-17) is equal to zero. Consequently, we must have inequalities (1.2-15) for $(x,t) \in \bar{\Omega}\times(0,\tau_1)$. Passing to the limit as $\varepsilon \to 0^+$, we obtain

$$(1.2\text{-}18) \qquad v_i(x,t) \le u_i(x,t) \le w_i(x,t)$$

for each i, $(x,t) \in \bar{\Omega}\times[0,\tau_1]$.

If $\tau_1 < T$, we redefine for $0 < \varepsilon < \delta/2$, $u_i^{\pm\varepsilon}(x,t) = u_i(x,t) \pm \varepsilon[1+3K\sqrt{n}$ $(t-\tau_1)]$ for $(x,t) \in \bar{\Omega}\times[\tau_1,T]$. Repeat the above arguments to deduce that (1.2-18) is true for each i, $(x,t) \in \bar{\Omega}\times[\tau_1,\tau_2]$, where $\tau_2 = \min\{T,\tau_1+\frac{1}{3K\sqrt{n}}\}$. Eventually, we obtain (1.2-18) in $\bar{\Omega}\times[0,T]$ for each i.

<u>Remark</u> 1.2-4. Theorem 1.2-5 is not in the most general form. If each f_i are independent of ∇u_i and the conditions $\alpha_i \ge 0$, $\beta_i > 0$ are replaced by: $\alpha_i \ge 0$, $\beta_i \ge 0$, $\alpha_i^2 + \beta_i^2 \ne 0$ for each $i = 1, \ldots, m$, on $\delta\Omega\times(0,T]$, the theorem remains

valid. For the proof of this case, we adapt the procedures in remark 1.2-1 and 1.2-2.

Remark 1.2-5. Suppose that in (1.2-8), (1.2-9) and (1.2-12) the parabolic operators L are assumed to be dependent on the ith equation. That is, we have $L = L_i$, which is parabolic at $(x,t) \in \Omega x(0,T]$ for each i, with all coefficients bounded as described in (1.2-1). Referring to the proof of Theorem 1.2-5 (cf. equation (1.2-16)), we see that the result of Theorem 1.2-5 remains valid.

(III) Non-Quasimonotone, Special Cases

Let f: $\Omega x(0,T] \times R^n \times R^m \to R^m$, where $f = (f_1, \ldots, f_m)$ be defined. Suppose that v(x,t) and w(x,t) are defined on $\bar{\Omega}x[0,T]$ with the same continuity and smoothness properties as described in the last subsection II, and $v_i \le w_i$ on $\bar{\Omega}x[0,T]$, $i = 1, \ldots, m$. Let Q and Q^δ, $\delta > 0$, be as described for Theorem 1.2-5. In the next theorem, we assume that for each $i = 1, \ldots, m$:

$$(1.2\text{-}19) \qquad L_i v_i + f_i(x,t,\nabla v_i,\tilde{u}) \ge 0$$

for all $(x,t,\tilde{u}) \in Q$ with $\tilde{u}_i = v_i$; and

$$(1.2\text{-}20) \qquad L_i w_i + f_i(x,t,\nabla w_i,\tilde{u}) \le 0$$

for alll $(x,t,\tilde{u}) \in Q$ with $\tilde{u}_i = w_i$. Here L_i are parabolic operators at $(x,t) \in \Omega x(0,T]$:

$$L_i \equiv \sum_{j,k=1}^{n} a^i_{jk}(x,t) \frac{\partial^2}{\partial x_j \partial x_k} + \sum_{j=1}^{n} b^i_j(x,t) \frac{\partial}{\partial x_j} - \frac{\partial}{\partial t},$$

where $a^i_{jk}(x,t)$ and $b^i_j(x,t)$ are all bounded functions in $\Omega x(0,T]$.

Theorem 1.2-6. *Let the conditions of the last paragraph be valid and* $\delta\Omega$ *be* C^2 *smooth. For each* i, $(x,t) \in \Omega x(0,T]$, $p \in R^n$, $(x,t,\tilde{u})$ *and* $(x,t,\hat{u})$ *in* Q^δ (δ *is a small positive constant*), *we assume that* f *satisfies*:

$$|f_i(x,t,p,\tilde{u}) - f_i(x,t,p,\hat{u})| \le K\sqrt{\sum_{i=1}^{m} (\tilde{u}_i - \hat{u}_i)^2}$$

for a positive constant K. *Suppose that* $u(x,t)$ *defined on* $\bar{\Omega}x[0,T]$ *satisfies all the properties as described in Theorem* 1.2-5 (i.e., *formulas* (1.2-12) *with* L *replaced by* L_i, (1.2-13) *and* (1.2-14)).

Then $\qquad v_i(x,t) \le u_i(x,t) \le w_i(x,t)$ *in* $\bar{\Omega}x[0,T]$.

Proof. The proof is exactly analogous to that of Theorem 1.2-5. In inequalities (1.2-16), we omit the step involving the term $f_j(x_1,t_1,\nabla v_j,v)$, and go directly to $f_j(x_1,t_1,\nabla u_j,\tilde{v})$. Details will be omitted.

In the study of competing-species interactions, one assumes that f has the property

$$(1.2\text{-}21) \qquad f_i(x,t,p,\tilde{u}) \le f_i(x,t,p,\hat{u})$$

for each i, any $p \in R^n$, $(x,t) \in \Omega x(0,T]$, whenever

$$(1.2\text{-}22) \qquad \tilde{u}_j \ge \hat{u}_j \ge 0,\ j \ne i \text{ and } \tilde{u}_i = \hat{u}_i.$$

Suppose that v, w are defined on $\bar{\Omega}x[0,T]$ with the same continuity and smoothness properties as in the last theorem, and for each i:

(1.2-23) $$0 \le v_i \le w_i \text{ on } \bar{\Omega}\times[0,T].$$

It is clear that (1.2-19) and (1.2-20) will be satisfied, if for each i:

(1.2-24) $$L_i v_i + f_i(x,t,\nabla v_i, w_1,\dots,w_{i-1},v_i,w_{i+1},\dots,w_m) \ge 0,$$

(1.2-25) $$L_i w_i + f_i(x,t,\nabla w_i, v_1,\dots,v_{i-1},w_i,v_{i+1},\dots,v_m) \le 0$$

for $(x,t) \in \Omega\times(0,T]$.

Theorem 1.2-7. Let $\delta\Omega$ be C^2 smooth, and f be as described in Theorem 1.2-6, with the additional hypotheses (1.2-21), (1.2-22) above. Let v, w be as in Theorem 1.2-6, with inequalities (1.2-19), (1.2-20) replaced by (1.2-24), (1.2-25) for each i, and (1.2-23) is satisfied. If u(x,t) is as described in Theorem 1.2-6, then

$$v_i(x,t) \le u_i(x,t) \le w_i(x,t) \quad \text{in} \quad \bar{\Omega}\times[0,T]$$

for each $i = 1, \dots, m$.

1.3 Basic Linear Theory and Fixed Point Theorems

In this section, we state some basic definitions and results in the classical theory of scalar partial differential equations of elliptic and parabolic type. The proofs will not be included here, because they are long and can be found in many other books, e.g. [126], [127].

Let V be a vector space over the real numbers R. A norm is a mapping from V into R (denoted hery by $\|x\|$ for each $x \in V$) satisfying

(i) $\|x\| \geq 0$ for all $x \in V$, $\|x\| = 0$ if and only if $x = 0$,

(ii) $\|\alpha x\| = |\alpha| \|x\|$ for all $\alpha \in R$, $x \varepsilon V$

(iii) $\|x+y\| \leq \|x\| + \|y\|$ for all $x, y \in V$ (triangle inequality)

The vector space equipped with a norm is called a normed linear space. If we define the distance between x and y by $\|x-y\|$, a normed linear space is a metric space. Consequently, we say a sequence $\{x_n\}$ in V converges to an element $x \in V$ if $\|x_n - x\| \to 0$ as $n \to \infty$. Also, $\{x_n\}$ is a Cauchy sequence if $\|x_n - x_m\| \to 0$, as $m, n \to \infty$. If V is complete, that is every Cauchy sequence converges to an element in V, then we say V is a Banach space.

Let Ω be a bounded domain in R^n, $x = (x, \ldots, x_n)$, $D_x^j u$ denotes any derivative of u(x) of order j with respect to any components of x. Let S be any subset of the closure of Ω (i.e. $\bar{\Omega}$), and $0 < \alpha < 1$. We say a function f on S is Hölder continuous in S with exponent α if the quantity:

(1.3-1) $$\sup_{x, y \in S, x \neq y} \frac{|f(x)-f(y)|}{|x-y|^{\alpha}}$$

is finite. Let k denote a nonnegative integer.

<u>Definition</u>. <u>The Hölder space</u> $H^{k+\alpha}(\bar{\Omega})$ <u>is the Banach space of continuous functions</u> u(x) <u>in</u> $\bar{\Omega}$, <u>having derivatives up to order</u> k <u>continuous in</u> $\bar{\Omega}$, <u>and have finite value for the norm:</u>

(1.3-2) $$|u|_{\Omega}^{(k+\alpha)} = \sum_{0 \leq j \leq k} \max_{\Omega} |D_x^j u| + \sum_{j=k} \sup_{x, x' \varepsilon \Omega, |x-x'| \leq \rho_0} \frac{|D_x^j u(x) - D_x^j u(x')|}{|x-x'|^{\alpha}} .$$

<u>The first sum is over all possible derivatives up to order</u> k <u>(including the</u>

function itself). The second sum is over all possible derivatives of order exactly k. $\rho_0 > 0$ is a fixed constant.

It can be readily shown that $|\cdot|_{\Omega}^{(k+\alpha)}$ is indeed a norm satisfying the properties (i) to (iii) above, and that $H^{k+\alpha}(\bar{\Omega})$ is indeed complete under this norm, and hence forms a Banach space. Let $T > 0$, $\bar{\Omega}_T = \bar{\Omega}\times[0,T]$. $\delta\Omega_T = \delta\Omega\times(0,T)$.

Definition. The Hölder space $H^{k+\alpha,(k+\alpha)/2}(\bar{\Omega}_T)$ is the Banach space of continuous functions $u(x,t)$ in $\bar{\Omega}_T$, having all derivatives of the form $D_t^r D_x^s u$, $0 \le 2r + s \le k$ continuous in $\bar{\Omega}_T$, and have finite value for the norm:

$$|u|_{\Omega_T}^{(k+\alpha)} = \sum_{0\le 2r+s\le k} \max_{\Omega_T} |D_t^r D_x^s u| + \sum_{2r+s=k} \sup_{\substack{(x,t),(x',t)\in\bar{\Omega}_T \\ |x-x'|\le\rho_0}} \frac{|D_t^r D_x^s u(x,t) - D_t^r D_x^s u(x',t)|}{|x-x'|^{\alpha}}$$

(1.3-3)

$$+ \sum_{k-1\le 2r+s\le k} \sup_{\substack{(x,t),(x,t')\in\bar{\Omega}_T \\ |t-t'|\le\rho_0}} \frac{|D_t^r D_x^s u(x,t) - D_t^r D_x^s u(x,t')|}{|t-t'|^{\frac{k+\alpha-(2r+s)}{2}}}$$

Let $x^0 = (x_1^0, \ldots, x_n^0)$ be any point on the boundary $\delta\Omega$ of a bounded domain Ω. We will call $(y_1, \ldots, y_n)$ a local Cartesian coordinate system with origin at x^0 if y and x are connected by the equations $y_i = \sum_{k=1}^{n} b_{ik}(x_k - x_k^0)$, $i = 1, \ldots, n$, where $\{b_{ik}\}$ forms an orthogonal constant matrix, and the y_n axis has the direction of the outward (with respect to Ω) normal to $\delta\Omega$ at x_0. In this section, we assume that at each point $\xi \in \delta\Omega$, there exists a tangent plane, with $\eta(\xi) = (\eta_1, \ldots, \eta_n)$ denoting the unit vector of the outward normal to $\delta\Omega$ at the point ξ.

We will say that a boundary surface of $\delta\Omega$ belongs to the class H^{ℓ}, $\ell > 1$, if it has the following property: There exists a number $\rho > 0$ such that the intersection of $\delta\Omega$ with a ball B_ρ of radius ρ with center at an arbitrary point $x'' \in \delta\Omega$ is a connected surface, the equation of which in the local Cartesian coordinate system $(y_1, \ldots, y_n)$ with the origin at x^0 has the form $y_n = \omega(y_1, \ldots, y_{n-1})$, which is a function of class $H^{\ell}(\bar{J})$. (Here, J is the projection of B_ρ on the surface $y_n = 0$).

Suppose a function $\phi(s)$ is given on a surface $\delta\Omega$ of class H^{ℓ_1}, $\ell_1 > 1$. We will say that $\phi(s)$ is a function of class $H^{\ell}(\delta\Omega)$, $\ell \leq \ell_1$, if as a function of $y_1, \ldots, y_{n-1}$ it is in the function space $H^{\ell}(\bar{J})$ at each $x^0 \in \delta\Omega$. We define the norm $|\phi|_{\delta\Omega}^{(\ell)}$ to be the largest of the norms $|\phi(y)|_J^{(\ell)}$ calculated for all points x^0 of $\delta\Omega$. Suppose ϕ is given on $\bar{\Omega}$ and $\phi(x)$ is in $H^{\ell}(\bar{\Omega})$. If $\delta\Omega$ is in H^{ℓ_1} with $\ell_1 \geq \max\{1, \ell\}$, then the restriction of ϕ to the boundary, $\phi(s)|_{s\in\delta\Omega}$, belongs to the class $H^{\ell}(\delta\Omega)$. Conversely, suppose $\phi(s)$ is in the class $H^{\ell}(\delta\Omega)$ and $\delta\Omega$ belongs to H^{ℓ}, $\ell > 1$. Then $\phi(s)$ can be extended to the whole domain Ω in such a way that the extended function $\phi(x)$ belongs to $H^{\ell}(\bar{\Omega})$. Moreover, this extension can be made for all functions $\phi(s)$ in $H^{\ell}(\delta\Omega)$ in exactly the same manner, so that the norms of $|\phi(s)|_{\delta\Omega}^{(\ell)}$ and $|\phi(x)|_{\Omega}^{(\ell)}$ are equal. For a function ϕ defined on $\delta\Omega x[0,T]$, we say that it is in $H^{\ell,\ell/2}(\overline{\delta\Omega_T})$ if at each $(x_0, t_0) \in \delta\Omega x[0,T]$, in the local Cartesian coordinates $y_n = \omega(y_1, \ldots, y_{n-1}, t) = \omega(y_1, \ldots, y_{n-1})$, ϕ is in $H^{\ell,\ell/2}(\bar{J}xI_{t_0})$ as a function of $(y_1, \ldots, y_{n-1}, t)$. Here I_{t_0} is the intersection of $[0,T]$ with a closed interval containing t_0 in its interior, and J is a projection as described in the last paragraph. We also define the $|\phi|_{\delta\Omega_T}^{(\ell)}$ as the largest of the norms $|\phi|_{\bar{J}xI_{t_0}}^{(\ell)}$ calculated for all $(x_0, t_0) \in \delta\Omega x[0,T]$. Other proeprties concerning extensions and restrictions to the boundary etc. carries over in a natural way from the situation without the variable t.

(I) Parabolic Equations

We denote the elliptic operator E on $\bar{\Omega}_T$ by:

$$E \equiv \sum_{i,j=1}^{n} a_{ij}(x,t) \frac{\partial^2}{\partial x_i \partial x_j} + \sum b_i(x,t) \frac{\partial}{\partial x_i} + c(x,t)$$

with $a_{ij}(x,t)$, $b_i(x,t)$, $c(x,t)$ defined on $\bar{\Omega}_T$ where

$$\mu \sum_{i=1}^{n} \xi_i^2 \leq \sum_{i,j=1}^{n} a_{ij}(x,t)\, \xi_i \xi_j \leq \nu \sum_{i=1}^{n} \xi_i^2$$

is satisfied for all $(\xi_1, \ldots, \xi_n) \in R^n$ and $(x,t) \in \bar{\Omega}_T$ (μ, ν are fixed positive constants.)

We consider the initial-boundary value problem for the parabolic equation:

$$(1.3\text{-}4) \qquad \begin{cases} \dfrac{\partial u}{\partial t} - E(x,t)u = f(x,t) & \text{for} \quad (x,t) \in \Omega x (0,T) \\ u(x,0) = g(x) & \text{for} \quad x \in \bar{\Omega} \\ u(x,t) = h(x,t) & \text{for} \quad (x,t) \in \delta\Omega x\ [0,T). \end{cases}$$

We say compatibility condition of order 1 is satisfied if:

$$(1.3\text{-}5) \qquad \begin{aligned} & g(x) = h(x,0) && \text{for all } x \in \partial\Omega, \text{ and} \\ & E(x,0)g(x) + f(x,0) = \frac{\partial h}{\partial t}(x,0) && \text{for all } x \in \partial\Omega. \end{aligned}$$

These are conditions on the prescribed "Dirichlet" data $h(x,t)$ on the boundary $\partial\Omega$ at $t = 0$, in relation to the initial data $g(x)$. The following is an existence theorem for problem (1.3-4) together with an estimate on the Hölder norm (1.3-3) with $k = 2$, for the solution in terms of those norms for data functions f, g and h.

<u>Theorem</u> 1.3-1. <u>Let</u> $0 < \alpha < 1$, <u>the coefficients</u> $a_{ij}(x,t)$, $b_i(x,t)$, $c(x,t)$ <u>belong to</u> $H^{\alpha,\alpha/2}(\bar{\Omega}_T)$, <u>and the boundary</u> $\delta\Omega$ <u>belongs to</u> $H^{2+\alpha}$. <u>Suppose that</u> $f \in H^{\alpha,\alpha/2}(\bar{\Omega}_T)$, $g \in H^{2+\alpha}(\bar{\Omega})$, $h \in H^{2+\alpha,\frac{2+\alpha}{2}}(\overline{\delta\Omega}_T)$, <u>and the compatibility condition of order</u> 1 <u>is satisfied</u>. <u>Then the initial-boundary value problem</u> (1.3-4) <u>has a unique solution</u> u <u>in the function space</u> $H^{2+\alpha,\frac{2+\alpha}{2}}(\bar{\Omega}_T)$. <u>Moreover, the following inequality is satisfied</u>

$$|u|^{(2+\alpha)}_{\Omega_T} \le K[|f|^{(\alpha)}_{\Omega_T} + |g|^{(2+\alpha)}_{\Omega} + |h|^{(2+\alpha)}_{\delta\Omega_T}]$$

<u>for a positive constant</u> K (<u>independent of</u> f, g <u>and</u> h)

For more general boundary conditions on $\delta\Omega x[0,T]$, we assume that $\alpha_1(x,t)$, ..., $\alpha_n(x,t)$ and $\beta(x,t)$ are defined on $\overline{\delta\Omega}_T$ and

$$\left|\sum_{i=1}^{n} \alpha_i(x,t)\eta_i(x)\right| \ge \delta > 0 \quad (\delta \text{ a constant})$$

for $(x,t) \in \delta\Omega_T$. We consider the problem:

$$\text{(1.3-6)} \quad \begin{aligned} &\frac{\partial u}{\partial t} - E(x,t)u = f(x,t) && \text{for} \quad (x,t) \in \Omega x(0,T) \\ &u(x,0) = g(x) && \text{for} \quad x \in \bar{\Omega} \\ &\sum_{i=1}^{n} \alpha_i(x,t)\frac{\partial u}{\partial x_i} + \beta(x,t)u = h(x,t) && \text{for} \quad (x,t) \in \delta\Omega x[0,T]. \end{aligned}$$

We say compatibility condition of order 0 is satisfied if

$$\sum_{i=1}^{n} \alpha_i(x,0)\frac{\partial g}{\partial x_i} + \beta(x,0)g = h(x,0) \quad \text{for all } x \in \delta\Omega.$$

<u>Theorem</u> 1.3-2. <u>Let</u> α, a_{ij}, b_i, c <u>and</u> $\delta\Omega$ <u>be as described in Theorem</u> 1.3-1, <u>and</u>

$\alpha_i(x,t)$, $\beta(x,t)$ be functions in $H^{1+\alpha,\frac{1+\alpha}{2}}(\overline{\delta\Omega_T})$. Suppose that $f \in H^{\alpha,\alpha/2}(\overline{\Omega}_T)$, $g \in H^{2+\alpha}(\overline{\Omega})$, $h \in H^{1+\alpha,\frac{1+\alpha}{2}}(\overline{\delta\Omega_T})$, and the compatibility condition of order 0 is satisfied. Then the initial value problem (1.3-6) has a unique solution u in the function space $H^{2+\alpha,\frac{2+\alpha}{2}}(\overline{\Omega}_T)$. Moreover

$$|u|^{(2+\alpha)}_{\Omega_T} \le K[|f|^{(\alpha)}_{\Omega_T} + |g|^{(2+\alpha)}_{\Omega} + |h|^{(1+\alpha)}_{\delta\Omega_T}]$$

for a positive constant K (independent of f, g and h).

Remark 1.3-1. Theorems 1.3-1 and 1.3-2 are valid if α are replaced everywhere by $k + \alpha$, $k > 0$ integer, and compatibility conditions of higher order are satisfied. For reference, see [127, p. 320] for details.

(II) Elliptic Equations

We consider a bounded domain $\Omega \subset R^n$. Let

$$L \equiv \sum_{i,j=1}^{n} a_{ij}(x) \frac{\partial^2}{\partial x_i \partial x_j} + \sum_{i=1}^{n} b_i(x) \frac{\partial}{\partial x_i} + c(x) \quad \text{in } \overline{\Omega},$$

where $\sum_{i,j=1}^{n} a_{ij}(x)\xi_i\xi_j \ge \mu_0 \sum_{i=1}^{n} \xi_i^2$, $\mu_0 > 0$, for all $x \in \overline{\Omega}$, all $(\xi_1, \ldots, \xi_n) \in R^n$. We consider the Dirichlet boundary value problem:

(1.3-7) $$Lu = f(x) \quad \text{in} \quad \Omega, \quad u = g(x) \quad \text{on} \quad \delta\Omega.$$

Theorem 1.3-3. Let $0 < \alpha < 1$, the coefficients $a_{ij}(x)$, $b_i(x)$, $c(x)$ belong to $H^\alpha(\overline{\Omega})$, $c(x) \le 0$, and the boundary $\delta\Omega$ belongs to $H^{2+\alpha}$. Suppose that $f \in H^\alpha(\overline{\Omega})$,

$g \in H^{2+\alpha}(\delta\Omega)$, *then the Dirichlet problem* (1.3-7) *has a unique solution* u *in the class* $H^{2+\alpha}(\bar{\Omega})$. *Moreover,*

$$|u|_{\Omega}^{(2+\alpha)} \le K[|f|_{\Omega}^{(\alpha)} + |g|_{\delta\Omega}^{(2+\alpha)}]$$

for a positive constant K (*independent of* f *and* g).

Remark 1.3-2. Note that $c(x) \le 0$ is assumed here so that the problem $Lu + \lambda u = 0$ in Ω, $u = 0$ in $\delta\Omega$ has $u \equiv 0$ in $\bar{\Omega}$ as the only solution when $\lambda = 0$; that is, $\lambda = 0$ is not an eigenvalue. More related properties are given below.

For more general boundary conditions, we consider

$$\text{(1.3-8)} \qquad Lu = f(x) \text{ in } \Omega, \ \frac{\partial u}{\partial \eta} + \beta(x)u = g \quad \text{on } \delta\Omega.$$

Theorem 1.3-4. *Let* α, a_{ij}, b_i, c *and* $\delta\Omega$ *be as described in Theorem* 1.3-3 (*note that* $c(x) \le 0$) *and* $\beta(x)$ *belongs to* $H^{1+\alpha}(\delta\Omega)$, $\beta(x) \ge \nu_0 > 0$ *on* $\delta\Omega$. *Suppose that* $f \in H^{\alpha}(\bar{\Omega})$, $g \in H^{1+\alpha}(\delta\Omega)$, *then the problem* (1.3-8) *has a unique solution* u *in the space* $H^{2+\alpha}(\bar{\Omega})$. *Moreover*

$$|u|_{\Omega}^{(2+\alpha)} \le K[|f|_{\Omega}^{(\alpha)} + |g|_{\delta\Omega}^{(1+\alpha)}]$$

for a positive constant K (independent of f, g).

Remark 1.3-3. For references see e.g. [89, Sections 6.3, 6.7] or [126, Chapter 3].

We now describe a few basic properties concerning the eigenvalue problem mentioned in Remark 1.3-2. For convenience, we only consider a simple elliptic operator:

$$\tilde{L} \equiv \operatorname{div}(p(x)\nabla) + c(x) \qquad \text{in } \bar{\Omega}$$

where $p(x) \in H^{1+\alpha}(\bar{\Omega})$, with $p(x) > 0$ in $\bar{\Omega}$, and $c(x) \in H^{\alpha}(\bar{\Omega})$, with $c(x) \le 0$ in $\bar{\Omega}$. Here div and ∇ denotes respectively the divergence and gradient. The eigenvalue problem

$$(1.3\text{-}9) \qquad \tilde{L}u + \lambda u = 0 \quad \text{in} \quad \Omega, \; u = 0 \quad \text{on } \delta\Omega$$

has a countable sequence of eigenvalues $\lambda_1 \le \lambda_2 \le \lambda_3 \le \ldots$ for which (1.3-9) has nontrivial solutions. The first eigenvalue λ_1 has the characterization

$$(1.3\text{-}10) \qquad \lambda_1 = \min_{\phi} \{[\int_{\Omega} p|\nabla\phi|^2 - c\phi^2 dx][\int_{\Omega} \phi^2 dx]^{-1}\} \quad , \; \phi \ne 0,$$

where the minimization is over a certain class of functions which has square integrable "strong first derivatives" in Ω and vanishes on the boundary in a certain sense (see [227, Chapter 3]). From the assumption $p > 0$, $c \le 0$ and (1.3-10), we conclude that $\lambda_1 > 0$. From the theories in [89, Chapter 7,8], and the assumption that $\delta\Omega$ is in $H^{2+\alpha}$, we deduce that the eigenfunction $u = \omega(x)$ for $\lambda = \lambda_1$ has to be in $H^{\alpha}(\bar{\Omega})$. From Theorem 1.3-3 above, using $-\lambda_1\omega$ as f, we obtain that $\omega(x)$ is in $H^{2+\alpha}(\bar{\Omega})$. Finally, from the characterization (1.3-10), which asserts that λ_1 is attained at $\phi = \omega(x)$, one deduces that $\omega(x)$ cannot change sign. Thus, from the maximum principles (Theorems 1.1-1 and 1.1-3), we obtain the result:

There is a principal eigenfunction $u = \omega(x)$ corresponding to the first eigenvalue $\lambda = \lambda_1 > 0$ of problem (1.3-9) with the property that $\omega(x) > 0$ for all $x \in \Omega$, and $\frac{\partial\omega}{\partial\eta} < 0$ on $\partial\Omega$. Under the conditions on p, c and $\partial\Omega$ above, the function ω is in $H^{2+\alpha}(\bar{\Omega})$.

Further details of the above arguments will not be included here, since they are not widely used in the rest of the book.

Many existence theorems for nonlinear differential equations and systems in the following chapters will be proved as fixed points for mappings between function spaces described above. Let V_1 and V_2 be normed linear spaces. A mapping $T: V_1 \to V_2$ is called compact if T maps every bounded sequence in V_1 into sequences in V_2 which contain convergent subsequence.

Theorem 1.3-5 (Schauder fixed point theorem). *Let B be a normed linear space and K be a non-empty convex compact subset of B. If $T: K \to K$ is continuous, then T has at least one fixed point in K (i.e., there exist a $u \in K$ so that $T(u) = u$).*

For direct proof of this theorem, see e.g. [17]. A variant of the theorem is the following corollary which is in more readily applicable form.

Theorem 1.3-6. *Let B be a Banach space, and C be a non-empty closed, bounded convex subset of B. If T is a compact and continuous map from B into itself such that $T: C \to C$, then T has a least one fixed point in C.*

For details of proof, see e.g. [17]. We also look for fixed point of maps by means of a "homotopic invariance" type of approach. Roughly speaking, if two appropriate maps can be deformed into each other in a certain continuous way, they would have the same "number" of zeros.

Theorem 1.3-7 (Leray-Schauder). *Let B be a Banach space, and O be a bounded open neighborhood of $p \in B$. Let $T: \bar{O} \to B$ be a continuous and compact map (i.e, maps every sequence in $\bar{O}$ into a sequence which contains convergent subsequence). Define a map $H: \bar{O} \times [0,1] \to B$ by*

$$H(u,\lambda) = u - \lambda Tu, \text{ for } u \in \bar{O},\ \lambda \in -[0,1]. \tag{1.3-11}$$

Suppose that $H(u,\lambda) \neq p$ for all $\lambda \in [0,1]$, $u \in \delta O$. Then the equation:

$$H(u,1) = p \quad , \quad \text{i.e.} \quad Tu = u - p$$

has a solution $u \in O$.

For a reference to this theorem, see [133]. A variant of this theorem, in a more readily applicable form, is formulated as follows.

Theorem 1.3-8. Let B be a Banach space, and λ be a real parameter in a bounded interval $a \le \lambda \le b$. Suppose $T: B\times[a,b] \to B$ is a transformation defined for all $u \in B$, $\lambda \in [a,b]$ with the following properties:

(i) For any fixed λ, $T(u,\lambda)$ is continuous in B.

(ii) For u in bounded sets of B, $T(u,\lambda)$ is uniformly continuous in λ, i.e., for any bounded $B_0 \subset B$ and any $\varepsilon > 0$, there exists a $\delta > 0$ such that if $u \in B_0$,
$|\lambda_1-\lambda_2| < \delta$, $a \le \lambda_1, \lambda_2 \le b$, then $\|T(u,\lambda_1) - T(u,\lambda_2)\| < \varepsilon$.

(iii) For any fixed λ, $T(u,\lambda)$ is a compact map from B into B.

(iv) There exists a constant M such that every possible solution of

$$u - T(u,\lambda) = 0 \ (\text{with } u \in B,\ \lambda \in [a,b]) \text{ satisfies:} \quad \|u\| \le M.$$

(v) There exists a unique solution of the equation $u - T(u,a) = 0$ in B.

Then, there exists a solution of the equation

$$u - T(u,b) = 0 \qquad \text{in } B.$$

The theorem is valid if $T(u,\lambda)$ is only defined for $\|u\| \le M'$, $\lambda \in [a,b]$ for some $M' > M$, and the assumptions (i) to (v) are modified accordingly.

1.4. An Existence Theorem for Semilinear Elliptic Systems

Existence of solution to scalar elliptic boundary value problem of the form $Lu + f(x,u) = 0$ in Ω, $u = g$ on $\delta\Omega$, (where L is elliptic), can be proved by means of monotone iteration scheme. Under appropriate conditions, one starts iterating from an "upper" or "lower" solution to the problem, and constructs a sequence of functions through solving linear problems. The sequence of functions will converge monotonically to a solution of the original problem. Such method is explained in Chapter 5; it can be adapted to numerical computations and to the study of systems under more restrictive conditions.

In this section, we discuss a different method of finding solutions. The method can readily treat general systems without having to deal with monotonicity which is more restrictive for the case of systems. The proof uses the Leray-Schauder's technique (Theorem 1.3-7) as described in the last section, and it is thus not constructive as in the monotone method. We present here theorems given in [220], which are used extensively in the next three chapters. More general versions had been obtained in e.g. [220], but will not be stated here.

In this section, we let Ω be a bounded domain in R^n, $x = (x_1, \ldots, x_n)$. Let $\delta\Omega \in H^{2+\alpha}$, $0 < \alpha < 1$, and

$$L \equiv \sum_{j,k=1}^{n} a_{jk}(x) \frac{\partial^2}{\partial x_j \partial x_k} + \sum_{j=1}^{n} b_j(x) \frac{\partial}{\partial x_j} + c(x) \quad \text{in } \bar{\Omega},$$

where a_{jk}, b_j, c are all in $H^{\alpha}(\bar{\Omega})$, with $c(x) \le 0$ in $\bar{\Omega}$; and

$$\mu \sum_{i=1}^{n} \xi_i^2 \le \sum_{j,k=1}^{n} a_{jk}(x)\xi_j\xi_k \le \nu \sum_{i=1}^{n} \xi_i^2$$

for all $x \in \bar{\Omega}$, $\xi = (\xi_1, \ldots, \xi_n) \in R^n$, where μ, ν are fixed positive constants.

We consider the boundary value problem

(1.4-1) $\quad Lu = f(x,u)$ in Ω, $u = \phi(x)$ on $\delta\Omega$,

with $u = (u_1, \ldots, u_m)$, $f = (f_1, \ldots, f_m)$, $\phi = (\phi_1, \ldots, \phi_m)$ and the operator L is applied componentwise. For each i, we assume that $f_i \in C^1(\bar{\Omega}\times R^m)$ and $\phi_i \in H^{2+\alpha}(\delta\Omega)$.

All the assumptions listed above will be made in this entire section, unless otherwise stated. The following lemma concerning a bound on the gradient of any possible twice continuously differentiable solutions in terms of bounds on f and its derivatives will be used in the proof of the existence theorem.

Lemma 1.4-1. For every $P > 0$, there is a constant $Q > 0$ such that if $u \in C^2(\bar{\Omega})$ is a solution of (1.4-1) with $|u(x)| \leq P$ in $\bar{\Omega}$, then $|\nabla u(x)| \leq Q$ in $\bar{\Omega}$. The constant Q depends only on P, the bounds of each $|f_i|$ on $\bar{\Omega}\times\{u: |u| \leq P\}$, and the maximums of the absolute values of the extensions $\tilde{\phi}_i$ of ϕ_i to $\bar{\Omega}$ and their first and second derivatives. (Here $|\nabla u|^2 = \sum_{i=1}^{m}\sum_{j=1}^{n}(u_i)^2_{x_j}$).

The proof of the lemma can be found in e.g. [126, p. 417]. If we have a family of f in (1.4-1) and a family of solutions u satisfying $|u(x)| \leq P$ in $\bar{\Omega}$, with the functions f being uniformly bounded in $\bar{\Omega}\times\{u: |u| \leq P\}$, then the family of solutions u will satisfy $|\nabla u(x)| \leq Q$ in $\bar{\Omega}$.

Theorem 1.4-1. Suppose that there exist constant vectors $\hat{\alpha} = (\alpha_1, \ldots, \alpha_m)$ and $\hat{\beta} = (\beta_1, \ldots, \beta_m)$, $\alpha_i < 0 < \beta_i$, $i = 1, \ldots, m$ such that, for each i:

(1.4-2)

$$f_i(x,u_1,\ldots,u_{i-1},\alpha_i,u_{i+1},\ldots,u_m) \leq 0 \leq f_i(x,u_1,\ldots,u_{i-1},\beta_i,u_{i+1},\ldots,u_m)$$

for all $x \in \bar{\Omega}$, all u_j satisfying $\alpha_j \le u_j \le \beta_j$, $j \ne i$, and

(1.4-3) $$\alpha_i - \phi_i(x) \le 0 \le \beta_i - \phi_i(x)$$

for all $x \in \delta\Omega$. Then the boundary value problem (1.4-1) has a solution u with $u_i \in H^{2+\alpha}(\bar{\Omega})$ and $\alpha_i \le u_i(x) \le \beta_i$ in $\bar{\Omega}$, each $i = 1, \ldots, m$.

Proof. Let $\mathcal{J} = \{u \in C^1(\bar{\Omega}; R^m): \alpha_i < u_i(x) < \beta_i,\ i = 1, \ldots, m;\ |\nabla u(x)| < M + 1,\ x \in \bar{\Omega}\}$, where M is a constant to be determined later. $\mathcal{J}$ is a bounded open neighborhood of 0 in the Banach space $C^1(\bar{\Omega}; R^m)$. Let $q > 0$ be arbitrary. For any $u \in \bar{\mathcal{J}}$, there exists a unique $w(x) = (w_1(x), \ldots, w_m(x))$ satisfying

$$(Lw-qw)(x) = f(x,u(x)) - qu(x) \qquad \text{in } \Omega,$$

(1.4-4)

$$w(x) = \phi(x) \qquad \text{on } \delta\Omega.$$

(Note that $f_i(x,u(x)) - qu_i(x)$ is in $H^{\alpha}(\bar{\Omega})$, and by Theorem 1.3-3 one has $w_i \in H^{2+\alpha}(\bar{\Omega}) \subset C^1(\bar{\Omega}; R))$. Define $Tu = w$, and we can show that T is a compact and continuous map from $\bar{\mathcal{J}}$ into $C^1(\bar{\Omega}, R^m)$. Let $H(u,\lambda) = u - \lambda Tu$ for $u \in \bar{\mathcal{J}}$, $\lambda \in [0,1]$, then $H: \bar{\mathcal{J}} \times [0,1] \to C^1(\Omega; R^m)$ is of the form as shown in (1.3-11). Since $0 \in \mathcal{J}$, it is clear that $H(u,\lambda) \ne 0$ when $u \in \delta\mathcal{J}$ and $\lambda = 0$. For $\lambda = 1$, if $H(u,\lambda) = 0$ for some $u \in \delta\mathcal{J}$, then we already have a solution to (1.4-1) with the properties described in the theorem, with u being a fixed point of T. Hence, we assume here that $H(u,1) \ne 0$ for $u \in \delta\mathcal{J}$. It remains to show that $H(u,\lambda) \ne 0$ for $u \in \delta\mathcal{J}$, $0 < \lambda < 1$, before applying Theorem 1.3-7. Suppose not, then there is a $\tilde{\lambda} \in (0,1)$, $\tilde{u} \in \delta\mathcal{J}$ with $\alpha_i \le \tilde{u}_i(x) \le \beta_i$, $|\nabla\tilde{u}(x)| \le M + 1$ for all $x \in \bar{\Omega}$, and $\tilde{u}$ satisfies:

$$L\tilde{u} = \tilde{\lambda} f(x,\tilde{u}) + (1-\tilde{\lambda})q\tilde{u} \quad \text{in } \Omega,$$

(1.4-5)
$$\tilde{u} = \tilde{\lambda}\phi \qquad \text{on } \delta\Omega,$$

(because $\tilde{u} = \tilde{\lambda}T\tilde{u}$). Because of the form of the right hand side of (1.4-5) and α_i, β_i are fixed, we can choose by means of Lemma 1.4-1 M large enough so that $|\nabla\tilde{u}(x)| \le M < M + 1$, for all $x \in \bar{\Omega}$. Since $\tilde{u} \in \delta\mathcal{J}$, we must then have $\tilde{u}_j(x_0) = \alpha_j$ or $\tilde{u}_j(x_0) = \beta_j$, for some $x_0 \in \bar{\Omega}$, some j, $1 \le j \le m$. Suppose that $\tilde{u}_j(x_0) = \beta_j$ (the other case can be proved in the same way), and $x_0 \in \Omega$. Let $R(x) = \tilde{u}_j - \beta_j$ in $\bar{\Omega}$. Then R(x) has an interior maximum at x_0, where $R(x_0) = 0$, $\frac{\partial R}{\partial x_i}(x_0) = 0$ and $\sum_{i,j=1}^{n} a_{ij}(x_0) \frac{\partial^2 R}{\partial x_i \partial x_j}(x_0) \le 0$. Consequently $LR(x_0) \le 0$. On the other hand

(1.4-6)
$$\begin{aligned} LR(x_0) &= L\tilde{u}_j(x_0) - L\beta_j \\ &= \tilde{\lambda}f_j(x_0, \tilde{u}_1(x_0), \ldots, \tilde{u}_{j-1}(x_0), \beta_j, \tilde{u}_{j+1}(x_0), \ldots, \tilde{u}_m(x_0)) \\ &> 0 \end{aligned}$$

because of (1.4-2) and $q > 0$, $\beta_j > 0$, leading to a contradiction. Thus x_0 must be on $\delta\Omega$. At $x_0 \in \delta\Omega$, we have from (1.4-5)

(1.4-7)
$$\tilde{u}_j(x_0) = \tilde{\lambda}\phi_j(x_0) \quad ; \quad \text{and}$$
$$\tilde{u}(x_0) = \beta_j \ge \phi_j(x_0)$$

from (1.4-3). Thus (1.4-7) leads to $\tilde{\lambda}\phi_j(x_0) \ge \phi_j(x_0)$ and $\phi_j(x_0) \le 0$. However $\tilde{\phi}_j(x_0) = \dfrac{\tilde{u}_j(x_0)}{\tilde{\lambda}} = \beta_j/\tilde{\lambda} > 0$; this again leads to a contradiction.

We have thus shown that $H(u,\lambda) \ne 0$ for all $u \in \delta\mathcal{J}$, $\lambda \in [0,1]$. By Theorem 1.3-7, there is a $u \in \mathcal{J}$ satisfying $Tu = u$, and u is a solution of (1.4-1). The functions $f_i(x,u(x))$ as functions of x are in $H^{\alpha}(\bar{\Omega})$, consequently from Theorem 1.3-4 we have $u_i \in H^{2+\alpha}(\bar{\Omega})$, $i = 1, \ldots, m$.

Theorem 1.4-2. *Let* L, f *and* ϕ *be as described in the beginning of this section for Theorem* 1.4-1. *Suppose that there exist* $\alpha(x) = (\alpha_1(x), \ldots, \alpha_m(x))$ *and* $\beta(x) = (\beta_1(x), \ldots, \beta_m(x))$, $\alpha_i(x)$, $\beta_i(x)$ in $C^2(\bar{\Omega})$, $i = 1, \ldots, m$, *such that for each* i, $\alpha_i(x) < \beta_i(x)$ *in* $\bar{\Omega}$,

$$(1.4\text{-}8)\qquad \begin{aligned} 0 &\le L\alpha_i(x) - f_i(x,u_1,\ldots,u_{i-1},\alpha_i(x),u_{i+1},\ldots,u_m), \\ 0 &\ge L\beta_i(x) - f_i(x,u_1,\ldots,u_{i-1},\beta_i(x),u_{i+1},\ldots,u_m) \end{aligned}$$

for all $x \in \bar{\Omega}$, $\alpha_j(x) \le u_j \le \beta_j(x)$, $j \ne i$; *and*

$$(1.4\text{-}9)\qquad \alpha_i(x) \le \phi_i(x) \le \beta_i(x) \quad \text{on } \delta\Omega.$$

Then the boundary value problem (1.4-1) *has a solution* u *with* $u_i \in H^{2+\alpha}(\bar{\Omega})$ *and* $\alpha_i(x) \le u_i(x) \le \beta_i(x)$ *in* $\bar{\Omega}$, *each* $i = 1, \ldots, m$.

Proof. Let $\mathcal{J} = \{u \in C^1(\bar{\Omega};R^m)$: $\alpha_i(x) < u_i(x) < \beta_i(x)$, $i = 1, \ldots, m$; $|\nabla u(x)| < M + 1$, $x \in \bar{\Omega}\}$ where M is to be determined later. Let $r_i(x) = \frac{1}{2}[\alpha_i(x)+\beta_i(x)]$ and $k > 0$ be a large enough constant so that

$$(1.4\text{-}10)\qquad \begin{aligned} &-f_i(x,u_1,\ldots,u_{i-1},\beta_i(x),u_{i+1},\ldots,u_m) + Lr_i(x) + k[\beta_i(x)-r_i(x)] > 0, \\ &-f_i(x,u_1,\ldots,u_{i-1},\alpha_i(x),u_{i+1},\ldots,u_m) + Lr_i(x) + k[\alpha_i(x)-r_i(x)] < 0 \end{aligned}$$

for all $x \in \bar{\Omega}$, $\alpha_j(x) \le u_j \le \beta_j(x)$, $j \ne i$, each $i = 1, \ldots, m$. For each $u \in \bar{\mathcal{J}}$, there exists a unique $w(x)$ satisfying

$$(1.4\text{-}11)\qquad Lw - kw = f(x,u(x)) - ku(x) \text{ in } \Omega, \quad w = \phi(x) \text{ on } \delta\Omega.$$

Define $Tu = w - r$ and $H(u,\lambda) = u - \lambda Tu$ for $u \in \bar{\mathcal{J}}$, $\lambda \in [0,1]$, $r = (r_1(x),\ldots, r_m(x))$. Hence H is of the form (1.3-11) in Theorem 1.3-7 from $\bar{\mathcal{J}}x[0,1]$ into $C^1(\bar{\Omega};R^m)$. For large enough M, we have $r \in \mathcal{J}$, and clearly $H(u,0) \neq r$ for $u \in \delta\mathcal{J}$. For $\lambda = 1$, if $H(u,1) = r$, then we have a solution to (1.4-1). Hence, we assume that $H(u,1) \neq r$ for $u \in \delta\mathcal{J}$. We need only to show that $H(u,\lambda) \neq r$ for $u \in \delta\mathcal{J}$, $0 < \lambda < 1$. Suppose not, then there is $\tilde{\lambda} \in (0,1)$, $\tilde{u} \in \delta\mathcal{J}$ with $\alpha_i(x) \le \tilde{u}_i(x) \le \beta_i(x)$, $|\nabla\tilde{u}(x)| \le M + 1$ in $\bar{\Omega}$ and $\tilde{u}$ satisfies $\tilde{u} = \tilde{\lambda}\tilde{w} + (1-\tilde{\lambda})r$ where

$$L\tilde{w} - k\tilde{w} = f(x,\tilde{u}(x)) - k\tilde{u}(x) \quad \text{in } \Omega, \ \tilde{w} = \phi(x) \text{ on } \delta\Omega.$$

This implies that

$$\begin{aligned} &L\tilde{u} = \tilde{\lambda}f(x,\tilde{u}) + (1-\tilde{\lambda})k(\tilde{u}-r)+(1-\tilde{\lambda})Lr \quad &&\text{in } \Omega, \\ &\tilde{u} - r = \tilde{\lambda}(\phi-r) &&\text{on } \delta\Omega. \end{aligned} \tag{1.4-12}$$

From Lemma 1.4-1 and the comments following it, we must have $|\nabla\tilde{u}| < M + 1$ in $\bar{\Omega}$ for M chosen large enough. Thus there is a $x_0 \in \bar{\Omega}$ where $\tilde{u}_j(x_0) = \alpha_j(x_0)$ or $\tilde{u}_j(x_0) = \beta_j(x_0)$, for some j, $1 \le j \le m$. Suppose $\tilde{u}_j(x_0) = \beta(x_0)$ and $x_0 \in \Omega$. Let $R(x) = \tilde{u}_j(x) - \beta_j(x)$ in $\bar{\Omega}$. Then $LR(x_0) \le 0$, since $R(x_0)$ is a maximum. On the other hand

$$\begin{aligned} LR(x_0) &= L\tilde{u}_j(x_0) - L\beta_j(x_0) \\ &\ge \tilde{\lambda}f_j(x_0,\tilde{u}_1(x_0),\ldots,\tilde{u}_{j-1}(x_0),\beta_j(x_0), \tilde{u}_{j+1}(x_0),\ldots,\tilde{u}_m(x_0)) \\ &\quad + (1-\tilde{\lambda})k(\tilde{u}_j(x_0)-r_j(x_0)) + (1-\tilde{\lambda})Lr_j(x_0) \\ &\quad - f_j(x_0,\tilde{u}_1(x_0),\ldots,\tilde{u}_{j-1}(x_0),\beta_j(x_0),\tilde{u}_{j+1}(x_0),\ldots,\tilde{u}_m(x_0)) \\ &= (-f_j(x_0,\ldots,\beta_j(x_0),\ldots,\tilde{u}_m(x_0)) + Lr_j(x_0) + k[\beta_j(x_0)-r_j(x_0)]). \\ &\qquad (1-\tilde{\lambda}) > 0 \end{aligned} \tag{1.4-13}$$

by the choice of k in (1.4-10). Hence x_0 must be on $\delta\Omega$. From (1.4-12), $\tilde{\lambda}(\phi_j(x_0)-r_j(x_0)) = \tilde{u}_j(x_0) - r_j(x_0) = \beta_j(x_0) - r_j(x_0) \geq \phi_j(x_0) - r_j(x_0)$; this implies that $\phi_j(x_0) - r_j(x_0) \leq 0$. However, $\phi(x_0) - r_j(x_0) = \frac{1}{\tilde{\lambda}}[\beta_j(x_0)-r_j(x_0)] > 0$, which leads to a contradiction. This completes the proof of the theorem.

Remark 1.4-1. Theorem 1.4-2 is true under more general conditions concerning f, boundary conditions and smoothness conditions. For example f may depend on ∇u with certain growth limitations. For more details, see [220].

We finally introduce a theorem which is used repeatedly in the following chapters. It is a "sweeping" principle which helps to obtain improved bounds for solutions of elliptic equations.

Theorem 1.4-3. *Let L be an elliptic operator with smoothness of its coefficients on $\bar{\Omega}$ as described in the beginning of this section, and $\delta\Omega \in H^{2+\alpha}$, $0 < \alpha < 1$. Let $u \in C^2(\bar{\Omega})$ be a solution of the scalar equations:*

$$(1.4\text{-}14) \qquad Lu + h(x,u) = 0 \quad \text{in} \quad \Omega, \quad u = g \text{ on } \delta\Omega .$$

Suppose $w_\lambda = w(x,\lambda)$ is a family of upper solutions of (1.4-14) which is increasing in λ, $a \leq \lambda \leq b$. (That is, w is C^2 in x in $\bar{\Omega}$, w and all its first partial derivatives with respect to x_i are continuous in $\lambda \in [a,b]$ uniformly for $x \in \bar{\Omega}$; $w(x,\lambda_1) \leq w(x,\lambda_2)$ for all $x \in \bar{\Omega}$, $a \leq \lambda_1 < \lambda_2 \leq b$ and

$$(1.4\text{-}15) \qquad Lw_\lambda + h(x,w_\lambda) \leq 0 \text{ in } \Omega, \quad w_\lambda \geq g \text{ on } \delta\Omega$$

for each $\lambda \in [a,b]$). Let $u(x) \leq w(x,b)$ for $x \in \bar{\Omega}$, and $u(x) \not\equiv w(x,\lambda)$, $x \in \bar{\Omega}$, for any $\lambda \in [a,b]$. Then

(1.4-16) $\qquad u(x) < w(x,a)$ <u>for all</u> $x \in \Omega$,

<u>provided that</u> $\frac{\partial h}{\partial v}(x,v)$ <u>exists and satisfies</u> $|\frac{\partial h}{\partial v}| \le K < \infty$ <u>for all</u> $x \in \Omega$, $\min_{x\in\bar\Omega} w(x,a) \le v \le \max_{x\in\bar\Omega} w(x,b)$.

<u>Proof</u>. The function $h(x,v) + Kv$ is a nondecreasing function of v for $\min_{x\in\bar\Omega} w(x,a) \le v \le \max_{x\in\bar\Omega} w(x,b)$, each fixed $x \in \Omega$. The function $u - w_b$ satisfies

$$(1.4\text{-}17) \qquad \begin{aligned} &(L-K)(u-w_b) \ge (h(x,w_b)+Kw_b) - (h(x,u) + Ku) \ge 0, \quad x \in \Omega, \\ &u - w_b \le 0 \quad \text{for} \quad x \in \delta\Omega. \end{aligned}$$

If $u(x) - w(x,b) = 0$ for some $x \in \Omega$, then by Theorem 1.1-2 we have $u(x) \equiv w(x,b)$, for all $x \in \bar\Omega$, contradicting our assumption. Hence $u(x) < w(x,b)$ for all $x \in \Omega$. Let

$$\mathcal{S} = \{a \le s \le b: \;\; u(x) < w(x,\lambda) \text{ for all } s \le \lambda \le b, \; x \in \Omega\}.$$

Suppose that the g.l.b. of $\mathcal{S}$ is $\bar\lambda > a$. We now show that there must be a point $\bar x \in \Omega$, where $u(\bar x) = w(\bar x,\bar\lambda)$. Suppose not, we deduce $(L-K)(u-w_{\bar\lambda}) \ge 0$ in Ω, $u - w_{\bar\lambda} \le 0$ for $x \in \delta\Omega$ as in (1.4-17); then, using $u - w_{\bar\lambda} \le 0$ in $\bar\Omega$ and Theorem 1.1-4 we find that $\frac{\partial}{\partial\eta}(u-w_{\bar\lambda}) > 0$ at those points at the boundary where $u = w_{\bar\lambda}$ (if such points exist). Consequently, for sufficiently small $\varepsilon > 0$, we have $u(x) < w(x,\bar\lambda-\varepsilon)$ for all $x \in \Omega$. This violates the definition of $\bar\lambda$, and we conclude that there must be a point $\bar x \in \Omega$ where $u(\bar x) = w(\bar x,\bar\lambda)$. Again, we use $(L-K)(u-w_{\bar\lambda}) \ge 0$ in Ω and $u - w_{\bar\lambda}$ attains a nonnegative maximum value at $\bar x \in \bar\Omega$ to conclude that $u(x) \equiv w(x,\bar\lambda)$ in $\bar\Omega$ by using Theorem 1.1-2. However this contradicts the hypothesis $u \not\equiv w_{\bar\lambda}$. Consequently, the g.l.b. of $\mathcal{S}$ must be the number a. Moreover, as before, in order not to contradicts $u \not\equiv w_a$ in $\bar\Omega$, we must have (1.4-16) valid for $x \in \Omega$.

Remark 1.4-2. In the last theorem, suppose that the inequality (1.4-15) is strict in Ω for each $\lambda \in [a,b]$, then there is no need for the hypothesis that $u(x) \not\equiv w(x,\lambda)$, $x \in \bar{\Omega}$ for any λ in $[a,b]$. The theorem remains valid because one would have equality in (1.4-15) for $w_{\hat{\lambda}}$ if $u \equiv w_{\hat{\lambda}}$ in $\bar{\Omega}$.

Remark 1.4-3. A more general version of the sweeping principle is also valid for systems. It is used in Theorem 4.2-1 of Chapter 4, and the proof is included in there.

Remark 1.4-4. In Theorem 1.4-3, if the increasing family of upper solution w_λ is replaced by an increasing family of lower solution v_λ. Then if $u(x) \geq v(x,a)$ for $x \in \bar{\Omega}$. $u \not\equiv v_\lambda$ in $\bar{\Omega}$ for any $\lambda \in [a,b]$, we can conclude $u(x) > v(x,b)$ for all $x \in \Omega$ provided that the other hypotheses hold.

Here, we say a function $w \in C^2(\bar{\Omega})$ is an upper solution of (1.4-14) if

$$Lw + h(x,w) \leq 0 \text{ in } \Omega, \text{ and } w \geq g \text{ on } \partial\Omega.$$

If both inequalities above are reversed, we say it is a lower solution. For more general theories, the reader is referred to Chapter 5.

Remark 1.4-5. In most of the theories in the following chapters, the uniform elliptic operators in Ω are restricted to the Laplacian operator, $\Delta \equiv \sum_{i=1}^{n} \partial^2/\partial x_i^2$, for convenience and simplicity.

Notes

The maximum principles for elliptic equations, Theorems 1.1 and 1.2, in which no assumption on the continuity of the coefficients is made, are obtained by E. Hopf [107]. Theorems 1.3 and 1.4 concerning the outward directional derivatives at the boundary for solutions of elliptic equations are due to E. Hopf [108] and Oleĭnik [174]. For parabolic equations, the maximum principles, Theorem 1.1-5 and 1.1-6, are due to Nirenberg [170]. The version concerning outward directional derivatives at the boundary, Theorem 1.1-7, is the result of Friedman [81]. For more references concerning these principles, the reader is referred to the books by Protter and Weinberger [188] and Walter [222]. The comparison results of Theorems 1.2-1 and 1.2-2 can be found in Nagumo and Simoda [165] and Westphal [228]. The various comparsion Theorems 1.2-3 to 1.2-7 for parabolic equations and systems are adapted from Fife and Tang [75]. Many similar results are collected in the books by Szarski [214], Walter [222], and Lakshmikantham and Leela [219]. Theorems 1.3-1 and 1.3-2 concerning classical linear theory for parabolic equations are found in Ladyženskaja, Solonnikov and Ural'ceva [127]. Similar theorems developed by another method (using potential theory) are described in Friedman [79], [80]. For classical linear elliptic theory, Theorems 1.3-3 and 1.3-4 are due to Schauder [203], [204] and Miranda [161] respectively. The fixed point Theorems 1.3-5 and 1.3-6 are the results of Schauder [202]. The fixed point Theorems 1.3-7 and 1.3-8 involving homotopic invariance arguments are due to Leray and Schauder [133]. The existence theorems for nonlinear elliptic systems, Theorems 1.4-1 and 1.4-2, are obtained by Tsai [220]; the method of proof is analogous to that for the existence of periodic solutions for systems of ordinary differential equations by Bebernes and Schmitt [20]. The sweeping principle, Theorem 1.4-3, is due to Serrin, as described in the book of Sattinger [201].

CHAPTER II

Interacting Population Reaction-Diffusion Systems, Dirichlet Conditions

2.1 Introduction

We will use the techniques described in the last chapter to study reaction-diffusion systems related to ecology. We consider steady states and stabilities for prey-predator and competing-species systems. In this chapter, we are primarily concerned with the case when values for the species are prescribed on the boundary (i.e., Dirichlet boundary conditions). In the next chapter, more elaborate problems and other boundary conditions are treated, together with certain asymptotic approximations. The special case of zero-flux boundary condition (i.e. homogeneous Neumann condition) is studied in Chapter 7. Numerical approximations and calculations by finite difference is presented in Chapter 6.

In this chapter, Ω is always assumed to be a bounded domain in R^n, with its boundary $\delta\Omega \in H^{2+\alpha}$, $0 < \alpha < 1$ (unless otherwise stated). We first study an existence theorem for an initial-boundary value problem describing interacting populations. A bound for the growth rates for the solutions is also contained in the proof. We consider the problem:

$$\frac{\partial u_i}{\partial t} = \sigma_i \Delta u_i + u_i R_i(u_1, u_2, \ldots, u_m) \quad \text{for } (x,t) \in \Omega \times (0,T],$$

$$(2.1\text{-}1) \qquad u_i(x,0) = \theta_i(x), \text{ for } x \in \bar{\Omega}$$

$$u_i(x,t) = g_i(x,t)\ , \text{ for } (x,t) \in \delta\Omega \times [0,T],$$

$i = 1, \ldots, m$. Here, we interpret $u_i(x,t)$ as concentrations of ecological species at position x and time t. The paremeters σ_i are positive constants describing diffusion rates. The functions θ_i and g_i are prescribed initial and boundary data. The functions R_i describe the growth rate of the ith species, and they are dependent on the concentrations of other species. We assume that $R_i: R^m \to R$ have continuous partial derivations up to second order, $1 \le i \le m$; and they satisfy the following food pyramid condition:

(2.1-2) For every $K > 0$ and $1 \le i \le m$, there exists a $r_i(K) > 0$ such that: if $u_1 \ge 0, u_2 \ge 0, \ldots, u_m \ge 0$ and $u_1 \le K, u_2 \le K, \ldots, u_{i-1} \le K$ then $R_i(u_1, \ldots, u_m) < r_i(K)$.

Such food pyramid condition is satisfied, for example, if each ith species is a prey for all the jth species with $j > i$, and overpopulation of each species limits its own growth rate. It includes the prey-predator and competing species interactions we will consider in detail in later sections. For convenience, we define $\Omega_T \equiv \Omega \times (0,T)$, with closure $\bar{\Omega}_T$, $\delta\Omega_T \equiv \delta\Omega \times (0,T)$ and $\delta\bar{\Omega}_T = \delta\Omega \times [0,T]$, where T is any positive number, $0 < \alpha < 1$.

Theorem 2.1-1. Let $\theta_i \in H^{2+\alpha}(\bar{\Omega})$, $\theta_i \ge 0$ in $\bar{\Omega}$ and $g_i \in H^{2+\alpha,1+\alpha/2}(\delta\bar{\Omega}_T)$, $g_i \ge 0$ in $\delta\bar{\Omega}_T$ be prescribed functions satisfying compatibility conditions: $\theta_i(x) = g_i(x,0)$ for all $x \in \delta\Omega$, $\frac{\partial g_i}{\partial t}(x,0) = \sigma_i \Delta\theta_i(x) + \theta_i R_i(\theta_i, \ldots, \theta_m)$ for all $x \in \delta\Omega$, $i = 1, \ldots, m$; and g_i, θ_i are restrictions of some $\psi_i(x,t) \in H^{2+\alpha,1+\alpha/2}(\bar{\Omega}_T)$, with $\psi_i(x,0) = \theta_i(x)$ in $\bar{\Omega}$ and $\psi_i(x,t) = g_i(x,t)$ in $\delta\Omega \times [0,T]$, where $\frac{\partial}{\partial t}(\frac{\partial\psi_i}{\partial t} - \sigma_i\Delta\psi_i)$ is continuous in $\bar{\Omega}_T$. Suppose that the functions R_i satisfy the smoothness and food pyramid condition (2.1-2) above. Then the initial-boundary value problem (2.1-1) has a solution $(u_1(x,t), \ldots, u_m(x,t))$ with $u_i \in H^{2+\alpha,1+\alpha/2}(\bar{\Omega}_T)$, $u_i(x,t) \ge 0$ in $\bar{\Omega}_T$, $i = 1, \ldots, m$. Furthermore, if no θ_i is identically zero, then each $u_i(x,t)$ is strictly positive for $(x,t) \in \Omega \times (0,T]$.

Proof. For $i = 1, \ldots, m$, let $a_i > \max\{|\psi_i|: (x,t) \in \bar{\Omega}_T\}$. We now define numbers $r_1, \ldots, r_m$ and $d_1, \ldots, d_m$ by induction. Let $r_1 > 0$, $d_1 > 0$ be constants such that $R_1(u_1, \ldots, u_m) < r_1$ for all $u_1 \ge 0, \ldots, u_m \ge 0$ (such r_1 exists by conditions (2.1-2)), and $d_1 > a_1^{-1}[r_1 \max_{\bar{\Omega}_T} |\psi_1| + \max_{\bar{\Omega}_T} |\frac{\partial\psi_1}{\partial t} - \sigma_1\Delta\psi_1|]$. Suppose that $r_1 > 0, \ldots, r_{i-1} > 0$ and $d_1 > 0, \ldots, d_{i-1} > 0$ had been chosen. Let

$K_i = \max\{a_j e^{(r_j+d_j)T} : 1 \leq j \leq i\} + \max\{\max_{\overline{\Omega}_T} |\psi_j| : 1 \leq j \leq i\}$ and define r_i to be the constant so that $u_1 \geq 0, \ldots, u_m \geq 0$ and, $u_1 \leq K_i, \ldots, u_{i-1} \leq K_i$ imply $R_i(u_1, \ldots, u_m) \leq r_i$. Also, set $d_i > a_i^{-1}[r_i \max_{\overline{\Omega}_T} |\psi_i| + \max_{\overline{\Omega}_T} |\frac{\partial \psi_i}{\partial t} - \sigma_i \Delta \psi_i|]$.

We define a Banach space B_δ, $0 < \delta < 1$, of functions $w(x,t) = (w_1(x,t), \ldots, w_m(x,t))$ whose components are continuous together with their derivatives with respect to x_i in $\overline{\Omega}_T$, $w_i(x,t) = 0$ for $(x,t) \in (\overline{\Omega} \times \{0\}) \cup (\partial\Omega \times [0,T])$, $i = 1, \ldots, m$, and have finite value for the norm

$$|w|_{B_\delta} = \sum_{i=1}^{m} |w_i|_{\Omega_T}^{(\delta)} + \sum_{i=1}^{m} \sum_{j=1}^{n} |\frac{\partial w_i}{\partial x_j}|_{\Omega_T}^{(\delta)}$$

(Here $| \ |_{\Omega_T}^{(\delta)}$ are Hölder norms as defined in section 1.3). For each $w \in B_\delta$, $\lambda \in [0,1]$, define $v = \Phi(w,\lambda)$ where $v = (v_1, \ldots, v_m)$ is the solution to the linear problem:

$$\frac{\partial v_i}{\partial t} = \sigma_i \Delta v_i + \lambda[w_i + \psi_i] R_i(w_1 + \psi_1, \ldots, w_m + \psi_m) - \lambda[\frac{\partial \psi_i}{\partial t} - \sigma_i \Delta \psi_i]$$

(2.1-3)

$$v_i(x,t) = 0 \qquad \text{for } (x,t) \in (\overline{\Omega}\times\{0\}) \cup (\delta\Omega\times[0,T]).$$

for $i = 1, \ldots, m$. By Theorem 1.3-1, v is uniquely defined in $H^{2+\delta,1+\delta/2}(\overline{\Omega}_T)$; moreover, if $|w|_{B_\delta} \leq C$, for some C, then $v = \Phi(w,\lambda)$, $0 \leq \lambda \leq 1$ will have uniform bound in $| \ |_{\Omega_T}^{(2+\delta)}$. Hence, by means of Ascoli's lemma, for fixed $\lambda \in [0,1]$, the mapping $\Phi: B_\delta \to B_\delta$ is compact, i.e. it maps bounded sequences in B_δ into sequences which contain convergent subsequence in B_δ. We now proceed to verify the remaining hypotheses for application of Leray-Schauder's fixed point theorem 1.3-8 in order to obtain a solution to our problem corresponding to $\lambda = 1$, $(u_1, \ldots, u_m) = \Phi(w,1) + \psi$ at a fixed point for Φ. Let $\hat{w}$, $\tilde{w}$ be in B_δ, $\lambda \in [0,1]$, then $z = \Phi(\hat{w},\lambda) - \Phi(\tilde{w},\lambda)$ satisfies

(2.1-4)

$$\frac{\partial z_i}{\partial t} - \sigma_i \Delta z_i = \lambda[\hat{w}_i+\psi_i]\, R_i(\hat{w}_1+\psi_1, \ldots, \hat{w}_m+\psi_m) - \lambda[\tilde{w}_i+\psi_i]\, R_i(\tilde{w}_1+\psi_1, \ldots, \tilde{w}_m+\psi_m) \quad \text{in } \Omega\times(0,T],$$

$$z_i(x,t) = 0 \quad \text{for } (x,t) \in (\bar{\Omega}\times\{0\}) \cup (\delta\Omega\times[0,T]),$$

for $i = 1, \ldots, m$. From the smoothness of R_i, one can see that if $|\hat{w}-\tilde{w}|_{B_\delta}$ is small, the right side of (2.1-4) is small as a function of (x,t) in the norm of $H^{\delta,\delta/2}(\bar{\Omega}_T)$. Hence, from Theorem 1.3-1, the norms, $|z_i|_{\Omega_T}^{(2+\delta)}$ are small. Consequently, $|z|_{B_\delta}$ is small, and one concludes that $\Phi(w,\lambda)$ is continuous in w for fixed λ. One can also verify analogously that for w in a bounded set in B_δ, $\Phi(w,\lambda)$ is uniformly continuous in λ for $\lambda \in [0,1]$. When $\lambda = 0$, equations (2.1-3) shows that $\Phi(w,0) = 0 = (0, \ldots, 0)$; therefore the equation $w - \Phi(w,0) = 0$ has the unique trivial solution in B_δ. To apply Leray-Schauder's theorem, it remains to show that there exists a constant M^* so that every possible solution $w = w^\lambda$ of $w - \Phi(w,\lambda) = 0$, $w \in B_\delta$, $\lambda \in [0,1]$ must satisfy $|w^\lambda|_{B_\delta} < M^*$.

Without loss of generality, we may assume that r_i in the first paragraph has the further property that: if $|u_1| < K_i, \ldots, |u_{i-1}| < K_i$, then $R_i(u_1, \ldots, u_m) < r_i$ for all $(u_1, \ldots, u_m)$ in R^m. This is permissible because we will eventually show that the solution to the initial value problem (2.1-1) with nonnegative initial boundary conditions will have $u_i(x,t) > 0$ in $\bar{\Omega}_T$, $i = 1, \ldots, m$. Consequently, modifying the equation by changing $R_i(u_i, \ldots, u_m)$ when some $u_j < 0$ does not really change the result of the theorem for the original equation concerning nonnegative solutions.

For convenience, let $L_i \equiv \sigma_i\Delta - \frac{\partial}{\partial t}$, $u_i = w_i + \psi_i$. The function $\tilde{\phi}_i \overset{(def)}{=} a_i e^{(r_i+d_i)t}$ on $\bar{\Omega}_T$ satisfies for $\lambda \in [0,1]$, $i = 1, \ldots, m$:

(2.1-5)

$$L_i\tilde{\phi}_i + \lambda[\tilde{\phi}_i+\psi_i]\, R_i(w_1+\psi_1, \ldots, w_{i-1}+\psi_{i-1}, \tilde{\phi}_i+\psi_i, w_{i+1} + \psi_{i+1}, \ldots, u_m) - \lambda\left[\frac{\partial\psi_i}{\partial t} - \sigma_i\Delta\psi_i\right]$$

$$\leq -a_i(r_i+d_i)e^{(r_i+d_i)t} + [a_i e^{(r_i+d_i)t} + \max_{\overline{\Omega}_T} |\psi_i|]\, r_i + \max_{\overline{\Omega}_T} |\frac{\partial \psi_i}{\partial t} - \sigma_i \Delta \psi_i|$$

$$\leq -a_i d_i + r_i \max_{\overline{\Omega}_T} |\psi_i| + \max_{\overline{\Omega}_T} |\frac{\partial \psi_i}{\partial t} - \sigma_i \Delta \psi_i| < 0 \text{ in } \Omega \times (0,T],$$

provided $|u_1| \leq K_1, \ldots, |u_{i-1}| \leq K_i$. Such condition will be satisfied if $|w_1| \leq \tilde{\phi}_1, \ldots |w_{i-1}| \leq \tilde{\phi}_{i-1}$ for $(x,t) \in \overline{\Omega}_T$. Note that $\tilde{\phi}_i + \psi_i > 0$ in $\overline{\Omega}_T$ by the choice of a_i. Moreover, we have $0 \leq \phi_i$, for $(x,t) \in (\overline{\Omega}\times\{0\}) \bigcup (\delta\Omega\times[0,T])$. On the other hand, define $\underset{\sim}{\phi}_i = -a_i e^{(r_i+d_i)t} = -\tilde{\phi}_i$ on $\overline{\Omega}_T$, $i = 1, \ldots, m$; they satisfy for $\lambda \in [0,1]$:

$$L_i \underset{\sim}{\phi}_i + \lambda[\underset{\sim}{\phi}_i + \psi_i]\, R_i(w_1+\psi_1, \ldots, w_{i-1}+\psi_{i-1}, \underset{\sim}{\phi}_i+\psi_i, w_{i+1}+\psi_{i+1}, \ldots, u_m)$$
$$-\lambda[\frac{\partial \psi_i}{\partial t} - \sigma_i \Delta \psi_i]$$

$$\text{(2.1-6)} \qquad \geq a_i(r_i+d_i)e^{(r_i+d_i)t} + [-a_i e^{(r_i+d_i)t} - \max_{\overline{\Omega}_T} |\psi_i|]\, r_i$$
$$- \max_{\overline{\Omega}_T} |\frac{\partial \psi_i}{\partial t} - \sigma_i \Delta \psi_i| > 0 \quad \text{in } \Omega \times (0,T].$$

provided $\underset{\sim}{\phi}_1 \leq w_1 \leq \tilde{\phi}_1, \ldots, \underset{\sim}{\phi}_{i-1} \leq w_{i-1} \leq \tilde{\phi}_{i-1}$ for $(x,t) \in \overline{\Omega}_T$. Also, we have $\underset{\sim}{\phi}_i \leq 0$, for $(x,t) \in (\overline{\Omega\times}\{0\}) \bigcup (\delta\Omega\times[0,T])$, $i = 1, \ldots, m$. Hence, by Theorem 1.2-6, formulas (2.1-3), (2.1-5), and (2.1-6), we conclude that any solution $w = w^\lambda = (w_1^\lambda, \ldots, w_m^\lambda)$ of $w - \Phi(w,\lambda) = 0$, $\lambda = [0,1]$, $w \in B_\delta$, must satisfy $\underset{\sim}{\phi}_i \leq w_i^\lambda \leq \tilde{\phi}_i$ in $\overline{\Omega}_T$, $i = 1, \ldots, m$. We have thus obtained an a-priori uniform bound for solutions of

$$\frac{\partial w_i^\lambda}{\partial t} = \sigma_i \Delta w_i^\lambda + \lambda[w_i^\lambda+\psi_i]\, R_i(w_1^\lambda+\psi_1, \ldots, w_m^\lambda+\psi_m) - \lambda[\frac{\partial \psi_i}{\partial t} - \sigma_i \Delta \psi_i] \text{ in } \Omega \times (0,T]$$

(2.1-7)

$$w_i^\lambda(x,t) = 0 \text{ for } (x,t) \in (\overline{\Omega}\times\{0\}) \bigcup (\delta\Omega\times[0,T])$$

$i = 1, \ldots, m$, $\lambda \in [0,1]$. By Theorem A1.1 in the appendix, the quantities $|\frac{\partial w_i^\lambda}{\partial x_j}|$ are uniformly bounded in $\overline{\Omega}_T$, $\lambda \in [0,1]$, by a constant depending on the a-priori bounds on $|w_k^\lambda|$ in $\bar{\Omega}_T$ and the constant $c_\lambda \overset{(def)}{=} \{\max(\sum_{j=1}^{n} [\frac{\partial w_k^\lambda}{\partial x_j}(x,t)]^2)^{1/2}: (x,t) \in (\delta\Omega x[0,T]) \bigcup (\bar{\Omega}x\{0\}),\ k = 1, \ldots, m\}$. Moreover, from remark A1.1, since the boundary condition is Dirichlet the quantity $c_\lambda, \lambda \in [0,1]$ can be bounded by a constant depending on the a-priori bounds on $|w_k^\lambda|$ in $\bar{\Omega}_T$ too. We have thus obtained uniform bound for $|\frac{\partial w_i^\lambda}{\partial x_j}|$ in $\bar{\Omega}_T$, $\lambda \in [0,1]$ $i = 1, \ldots, m$, $j = 1, \ldots, n$. By Theorem A1.2 in the appendix, as long as we have uniformly bounds for $|w_i^\lambda|$ and $|\frac{\partial w_i^\lambda}{\partial x_j}|$ in $\bar{\Omega}_T$, we can obtain uniform bound for $|\frac{\partial w_i^\lambda}{\partial t}|$ in Ω_T, $i = 1, \ldots, m$, $\lambda \in [0,1]$; and also this will lead to uniform bound for $|\frac{\partial w_i^\lambda}{\partial x_j}|_{\Omega_T}^{(\sigma)}$ for some $0 < \sigma <$ for all $\lambda \in [0,1]$.

Referring to equation (2.1-7), the term $P_i(w_1^\lambda(x,t), \ldots, w_m^\lambda(x,t))$ will therefore, as a function of (x,t), possess uniform bound for the norm $|\ \ |_{\Omega_T}^{(\sigma)}$, each $\lambda \in [0,1]$. Thus from the linear theory in Theorem 1.3-1 and equation (2.1-7), we obtain a uniform bound for the norm $|w^\lambda|_{\Omega_T}^{(2+\tilde{\alpha})}$ where $\tilde{\alpha} = \min\{\sigma,\alpha\}$. Consequently, we obtain uniform bound for $|\frac{\partial w_i^\lambda}{\partial x_j}|_{\Omega_T}^{(\delta)}$, all $\lambda \in [0,1]$, where $\delta \in (0,1)$ was chosen. We thus conclude that every possible solution $w = w^\lambda$ of (2.1-7) must satisfy: $|w^\lambda|_{B_\delta} < M^*$ for some large M^*. From Leray-Schauder's Theorem 1.3-8, the problem (2.1-7) has a solution for $\lambda = 1$, $w = (w_1^1(x,t), \ldots, w_m^1(x,t))$.

For each i, $u_i = w_i^1 + \psi_i$ satisfies the scalar problem:

$$
(2.1\text{-}8) \quad \begin{cases} \dfrac{\partial u_i}{\partial t} = \sigma_i \Delta u_i + u_i R_i(w_1^1+\psi_1, \ldots, w_m^1+\psi_m) \text{ in } \Omega x(0,T], \\ u_i(x,t) = \psi_i(x,t) \text{ on } (\bar{\Omega}x\{0\}) \bigcup (\delta\Omega x[0,T]). \end{cases}
$$

On the other hand $u \equiv 0$ satisfies (2.1-8) with the second "=" replaced by <. By the comparison theorem 1.2-3 for scalar equations, we conclude that $0 < w_i^1(x,t) + \psi_i(x,t)$ for $(x,t) \in \bar{\Omega}_T$. Consequently, the function $(z_1^1(x,t), \ldots, z_m^1(x,t))$, where $z_i^1 \overset{(def)}{=} w_i^1 + \psi_i$, $i = 1, \ldots, m$ is an actual solution of the original problem (2.1-1), without any modification of R_i as described before. To be consistent with the statement of the theorem, we now denote $(z_1^1(x,t), \ldots, z_m^1(x,t)) = (u_1(x,t), \ldots, u_m(x,t))$, $(x,t) \in \bar{\Omega}_T$.

From previous arguments, we have $u_i(x,t) = z_i^1(x,t)$ are in $H^{2+\alpha,1+\alpha/2}(\bar{\Omega}_T)$. Thus the function $R_i(u_1(x,t), \ldots, u_m(x,t))$ on the right on (2.1-1), as a function of (x,t), is in $H^{\alpha,\alpha/2}$. From linear theory given by Theorem 1.3-1, we have $u_i(x,t) \in H^{2+\alpha,1+\alpha/2}(\bar{\Omega}_T)$ since it satisfies (2.1-1), treating R_i as coefficient. Finally, we now show that the solution components satisfy $u_i(x,t) > 0$ in $\Omega \times (0,T]$ if no θ_i is identically zero. Let $Q_i(x,t) = -u_i(x,t)e^{-ht}$ for $(x,t) \in \bar{\Omega}_T$, where

$h = \max\{|R_i(y_1, \ldots, y_m)|: 1 \leq i \leq m,\ 0 \leq y_k \leq a_k e^{(r_k+d_k)T} + \max_{\bar{\Omega}_T}|\psi_k|$, for $1 \leq k \leq m\}$. The functions Q_i satisfies

$$\sigma_i \Delta Q_i - \frac{\partial Q_i}{\partial t} + Q_i[R_i(u_1, \ldots, u_m) - h] = 0 \quad \text{in } \Omega \times (0,T]$$

where $R_i(u_1, \ldots, u_m) - h \leq 0$ in $\Omega \times (0,T]$, (since we know $0 \leq u_i(x,t) = z_i^1(x,t) \leq a_i e^{(r_i+d_i)t} + \max_{\Omega_T}|\psi_i|$ on $\bar{\Omega}_T$). Suppose that $M_i \overset{(def)}{\equiv} \max_{\bar{\Omega}_T} Q_i(x,t) = 0$, and is attained at some point in $\Omega \times (0,T]$. By the maximum principle, Theorem 1.1-8, we conclude that $u_i(x,t) \equiv M_i = 0$, and hence $\theta_i(x) \equiv 0$. This completes the proof of Theorem 2.1-1.

Remarks: The solution as described in Theorem 2.1-1 is unique, and further discussions can be found in [229]. The theorem can be readily extended to include the case when the Laplacian Δ is replaced by uniformly elliptic operators in Ω. Theorem 2.1-1 is not absolutely necessary for the development of the remaining part of this chapter or book. Behavior of solutions as t changes can be studied if one presumes solution of (2.1-1) exist.

We conclude this section with a simple lemma which will be used in the next few sections.

Lemma 2.1-1. Let σ and c be positive constants. The boundary value problem: $\sigma\Delta w - cw = 0$ in Ω, $w = g(x) \geq 0, \not\equiv 0$ in $\partial\Omega$, where $g(x)$ is continuous, has a unique $C^2(\Omega)\cap C(\bar{\Omega})$ solution which is strictly positive in Ω.

Proof. Existence and uniqueness of classical $C^2(\Omega)\cap C(\bar{\Omega})$ solution for the scalar linear problem are standard results (also cf section 1.3). Let $u = -w$ in $\bar{\Omega}$. Suppose that u has a nonnegative maximum M at a point in Ω. By the maximum principle, Theorem 1.1-2, we have $u(x) \equiv M$, i.e., $w(x) \equiv -M \leq 0$ in $\bar{\Omega}$. This contradicts the fact that $w \geq 0, \not\equiv 0$ on $\partial\Omega$. Consequently, we must have the $u(x) < 0$ for all $x \in \Omega$. That is $w(x) > 0$ for $x \in \Omega$.

In all the sections in this chapter, $\lambda_1 > 0$ will denote the first (principal) eigenvalue for the Dirichlet problem $\Delta u + \lambda u = 0$ in Ω, $u = 0$ on $\delta\Omega$. The function $\omega(x)$ will denote the corresponding principal eigenfunction. (Note: $\omega(x)$ has the properties that $\omega(x) > 0$ in Ω, $\frac{\partial\omega}{\partial\eta} < 0$ on $\partial\Omega$ where η is the unit outward normal at $\partial\Omega$, as indicated in section 1.3).

2.2 Prey-Predator with Dirichlet Boundary Condition

We consider special cases of interacting population reaction-diffusion problem (2.1-1), when there are two species interacting with prey-predator relationship. We will obtain various types of steady states when there are coexistence or extinction of certain species. Large-time asymptotic behavior of the system is also discussed. Specifically, we discuss the system:

$$\begin{aligned} \frac{\partial u_1}{\partial t} &= \sigma_1\Delta u_1 + u_1[a + f_1(u_1,u_2)] \\ \frac{\partial u_2}{\partial t} &= \sigma_2\Delta u_2 + u_2[-r + f_2(u_1,u_2)] \end{aligned} \quad \text{for } x \in \Omega,\ t > 0 \tag{2.2-1}$$

where $u_i(x,t)$, $i = 1,2$ represent the concentration of prey and predator respectively at position x and time t. The parameters a, r, σ_1, σ_2 are positive constants with a and r representing growth and mortality rates when no interaction occurs, σ_1 and σ_2 representing diffusion rates. The functions $f_i: R^2 \to R$ have Hölder continuous partial derivatives up to second order in compact sets, $i = 1,2$. Further, we assume that

$$f_1(0,0) = f_2(0,0) = 0. \tag{2.2-2}$$

For (u_1,u_2) in the first closed quadrant, the first partial derivatives of f_1, f_2 satisfy:

$$\frac{\partial f_1}{\partial u_1} < 0,\ \frac{\partial f_1}{\partial u_2} < 0,\ \frac{\partial f_2}{\partial u_1} > 0,\ \frac{\partial f_2}{\partial u_2} < 0; \quad \text{and} \tag{2.2-3}$$

(2.2-4) for each $k > 0$, $-r + f_2(k,u_2) < 0$ for all sufficiently large $u_2 > 0$.

The inequalities (2.2-3) reflect the prey-predator relationship; and (2.2-4) describes the situation that for any fixed supply of prey, excessive predator crowding will cause its own population decrease.

We will see that the behavior of the solutions differ drastically according to $a < \sigma_1\lambda_1$ or $a > \sigma_1\lambda_1$. A steady state solution of (2.2-1) means a solution of

$$\left.\begin{aligned} &\sigma_1\Delta u_1 + u_1[a + f_1(u_1,u_2)] = 0 \\ &\sigma_2\Delta u_2 + u_2[-r + f_2(u_1,u_2)] = 0 \end{aligned}\right\} \quad \text{in } \Omega \tag{2.2-5}$$

Theorem 2.2-1 to Theorem 2.2-4 discuss the situation when the prey population is kept at zero at the boundary. Theorem 2.2-1 to Theorem 2.2-3 shows that with such boundary conditions, the prey become extinct if $a < \sigma_1\lambda_1$.

Theorem 2.2-1. *Suppose that* $a < \sigma_1\lambda_1$. *Let* $\bar{u}_i(x) \geq 0$, $i = 1, 2$, *in* $\bar{\Omega}$ *be functions in* $H^{2+\alpha}(\bar{\Omega})$, *with* $(\bar{u}_1(x),\bar{u}_2(x))$ *satisfying the steady state equations* (2.2-5). *If* $\bar{u}_1(x)|_{\partial\Omega} = 0$, *then* $\bar{u}_1(x) \equiv 0$ *for all* $x \in \Omega$.

<u>Proof</u>. For each $k > 0$, we have the inequality:

$$\begin{aligned}(2.2\text{-}6)\quad & \sigma_1 \Delta k\omega(x) + k\omega[a + f_1(k\omega,\bar{u}_2)] = k\omega[(a-\sigma_1\lambda_1) + f_1(k\omega,\bar{u}_2)] < 0 \text{ in } \Omega \\ & k\omega(x) > 0 \quad \text{for} \quad x \in \partial\Omega.\end{aligned}$$

That is, $u = k\omega$ is an upper solution (cf. section 1.4 and 5.1) for the boundary value problem:

$$\sigma_1 \Delta u + u[a + f_1(u,\bar{u}_2(x))] = 0 \text{ in } \Omega,\ u = 0 \quad \text{on } \partial\Omega.$$

Since $\frac{\partial\omega}{\partial\eta} < 0$ on $\partial\Omega$, for $\tilde{k} > 0$ sufficiently large, we have $\bar{u}_1(x) < \tilde{k}\omega(x)$ for $x \in \Omega$. The family of upper solutions $k\omega(x)$, $0 < k < \tilde{k}$ (with strict inequality in Ω in (2.2-6)) and the sweeping principle in Theorem 1.4-3, imply that $\bar{u}_1(x) < k\omega(x)$ in $\bar{\Omega}$ at $k = 0$. The nonnegative assumption on $\bar{u}_1(x)$ consequently implies that $\bar{u}_1(x) \equiv 0$ in $\bar{\Omega}$.

<u>Theorem</u> 2.2-2. <u>The boundary value problem</u> (2.2-5) <u>with boundary conditions</u>

$$(2.2\text{-}6)\qquad u_1 = 0\ ,\ u_2 = g(x)\quad ,\ \text{for}\quad x \in \partial\Omega,$$

<u>where</u> $g(x) > 0$, $\not\equiv 0$ <u>on</u> $\partial\Omega$ <u>and</u> g <u>has an extension</u> $\hat{g} \in H^{2+\alpha}(\bar{\Omega})$, <u>has a solution of the form</u> $(u_1(x),u_2(x)) = (0,\tilde{u}_2(x))$, <u>where</u> $\tilde{u}_2(x) > 0$ <u>in</u> Ω, <u>and</u> $\tilde{u}_2 \in H^{2+\alpha}(\bar{\Omega})$.

<u>Proof</u>. Consider the boundary value problem for the scalar function u satisfying: $\sigma_2 \Delta u + u[-r + f_2(0,u)] = 0$ in Ω, $u = g$ on $\partial\Omega$. The zero function is a lower solution, and a large positive constant function is an uppr solution (by (2.2-4)). Hence by a more restrictive version of theorem 1.4-2 (see also section 5.1), there exists a solution $u(x) \in H^{2+\alpha}(\bar{\Omega})$, with $0 < u(x) < K$, K sufficiently large.

We now show that u is positive in Ω. Suppose that $u(x_0) = 0$ at some point $x_0 \in \Omega$, then in a small neighborhood $\mathcal{O}$ of x_0 we have $\sigma_2 \Delta u + h(x)u = 0$ in $\mathcal{O}$, and $h(x) < 0$ in $\mathcal{O}$.

By the maximum (minimum) principle, Theorem 1.1-2, we have $u(x) \equiv 0$ in $\mathcal{O}$. Hence the set where $u = 0$ is both open and closed, and is therefore the whole Ω. However, the continuity of u in $\bar{\Omega}$ contradicts the fact that $g \not\equiv 0$ on $\partial\Omega$. Therefore we have $u(x) > 0$ in Ω. Letting $\tilde{u}_2(x) = u(x)$, the function $(0,\tilde{u}_2(x))$ clearly satisfies the properties described in the theorem.

In order to consider the time stability of the steady state solutions in the last two theorems, we will need the following comparison lemma involving differential equalities.

<u>Lemma</u> 2.2-1. <u>Let</u> $v_i(x,t)$, $w_i(x,t)$, $(x,t) \in \bar{\Omega}x[0,\infty)$, $i = 1,2$ <u>be functions in</u> $H^{2+\alpha,(2+\alpha)/2}(\bar{\Omega}_T)$, each $T > 0$, <u>satisfying the inequalities</u>:

$$(2.2\text{-}7)\qquad \begin{aligned} &0 < v_i < w_i \quad , \quad i = 1,2 \\ &\sigma_1 \Delta v_1 + v_1[a + f_1(v_1,w_2)] - \frac{\partial v_1}{\partial t} > 0 \\ &\sigma_1 \Delta w_1 + w_1[a + f_1(w_1,v_2)] - \frac{\partial w_1}{\partial t} < 0 \\ &\sigma_2 \Delta v_2 + v_2[-r + f_2(v_1,v_2)] - \frac{\partial v_2}{\partial t} > 0 \\ &\sigma_2 \Delta w_2 + w_2[-r + f_2(w_1,w_2)] - \frac{\partial w_2}{\partial t} < 0 \end{aligned}$$

<u>Let</u> $(u_1(x,t), u_2(x,t))$ <u>be a solution of</u> (2.2-1) <u>with</u> $u_i \in H^{2+\alpha,(2+\alpha)/2}(\bar{\Omega}_T)$, <u>and with initial boundary conditions such that</u>

$$(2.2\text{-}8)\qquad \begin{aligned} &v_i(x,0) < u_i(x,0) < w_i(x,0) \quad , \quad x \in \Omega \quad , \; i = 1,2 \\ &v_i(x,t) < u_i(x,t) < w_i(x,t) \quad , \quad (x,t) \in \partial\Omega x[0,\infty), \; i = 1,2. \end{aligned}$$

Then $(u_1(x,t), u_2(x,t))$ *will satisfy*

$$v_i(x,t) \le u_i(x,t) \le w_i(x,t), \quad (x,t) \in \bar{\Omega}\times[0,\infty) \tag{2.2-9}$$

Proof. The lemma is almost a special case of Theorem 1.2-6, except that the operator L is different for each $i = 1,2$. Note that inequalities (2.2-7) imply that inequalities (1.2-19) and (1.2-20) are valid. The proofs in Theorems 1.2-5 and 1.2-6 can thus be readily adapted to this lemma.

The next Theorem shows that if $a < \sigma_1\lambda_1$ and the prey concentration is held at zero at the boundary, then the prey will eventually become extinct in the entire domain Ω.

Theorem 2.2-3. *Let* $a < \sigma_1\lambda_1$. *Suppose* $\bar{u}_2(x) > 0$ *in* $\bar{\Omega}$ *is a solution of the boundary value problem*

$$(2.2\text{-}10) \qquad \begin{aligned} &\sigma_2\Delta u + u[-r + f_2(0,u)] = 0 && \text{in } \Omega, \\ &u = g(x) > 0 && \text{on } \partial\Omega, \end{aligned}$$

where g *has an extension* $\hat{g} \in H^{2+\alpha}(\bar{\Omega})$. *Let* $(u_1(x,t), u_2(x,t))$ *with* $u_i \in H^{2+\alpha,(2+\alpha)/2}(\bar{\Omega}_T)$, *each* $T > 0$, $i = 1,2$, *be a solution of* (2.2-1) *with initial-boundary conditions*

$$\left.\begin{aligned} u_1(x,0) &= \phi_1(x) \ge 0, \not\equiv 0 \\ u_2(x,0) &= \phi_2(x) \ge 0 \end{aligned}\right\} \quad \text{for } x \in \bar{\Omega}$$

$$\left.\begin{aligned} u_1(x,t) &= 0 \\ u_2(x,t) &= g(x) > 0 \end{aligned}\right\} \quad \text{for } (x,t) \in \partial\Omega\times[0,\infty)$$

where ϕ_1, ϕ_2, g *satisfy the compatibility condition of order* 1 *on* $\partial\Omega$ *at* $t = 0$ *as described in section* 1.3. *Then* $(u_1(x,t), u_2(x,t)) \to (0,\bar{u}_2(x))$ *as* $t \to +\infty$, $x \in \bar{\Omega}$.

Remarks. The existence of $(u_1(x,t),u_2(x,t))$ in the function spaces as described follows from Theorem 2.1-1, where the food pyramid condition etc. can be seen to be satisfied readily. The existence of $\bar{u}_2$ in $H^{2+\alpha}(\bar{\Omega})$ and strictly positive in $\bar{\Omega}$ follows from Theorem 2.2-2. An immediate consequence of Theorem 2.2-3 is that a positive solution of (2.2-10) is unique. One can describe the solution $(0,\bar{u}_2(x))$ as "globally stable" among solutions with nonnegative initial conditions.

Proof. The proof is by an application of Lemma 2.2-1 with appropriate choices of v_i, w_i, $i = 1,2$. We choose $v_1(x,t) \equiv 0$ for $(x,t) \in \bar{\Omega}x[0,\infty)$ and $w_1(x,t)$ to be the solution of the initial-boundary value problem

$$\sigma_1 \Delta w_1 + w_1[a + f_1(w_1,0)] - \frac{\partial w_1}{\partial t} = 0, \text{ for } (x,t) \in \Omega x(0,\infty),$$

$$w_1(x,0) = \phi_1(x) \geq 0, \not\equiv 0, \qquad \text{for } x \in \bar{\Omega},$$

$$w_1(x,t) = 0\ , \qquad \text{for } (x,t) \in \delta\Omega x[0,\infty)$$

(Existence of $w_1(x,t) > 0$, $(x,t) \in \Omega x(0,\infty)$ follows from Theorem 2.1-1 and its following remark). Now we show that w_1 decays exponentially as $t \to +\infty$. Let $a/\sigma_1 < \lambda_1' < \lambda_1$, and Ω' be a domain containing $\bar{\Omega}$. Let $\psi(x)$ be a function satisfying $\Delta\psi + \lambda_1'\psi = 0$ in Ω', $\psi|_{\delta\Omega'} = 0$, $\psi(x) > 0$ for $x \in \bar{\Omega}$ and $\sup_{x\in\Omega'} |\psi(x)| = 1$. (See for example [19], [73] or section 1.3 for the existence of such λ_1' and ψ). Define $\bar{z}(x,t)$ by $w_1(x,t) = \bar{z}(x,t)\psi(x)e^{-\alpha_1 t}$, α_1 to be chosen. The function $\bar{z}(x,t)$ satisfies in Ω:

$$\sigma_1 \Delta\bar{z} + 2\sigma_1 \frac{\nabla\psi}{\psi} \cdot \nabla\bar{z} + \bar{z}[(a-\sigma_1\lambda_1') + f_1(w_1,0) + \alpha_1] - \frac{\partial\bar{z}}{\partial t} = 0.$$

Consequently, if $\alpha_1 > 0$ is chosen so that $a - \sigma_1\lambda_1' + \alpha_1 < 0$, we have

$$\sigma_1 \Delta\bar{z} + 2\sigma_1 \frac{\nabla\psi}{\psi} \cdot \nabla\bar{z} - \frac{\partial\bar{z}}{\partial t} \geq 0 \quad \text{in } \Omega x(0,\infty),$$

$$\bar{z} = 0 \quad \text{on } \delta\Omega x[0,\infty),\ \bar{z}(x,0) \geq 0, \not\equiv 0 \text{ for } x \in \Omega.$$

The maximum principle, Th. 1.1-6, implies that $0 \le \bar{z}(x,t) \le \sup\{\phi_1(x)/\psi(x): x \in \bar{\Omega}$ for $(x,t) \in \bar{\Omega}\times[0,\infty)$. Hence, we have

$$0 \le w_1(x,t) \le Ke^{-\alpha_1 t}$$

for $(x,t) \in \bar{\Omega}\times[0,\infty)$ and some constant K.

Now take $v_2(x,t) \equiv 0$ and $w_2(x,t) \equiv C$ for some large constant. Applying Lemma 2.2-1, we see that $0 \le u_1(x,t) \le Ke^{-\alpha_1 t}$ and $0 \le u_2(x,t) \le C$ for $(x,t) \in \bar{\Omega}\times[0,\infty)$. Moreover, by Theorem 2.1-1 (or the maximum principle) we have $u_2(x,t) > 0$ for $(x,t) \in \Omega\times(0,\infty)$.

It is obvious that Lemma 2.2-1 may be generalized, so as to apply to solutions of (2.2-1) considered on domains of the form $\bar{\Omega}\times[T,\infty)$, where $T > 0$. The 0's in condition (2.2-8) are to be replaced by T's.

With this in mind, we improve the choice of $v_2(x,t)$ to be $\beta(t)\bar{u}_2(x)$, $t \ge 1$, $x \in \bar{\Omega}$, where $\beta(t) = 1-ke^{-\ell t}$, with k, ℓ to be chosen constants. We have

$$\begin{aligned} &\sigma_2\Delta v_2 + v_2[-r + f_2(0,v_2)] - \frac{\partial v_2}{\partial t} \\ (2.2\text{-}11)\qquad &= \sigma_2\beta\Delta\bar{u}_2 + \beta\bar{u}_2[-r + f_2(0,\beta\bar{u}_2)] - \beta'(t)\bar{u}_2 \\ &= \beta\bar{u}_2[f_2(0,\beta\bar{u}_2) - f_2(0,\bar{u}_2)] - k\ell e^{-\ell t}\,\bar{u}_2 \\ &= \beta\bar{u}_2\,\frac{\partial f_2}{\partial u_2}(0,\theta\bar{u}_2)(-ke^{-\ell t}\bar{u}_2) - k\ell e^{-\ell t}\bar{u}_2\,,\quad 1 - ke^{-\ell t} < \theta < 1. \end{aligned}$$

Let $k \overset{(def)}{=} \max\{0, e^{\ell}\max\{1 - \frac{u_2(x,1)}{\bar{u}_2(x)} : x \in \bar{\Omega}\}\} < e^{\ell}$ (with $\ell > 0$ still to be chose Note that $1-ke^{-\ell t} \ge 1-e^{\ell-\ell t} \ge 0$ for $t \in [1,\infty)$. With this choice of k, we have $v_2(x,1) = (1-ke^{-\ell})\bar{u}_2(x) \le (1-\max_{x\in\bar{\Omega}}\{1 - \frac{u_2(x,1)}{\bar{u}_2(x)}\})\,\bar{u}_2(x) \le (1-[1-\frac{u_2(x,1)}{\bar{u}_2(x)}])\,\bar{u}_2(x$ $u_2(x,1)$. Now, one can readily see that

$$(1-ke^{-\ell})\bar{u}_2(x)\,\frac{\partial f_2}{\partial u_2}(0,\theta\bar{u}_2) + \ell = \min\{1, \min\{\frac{u_2(x,1)}{\bar{u}_2(x)} : x\in\bar{\Omega}\}\}\,\bar{u}_2(x)\,\frac{\partial f_2}{\partial u_2}(0,\theta\bar{u}_2) + \ell$$

and thus we can choose $\ell > 0$ sufficiently small so that

$$(1-ke^{-\ell})\bar{u}_2(x) \frac{\partial f_2}{\partial u_2}(0,\theta\bar{u}_2) + \ell < 0 \quad \text{for all } x \in \bar{\Omega}.$$

With these choices of k and ℓ, the latter expressions (2.2-11) is > 0 for $(x,t) \in \bar{\Omega}x[1,\infty)$. By the variant of Lemma 2.2-1 described above, we therefore obtain $v_2(x,t) < u_2(x,t)$ for $(x,t) \in \bar{\Omega}x[1,\infty)$. (Note that $v_2(x,t) \to \bar{u}_2(x)$ for $x \in \bar{\Omega}$ as $t \to +\infty$).

We finally improve $w_2(x,t)$ to be $\alpha(t)\bar{u}_2(x)$, $t > T$, $x \in \bar{\Omega}$, where $\alpha(t) = 1 + pe^{-qt}$ with T, p, q > 0 to be chosen. We have

$$\sigma_2\Delta w_2 + w_2[-r + f_2(w_1,w_2)] - \frac{\partial w_2}{\partial t}$$

$$= \alpha\bar{u}_2[f_2(w_1,\alpha\bar{u}_2) - f_2(0,\bar{u}_2)] + pqe^{-qt}\,\bar{u}_2$$

$$(2.2\text{-}12) \qquad < \alpha\bar{u}_2[f_2(Ke^{-\alpha_1 t},\alpha\bar{u}_2) - f_2(0,\alpha\bar{u}_2)] + \alpha\bar{u}_2[f_2(0,\alpha\bar{u}_2) - f_2(0,\bar{u}_2)] + pqe^{-qt}\bar{u}_2$$

$$< \alpha\bar{u}_2[K_1e^{-\alpha_1 t} - K_2pe^{-qt}\bar{u}_2] + pqe^{-qt}\bar{u}_2$$

for certain positive constants K_1, K_2. (We will see that we can take $K_2 = \min\{ |\frac{\partial f_2}{\partial u_2}(0,s)|: \rho_1 < s < \frac{\rho_2}{\rho_1} C\}$, where $\rho_1 = \min_{x\in\bar{\Omega}} \bar{u}_2(x)$ and $\rho_2 = \max_{x\in\bar{\Omega}} \bar{u}_2(x)$).

We wish to show that the expression in (2.2-12) is < 0 in $\bar{\Omega}x[T,\infty)$ for proper choice of T, p, q. Since we know that $u_2(x,t) < C$, we impose the condition

$$(2.2\text{-}13) \qquad (1+pe^{-qT})\rho_1 = C$$

which implies that $w_2(x,T) > u_2(x,T)$ for all $x \in \bar{\Omega}$. Condition (2.2-13) also implies that $\alpha(t)\bar{u}_2(x) < \frac{C\rho_2}{\rho_1}$ for $(x,t) \in \bar{\Omega}x[T,\infty)$, and consequently we may assume K_2 be as described above. Thus, K_1 and K_2 are independent of the particular T, p, q as long as (2.2-13) is satisfied.

We choose $q < \alpha_1$ and T sufficiently large so that

$$K_1 e^{-\alpha_1 T} < \frac{1}{3} K_2(C-\rho_2).$$

It follows that $K_1 e^{-\alpha_1 T} < \frac{1}{3}K_2[(1+pe^{-qT})\overline{u}_2(x)-\overline{u}_2(x)] = \frac{1}{3}K_2 pe^{-qT}\overline{u}_2(x)$ for $x \in \overline{\Omega}$ and $K_1 e^{-\alpha_1 t} < \frac{1}{3}K_2 pe^{-qt}\overline{u}_2(x)$ for $(x,t) \in \overline{\Omega}x[T,\infty)$. Finally, we choose q smaller if necessary, so that

$$q < \frac{1}{3}K_2\rho_1$$

Having chosen T and q, we choose p to satisfy (2.2-13). Thus we have the differential inequalities (2.2-7) for $w_2(x,t)$, with $(x,t) \in \overline{\Omega}x[T,\infty)$.

Applying the general version of Lemma 2.2-1 as described for $(x,t) \in \overline{\Omega}x[T,\infty)$, we have $v_1(x,t) \equiv 0$, $w_1(x,t) \to 0$, $v_2(x,t) \to \overline{u}_2(x)$ and $w_2(x,t) \to \overline{u}_2(x)$ as $t \to +\infty$, $x \in \overline{\Omega}$. Consequently $(u_1(x,t), u_2(x,t)) \to (0,\overline{u}_2(x))$ as $t \to +\infty$, $x \in \overline{\Omega}$.

We next consider the same boundary condition for u_1 as before, i.e., $u_1(x,t) = 0$ for $x \in \delta\Omega$, $t > 0$, with the reverse assumption that $a > \sigma_1\lambda_1$. We will see that coexistence of prey and predator are possible when the mortality rate r of the predator is relatively large. (That is, there is a steady state at which neither u_1 nor u_2 is identically zero in Ω.)

Theorem 2.2-4. *Let* $a > \sigma_1\lambda_1$. *Let* $\delta > 0$ *be any number such that* $|f_1(0,\delta)| < a - \sigma_1\lambda_1$. *If there is a* $k > 0$ *such that* $f_1(k,0) < -a$ *and* r *satisfies*

$$r > f_2(k,\delta)$$

then the boundary value problem

$$\text{(2.2-14)} \qquad \begin{aligned} &\sigma_1\Delta u_1 + u_1[a + f_1(u_1,u_2)] = 0 \\ &\sigma_2\Delta u_2 + u_2[-r + f_2(u_1u_2)] = 0 \end{aligned} \quad \text{in } \Omega$$

$$\begin{aligned} u_1(x) &= 0 \\ u_2(x) &= \rho(x) < \delta \end{aligned} \quad \text{on } \delta\Omega$$

has a solution with $u_1(x) > 0$ for $x \in \Omega$, $0 < u_2(x) < \delta$, $\not\equiv 0$ in Ω. Here, $0 < \rho(x) < \delta$, $\rho(x) \not\equiv 0$ on $\delta\Omega$, and has an extension $\hat{\rho}(x)$, $x \in \bar{\Omega}$, where $\hat{\rho}(x) \in H^{2+\alpha}(\bar{\Omega})$. The solution components $u_1(x)$, $u_2(x)$ are in $H^{2+\alpha}(\bar{\Omega})$.

Proof. Let $\phi_1(x) = c\omega(x)$, $c > 0$, where $\omega(x)$ is the principal eigenfunction described before. We have $\sigma_1\Delta\phi_1(x) + \phi_1(x)[a + f_1(\phi_1(x),y_2)] = (a-\sigma_1\lambda_1)\phi_1(x) + \phi_1(x)f_1(c\omega(x),y_2) > 0$, for $0 < y_2 < \delta$ provided that c is small enough. That is, such $\phi_1(x)$ is a lower solution for $\sigma_1\Delta u + u[a + f_1(u,y_2)] = 0$ in Ω, $u|_{\delta\Omega} = 0$ for each $0 < y_2 < \delta$. The function $\theta_1(x) \equiv k$ satisfies $\sigma_1\Delta\theta_1 + \theta_1[a + f_1(k,y_2)] < 0$ for each $y_2 > 0$, by the assumption on k, and is therefore an upper solution for the problem just described. On the other hand, $\phi_2(x) \equiv 0$ and $\theta_2(x) \equiv \delta$ are respectively lower and upper soltuions for $\sigma_2\Delta u + u[-r + f_2(y_1,u)] = 0$ in Ω, $u|_{\delta\Omega} = \rho(x)$ for each $o < y_1 < k$, by the assumption $r > f_2(k,\delta)$. We now reduce $c > 0$ in the definition of ϕ_1, if necessary, such that $c\max_{\bar{\Omega}} \omega(x) < k$; then by Theorem 1.4-2, there exists a solution $(u_1(x), u_2(x))$ to the boundary value problem (2.2-14), with $c\omega(x) < u_1(x) < k$, $0 < u_2(x) < \delta$, $x \in \bar{\Omega}$. We have $u_2(x) \not\equiv 0$ in Ω because $\rho(x) \not\equiv 0$ on $\delta\Omega$.

Even when $a > \sigma_1\lambda_1$, the solution $(0,\tilde{u}_2(x))$ described in Theorem 2.2-2 still exists. One can therefore say that (u_1,u_2) bifurcates into more than one solution for the boundary value problem (2.2-14) as the parameter a crosses over a_0 for some $a_0 > \lambda_1\sigma_1$, because of Theorems 2.2-1, 2.2-2 and 2.2-4.

We next analyze other kinds of steady states while the boundary concentration for the predator is held at zero, rather than that for the prey. In Theorem 2.2-5, we find an estimate for the size of the predator mortality rate r, which will cause extinction of the predator. Coexistence of prey and predator is possible when r is not large.

Theorem 2.2-5. _Suppose that there is a positive number_ u_1^* _such that_ $a + f_1(u_1^*,0) = 0$. _Let_ K _be any number_ $> u_1^*$. _If_ $r > 0$ _is large enough so that_ $-r + f_2(K,0) < \sigma_2\lambda_1$, _then for any nonnegative functions_ $u_1(x,t)$, $u_2(x,t)$ _in_ $H^{2+\alpha,(2+\alpha)/2}(\bar{\Omega}_T)$, _each_ $T > 0$, _satisfying equation_ (2.2-1) _and initial boundary conditions_

$$0 \leq u_i(x,0) \leq K, \quad x \in \bar{\Omega}, \quad i = 1,2,$$

$$\left.\begin{array}{l} 0 \leq u_1(x,t) \leq K \\ u_2(x,t) = 0 \end{array}\right\} \quad (x,t) \in \delta\Omega x[0,\infty),$$

it must be true that $u_2(x,t) \to 0$ _as_ $t \to \infty$, $x \in \bar{\Omega}$.

Proof. We employ Lemma 2.2-1. First, we choose $v_1(x,t) \equiv 0$ and $w_1(x,t) \equiv K$. Clearly the derivatives of v_1 satisfy (2.2-7). Now

$$\sigma_1\Delta K + K[a + f_1(K,u_2(x,t))] \leq K[a + f_1(K,0)] \leq K[a + f_1(u_1^*,0)] = 0,$$

hence w_1 satisfies (2.2-7).

We take $v_2(x,t) \equiv 0$ and $w_2(x,t) = \alpha(t)\omega(x)$, where $\alpha(t) = Ce^{-\ell t}$, with C,ℓ to be chosen, and $\omega(x)$ is a principal eigenfunction. We have

$$\sigma_2\Delta w_2 + w_2[-r + f_2(w_1,w_2)] - \partial w_2/\partial t$$

$$= \sigma_2\alpha\Delta\omega + \alpha\omega[-r + f_2(K,\alpha\omega)] + \ell\alpha\omega$$

$$\leq \alpha\omega[-\sigma_2\lambda_1 - r + f_2(K,0) + \ell].$$

We may choose $\ell > 0$ sufficiently small so that the latter expression is negative in $\Omega x(0,\infty)$. Moreover, $\frac{\partial\omega}{\partial\eta} < k < 0$ for some constant k, all $x \in \delta\Omega$, and $u_2(x,0) =$ on $\partial\Omega$, therefore we may choose $C > 0$ sufficiently large so that $C\omega(x) \geq u_2(x,0)$ in $\bar{\Omega}$. Choosing such C and ℓ, we see that $w_2(x,t)$ satisfies (2.2-7) together with its part in (2.2-8). The other parts of (2.2-8) are seen to be readily satisfied by v_1, w_1, v_2. Applying Lemma 2.2-1 with this set of v_i, w_i, and since $w_2(x,t) \to 0$ as $t \to \infty$, $x \in \bar{\Omega}$, we obtain the desired result.

It is readily seen that one can have a steady state solution of (2.2-5) with $u_2 \equiv 0$ in $\bar{\Omega}$. Let K be as defined in Theorem 2.2-5, and suppose that $\phi(x) \in H^{2+\alpha}(\bar{\Omega})$, $0 < \phi(x) < K$ for $x \in \delta\Omega$. The boundary value problem $\sigma_1 \Delta u + u[a+f_1(u,0)] = 0$ in Ω, $u|_{\delta\Omega} = \phi|_{\delta\Omega}$, has 0 and K respectively as lower and upper solutions. Hence, there exists a solution $u = \hat{u}_1(x)$, $0 < \hat{u}_1(x) < K$, $x \in \bar{\Omega}$. Moreover, $(\hat{u}_1(x),0)$, $x \in \Omega$, is a solution of the steady-state equations (2.2-5) and boundary conditions $0 < u_1 < K$, $u_2 = 0$ on $\delta\Omega$.

When the mortality rate r of the predator is not large enough to satisfy the condition in Theorem 2.2-5, coexistence of prey and predator in equilibrium is possible under the same type of boundary conditions. More discussions of such nature can be found in [145].

A more intriguing problem is the possibility of coexistence when the boundary conditions of <u>both</u> prey and predator are set to be zero. Such problem has been studied for the Volterra-Lotka prey-predator system:

$$
\begin{aligned}
&\Delta u_1 + u_1[a-bu_1-cu_2] = 0 \\
(2.2\text{-}15) \qquad &\Delta u_2 + u_2[e+fu_1-gu_2] = 0 \qquad \text{in } \Omega \\
&u_1 = u_2 = 0 \quad \text{on } \partial\Omega
\end{aligned}
$$

where a, b, c, e, f and g are positive constants. The system satisfies all the hypotheses (2.2-2) to (2.2-4) with $f_1(u_1,u_2) = -bu_1-cu_2$, $f_2(u_1,u_2) = fu_1-gu_2$. The only significant difference is that the intrinsic rate for the predator u_2 is changed from negative (-r) to positive (e). This is certainly possible if u_2 has food supply other than u_1. It is shown in Chapter 5 that if

$$a > \lambda_1 , \; e > \lambda_1 ,$$
$$cf < gb , \text{ and } a > \frac{gb}{gb-cf}(\lambda_1 + \frac{ce}{g})$$

Then possible coexistence solution of (2.2-15) exist with both $u_1(x)$ and $u_2(x)$ being positive in Ω. Further discussion of such problem can also be found in Chapter 7, section 7.4. Recent contributions to this problem can be found in e.g [25] and [151].

2.3 Competing Species with Positive Dirichlet Conditions, Stability of Steady States

In this and the next section we consider a system of two competing species with Dirichlet boundary conditions. We study various possibilities of non-negative steady states and their stabilities. We first consider the situation when the Dirichlet conditions are positive, which is easier to analyze. The system of equations are:

$$\text{(2.3-1)} \qquad \begin{aligned} \frac{\partial u_1}{\partial t} &= \sigma_1 \Delta u_1 + u_1[a + f_1(u_1,u_2)] \\ \frac{\partial u_2}{\partial t} &= \sigma_2 \Delta u_2 + u_2[b + f_2(u_1,u_2)] \end{aligned} \qquad \text{for } x \in \Omega,\ t > 0,$$

where $u_i(x,t)$, $i = 1,2$ represent the concentration of two species at position x and time t. The parameters a, b, σ_1, σ_2 are positive constants, with a and b representing growth rates when no interaction occurs, σ_1 and σ_2 representing diffusion rates. The functions f_i: $R^2 \to R$ have Hölder continuous partial derivatives up to second order in compact sets, $i = 1,2$. Further, we assume that

$$\text{(2.3-2)} \qquad f_1(0,0) = f_2(0,0) = 0.$$

For (u_1,u_2) in the first open quadrant, the first partial derivatives of f_1, f_2 satisfy:

$$\frac{\partial f_i}{\partial u_j} < 0 \text{ for each } i,j = 1 \text{ or } 2. \tag{2.3-3}$$

Such condition reflects the competing nature of the two species, because the rise in concentration of one species would reduce the growth rate of the other. The above assumptions are always made in this and the next section.

The next lemma first establishes the fact that nonnegative steady state solutions with nontrivial nonnegative Dirichlet data must be positive in the interior.

__Lemma__ 2.3-1. __Let__ $(\bar{u}_1(x),\bar{u}_2(x))$, __with__ $\bar{u}_i(x) \in H^{2+\alpha}(\bar{\Omega})$, $i = 1,2$, __be a solution of the boundary value problem:__

$$\begin{aligned} &\sigma_1 \Delta u_1 + u_1[a + f_1(u_1,u_2)] = 0 \\ &\qquad\qquad\qquad\qquad\qquad\qquad\qquad x \in \Omega \\ &\sigma_2 \Delta u_2 + u_2[b + f_2(u_1,u_2)] = 0 \end{aligned} \tag{2.3-4}$$

$$u_i(x) = g_i(x) \geq 0, \not\equiv 0 \text{ on } \delta\Omega, \; i = 1,2.$$

__Suppose that__ $\bar{u}_i(x) \geq 0$ __in__ $\bar{\Omega}$, $i = 1,2$, __then__ $\bar{u}_i(x) > 0$ __in__ Ω, $i = 1,2$.

__Proof__. Let $w = -\bar{u}_1(x)e^{-at}$. Direct computation yields $\sigma_1 \Delta w - \frac{\partial w}{\partial t} + wf_1(\bar{u}_1,\bar{u}_2) = 0$. where $f_1(\bar{u}_1,\bar{u}_2) < 0$. If $\bar{u}_1(\hat{x}) = 0$, for $\hat{x} \in \Omega$, then $w(\hat{x},t) = 0$ for a positive t. The maximum principle (Theorem 1.1-8) therefore implies that $g_1(x) \equiv 0$, contradicting the assumption. Consequently, $\bar{u}_1(x) > 0$ in Ω. Similarly, we can prove that $\bar{u}_2(x) > 0$ in Ω.

The existence of positive coexistence steady states can be established by adding the following hypotheses (2.3-5) to (2.3-2) and (2.3-3).

Lemma 2.3-2. Consider the boundary value problem (2.3-4) described in Lemma 2.3-1, while we further assume that $g_i(x) > 0$ on $\delta\Omega$ and has an extension $\hat{g}_i \in H^{2+\ell}(\bar{\Omega})$, $i = 1,2$. Assume that there exists a positive constant C so that

$$(2.3\text{-}5) \qquad a + f_1(C,0) < 0, \quad b + f_2(0,C) < 0.$$

Then there exists a solution $(\bar{u}_1(x),\bar{u}_2(x))$ with $\bar{u}_i(x) \in H^{2+\ell}(\bar{\Omega})$, $0 < \bar{u}_i(x) < K_i$, $x \in \bar{\Omega}$, $i = 1,2$. (Here K_1, K_2 are positive constants).

Proof. By hypotheses (2.3-3) and (2.3-5), there exist positive constants K_1, K_2 so that $a + f_1(K_1,0) < 0$, $b + f_2(0,K_2) < 0$, $g_i(x) < K_i$ on $\partial\Omega$, $i = 1,2$. For each $0 < u_2 < K_2$, the functions $\phi_1(x) = 0$, $\psi_1(x) = K_1$ are respectively lower and upper solutions of the boundary value problem $\sigma_1\Delta u + u[a + f_1(u,u_2)] = 0$ in Ω, $u = g_1$ on $\delta\Omega$. Similarly, $\phi_2 = 0$, $\psi_2 = K_2$ are respectively lower and upper solutions for the problem $\sigma_2\Delta u + u[b + f_2(u_1,u)] = 0$ in Ω, $u = g_2$ on $\delta\Omega$, for each $0 < u_1 < K_1$. By Theorem 1.4-2, there exists a solution $(\bar{u}_1(x), \bar{u}_2(x))$ with $\bar{u}_i \in H^{2+\ell}(\bar{\Omega})$, $0 < \bar{u}_i(x) < K_i$, $i = 1,2$ to the boundary value problem described in the lemma. By Lemma 2.3-1, $0 < \bar{u}_i(x)$ in $\bar{\Omega}$.

Remark. The system (2.3-1) together with hypotheses (2.3-2), (2.3-3) and (2.3-5) are general assumptions for competing species including the classical Volterra-Lotka model with diffusion.

The following Theorem gives sufficient condition for asymptotic stability of the coexistence steady state solutions found in Lemma 2.3-2.

Theorem 2.3-1. Let $(\bar{u}_1(x),\bar{u}_2(x))$ be and equilibrium solution to (2.3-4) as described in Lemma 2.3-2 ($g_i > 0$, with extension $\hat{g}_i \in H^{2+\ell}(\bar{\Omega})$, $i = 1,2$). Suppose that:

$$(2.3\text{-}6) \qquad \left|\frac{\bar{u}_i(x)}{\bar{u}_j(x)} \cdot \frac{(\partial f_j/\partial u_i)(\bar{u}_1(x),\bar{u}_2(x))}{(\partial f_j/\partial u_j)(\bar{u}_1(x),\bar{u}_2(x))}\right| < \min_{x\in\bar{\Omega}} \left|\frac{\bar{u}_i(x)}{\bar{u}_j(x)} \frac{(\partial f_i/\partial u_i)(\bar{u}_1(x),\bar{u}_2(x))}{(\partial f_i/\partial u_j)(\bar{u}_1(x),\bar{u}_2(x))}\right|$$

for each $x \in \bar{\Omega}$, $i \neq j$, $1 < i, j < 2$, *then* $(\bar{u}_1(x), \bar{u}_2(x))$ *is asymptotically stable*. (*Here asymptotic stability is interpreted to mean that for any solution* $(u_1(x,t), u_2(x,t))$ *with* $u_i \in H^{2+\ell, 1+\ell/2}(\bar{\Omega}x[0,T])$, *each* $T > 0$, $i = 1,2$ *of the system* (2.3-1) *with boundary conditions* $u_i(x,t) = g_i(x)$ *and initial conditions* $u_i(x,0)$ *close enough to* $\bar{u}_i(x)$ *for all* $x \in \bar{\Omega}$, $i = 1,2$, *one has* $u_i(x,t) \to \bar{u}_i(x)$ *uniformly as* $t \to +\infty$, $i = 1,2$.)

Proof. Assumption (2.3-6) implies that there are ρ_1, ρ_2 close enough to 1 with $\rho_1 < 1 < \rho_2$ such that

$$\frac{\bar{u}_i(x)}{\bar{u}_j(x)} \frac{\max_{\rho_1 < s,\tau < \rho_2} |(\partial f_j/\partial u_i)(s\bar{u}_1(x), \tau\bar{u}_2(x))|}{\min_{\rho_1 < s < 1} |(\partial f_j/\partial u_j)(s\bar{u}_1(x), \bar{u}_2(x))|}$$

$$< \min_{x \in \bar{\Omega}} \left\{ \frac{\bar{u}_i(x)}{\bar{u}_j(x)} \frac{\min_{\rho_1 < s,\tau < \rho_2} |(\partial f_i/\partial u_i)(s\bar{u}_1(x), \tau\bar{u}_2(x))|}{\max_{\rho_1 < s < 1} |(\partial f_i/\partial u_j)(s\bar{u}_1(x), \bar{u}_2(x))|} \right\} - \bar{\varepsilon}_1$$

for each $x \in \bar{\Omega}$, $i \neq j$, $1 < i, j < 2$, where $\bar{\varepsilon}_1$ is a small positive number. (Recall that Lemma 2.3-1 implies that $\bar{u}_i(x) > 0$ in $\bar{\Omega}$, $i = 1,2$). We will construct appropriate lower and upper solutions v_i, w_i, $i = 1,2$, and apply Theorem 1.2-6. Let

$$G(x) = \bar{u}_2(x) \min_{\rho_1 < s,\tau < \rho_2} \left|\frac{\partial f_2}{\partial u_2}(s\bar{u}_1(x), \tau\bar{u}_2(x))\right|$$

$$\cdot \left(\bar{u}_1(x) \max_{\rho_1 < s < 1} \left|\frac{\partial f_2}{\partial u_1}(s\bar{u}_1(x), \bar{u}_2(x))\right|\right)^{-1},$$

and α be a number satisfying $1 < \alpha < \rho_2$, $(1-\rho_1)(\alpha-1)^{-1} > G(x)$ for all $x \in \bar{\Omega}$. Define $w_2 = [1 + (\alpha-1)e^{-nt}]\,\bar{u}_2(x)$ where $n > 0$ will be determined later so that w_2 becomes an upper solution. On the other hand define $v_1 = [1 - (1-\beta)e^{-nt}]\cdot \bar{u}_1(x)$, where $\beta = 1 - \min_{x\in\bar{\Omega}}(\alpha-1)G(x) + \bar{\varepsilon}_2(\alpha-1)$ where $0 < \bar{\varepsilon}_2 < \min_{x\in\bar{\Omega}} G(x)$, so $\rho_1 < \beta < 1$. We have

$$\sigma_2\Delta w_2 + w_2[b + f_2(v_1,w_2)] - \frac{\partial w_2}{\partial t}$$

$$= [1 + (\alpha-1)e^{-nt}]\bar{u}_2[f_2(v_1,w_2) - f_2(v_1,\bar{u}_2) + f_2(v_1,\bar{u}_2) - f_2(\bar{u}_1,\bar{u}_2)]$$

$$+ n(\alpha-1)e^{-nt}\bar{u}_2$$

$$< [1 + (\alpha-1)e^{-nt}]\bar{u}_2[\max_{1<\tau<\rho_2}\{\frac{\partial f_2}{\partial u_2}(v_1,\tau\bar{u}_2)\}\ (\alpha-1)\bar{u}_2e^{-nt}$$

$$- \min_{\rho_1<s<1}\{\frac{\partial f_2}{\partial u_1}(s\bar{u}_1,\bar{u}_2)\}\ (1-\beta)\bar{u}_1e^{-nt}] + n(\alpha-1)e^{-nt}\bar{u}_2$$

$$< e^{-nt}\bar{u}_2\{[1 + (\alpha-1)e^{-nt}]k + n(\alpha-1)\}$$

where k is a negative number, because

$$(1-\beta)\bar{u}_1(x)\ |\min_{\rho_1<s<1}\{\frac{\partial f_2}{\partial u_1}(s\bar{u}_1(x),\bar{u}_2(x))\}|$$

$$< [(\alpha-1)G(x) - \bar{\varepsilon}_2(\alpha-1)]\bar{u}_1(x)\max_{\rho_1<s<1}\ |\frac{\partial f_2}{\partial u_1}(s\bar{u}_1(x),\bar{u}_2(x))|$$

$$= (\alpha-1)\bar{u}_2(x)\ |\max_{\rho_1<s,\tau<\rho_2}\frac{\partial f_2}{\partial u_2}(s\bar{u}_1(x),\tau\bar{u}_2(x))\ | - \varepsilon_1(x)$$

$$< (\alpha-1)\bar{u}_2(x)\ |\max_{1<\tau<\rho_2}\frac{\partial f_2}{\partial u_2}(v_1(x,t),\tau\bar{u}_2(x))\ | - \varepsilon_1(x),$$

for each $x \in \bar{\Omega}$, $t > 0$,

(here $\varepsilon_1(x) = \bar{\varepsilon}_2(\alpha-1)\bar{u}_1(x)\max_{\rho_1<s<1}|(\partial f_2/\partial u_1)(s\bar{u}_1(x),\bar{u}_2(x))| > 0$, for $x \in \bar{\Omega}$).

Choosing n to satisfy $0 < n(\alpha-1) < -k$, we have w_2 as an upper solution.

For v_1, we have the differential inequality:

$$\sigma_1\Delta v_1 + v_1[a + f_1(v_1,w_2)] - \frac{\partial v_1}{\partial t}$$

$$= [1 - (1-\beta)e^{-nt}]\bar{u}_1[f_1(v_1,w_2) - f_1(v_1,\bar{u}_2) + f_1(v_1,\bar{u}_2) - f_1(\bar{u}_1,\bar{u}_2)]$$

$$- n(1-\beta)e^{-nt}\overline{u}_1$$

$$\geq [1 - (1-\beta)e^{-nt}]\overline{u}_1[\min_{1\leq\tau\leq\rho_2}\{\frac{\partial f_1}{\partial u_2}(v_1,\tau\overline{u}_2)\}(\alpha-1)\overline{u}_2e^{-nt}$$

$$- \max_{\rho_1\leq s\leq 1}\{\frac{\partial f_1}{\partial u_1}(s\overline{u}_1,\overline{u}_2)\}(1-\beta)\overline{u}_1e^{-nt}] - n(1-\beta)e^{-nt}\overline{u}_1$$

$$\geq e^{-nt}\overline{u}_1\{[1 - (1-\beta)e^{-nt}]p - n(1-\beta)\},$$

where p is a positive number, because

$$\overline{u}_2(x)(\alpha-1)\ |\min_{1\leq\tau\leq\rho_2}\{\frac{\partial f_1}{\partial u_2}(v_1(x,t),\tau\overline{u}_2(x))\}|$$

$$= (\alpha-1)\overline{u}_2(x)\max_{1\leq\tau\leq\rho_2}|\frac{\partial f_1}{\partial u_2}(v_1(x,t),\tau\overline{u}_2(x))|$$

$$\leq (\alpha-1)\overline{u}_2(x)\max_{\rho_1\leq s,\tau\leq\rho_2}|\frac{\partial f_1}{\partial u_2}(s\overline{u}_1(x),\tau\overline{u}_2(x))|$$

$$< (\alpha-1)\ [\min_{x\in\overline{\Omega}}\{\frac{\overline{u}_2(x)}{\overline{u}_1(x)}\ \frac{\min_{\rho_1\leq s,\tau\leq\rho_2}|\frac{\partial f_2}{\partial u_2}(s\overline{u}_1(x),\tau\overline{u}_2(x))|}{\max_{\rho_1\leq s\leq 1}|\frac{\partial f_2}{\partial u_1}(s\overline{u}_1(x),\overline{u}_2(x))|} - \overline{\varepsilon}_1\ ]$$

$$\cdot\overline{u}_1(x)\min_{\rho_1\leq s\leq 1}|\frac{\partial f_1}{\partial u_1}(s\overline{u}_1(x),\overline{u}_2(x))|$$

$$= [\min_{x\in\overline{\Omega}}(\alpha-1)G(x) - (\alpha-1)\overline{\varepsilon}_1]\overline{u}_1(x)\ |\max_{\rho_1\leq s\leq 1}\frac{\partial f_1}{\partial u_1}(s\overline{u}_1(x),\overline{u}_2(x))|$$

$$= (1-\beta)\overline{u}_1(x)\ |\max_{\rho_1\leq s\leq 1}\frac{\partial f_1}{\partial u_1}(s\overline{u}_1(x),\overline{u}_2(x))\ | + \varepsilon_2(x),$$

for each $x \in \overline{\Omega}$, $t > 0$,

(here $\varepsilon_2(x) = (\overline{\varepsilon}_2-\overline{\varepsilon}_1)(\alpha-1)|\max_{\rho_1\leq s\leq 1}(\partial f_1/\partial u_1)(s\overline{u}_1(x),\overline{u}_2(x))|\ \overline{u}_1(x) < 0$, for $x \in \overline{\Omega}$, since we may reduce $\overline{\varepsilon}_2$ so that $0 < \overline{\varepsilon}_2 < \overline{\varepsilon}_1$). Reducing the choice of n if necessary, so that $n(1-\beta) < \beta p$, we have v_1 as an lower solution.

Since all the first partial derivatives of f_1 and f_2 have the same sign, we can interchange the role of $\bar{u}_1$, f_1 with $\bar{u}_2$, f_2 respectively and construct lower and upper solutions v_2, w_1 in exactly the same manner as before. Here v_2, w_1 are of the form $v_2 = [1 - (1-\tilde{\beta})e^{-mt}]\bar{u}_2(x)$, $w_1 = [1 + (\tilde{\alpha}-1)e^{-mt}]\bar{u}_1(x)$ where $\tilde{\beta}$, $\tilde{\alpha}$, m are chosen constants with $\rho_1 < \tilde{\beta} < 1$, $1 < \tilde{\alpha} < \rho_2$, $m > 0$.

Finally, we observe that $v_1(x,t) \to \bar{u}_1(x)$, $v_2(x,t) \to \bar{u}_2(x)$ from below as $t \to +\infty$, and $w_1(x,t) \to \bar{u}_1(x)$, $w_2(x,t) \to \bar{u}_2(x)$ from above as $t \to +\infty$. Applying Theorem 1.2-6, we clearly have $(\bar{u}_1(x),\bar{u}_2(x))$ as an asymptotically stable solution as described in the theorem.

Remarks. (i) In Theorem 2.3-1, the existence of $(\bar{u}_1(x),\bar{u}_2(x))$ satisfying (2.3-6) is assumed. Consequently, one does not have to assume hypothesis (2.3-5), which leads to the existence of $(\bar{u}_1(x),\bar{u}_2(x))$ which may not even satisfying (2.3-6).

(ii) For the existence of a solution $(u_1(x,t),u_2(x,t))$ of the initial boundary value problem for equations (2.3-1), see Theorem 2.1-1.

As an application of Theorem 2.3-1, we consider the following boundary value problem:

Example (i)

$$\text{(2.3-7)}\quad \begin{aligned} \frac{\partial u_1}{\partial t} &= \sigma_1 \Delta u_1 + u_1[10 - 4u_1 - u_2] \\ \frac{\partial u_2}{\partial t} &= \sigma_2 \Delta u_2 + u_2[10 - u_1 - 4u_2] \end{aligned} \qquad (x,t) \in \Omega x(0,\infty),$$

$$u_i = g_i(x), \quad \text{for } x \in \delta\Omega,\ t > 0,\ i = 1,2,$$

where $1.5 < g_i(x) < 4$. Let $\phi_1(x) \equiv 1.5$, $\psi_1(x) \equiv 4$. We have $\sigma_1\Delta\phi_1 + \phi_1[10 - 4\phi_1 - u_2] = 1.5[4 - u_2] > 0$ for each $1.5 < u_2 < 4$, and $\sigma_1\Delta\psi_1 + \psi_1[10 - 4\psi_1 - u_2] = 4[-6 - u_2] < 0$ for each $1.5 < u_2 < 4$. Similarly, we let $\phi_2(x) \equiv 1.5$, $\psi_2(x) \equiv 4$ and prove as in Lemma 2.3-2 that there is an equilibrium $(\tilde{u}_1(x),\tilde{u}_2(x))$, with

$\tilde{u}_i(x) \in H^{2+\ell}(\bar{\Omega})$, $1.5 < \tilde{u}_i(x) < 4$, for $x \in \bar{\Omega}$, $i = 1,2$. For (2.3-7), $\partial f_1/\partial u_1 = -4$, $\partial f_1/\partial u_2 = -1$, $\partial f_2/\partial u_1 = -1$, $\partial f_2/\partial u_2 = -4$. Therefore

$$|\tilde{u}_1(x) \left(\frac{\partial f_1}{\partial u_1}\right)(\tilde{u}_1(x),\tilde{u}_2(x))| \cdot |[\tilde{u}_2(x) \left(\frac{\partial f_1}{\partial u_2}\right) (\tilde{u}_1(x),\tilde{u}_2(x))]^{-1}|$$

$$> \frac{(1.5)(4)}{(4)(1)} = 1.5,$$

and

$$| \tilde{u}_1(x) \left(\frac{\partial f_2}{\partial u_1}\right) (\tilde{u}_1(x),\tilde{u}_2(x))| \cdot |[\tilde{u}_2(x) \left(\frac{\partial f_2}{\partial u_2}\right) (\tilde{u}_1(x),\tilde{u}_2(x))]^{-1}|$$

$$< \frac{(4)(1)}{(1.5)(4)} = \frac{2}{3} .$$

We see that (2.3-6) is satisfied for $i = 1$, $j = 2$. Similarly, one checks that (2.3-6) is also satisfied for $i = 2$, $j = 1$. Theorem 2.3-1 then implies that $(\tilde{u}_1(x), \tilde{u}_2(x))$ is asymptotically stable.

Example (ii). For a less restrictive situation, consider equations (2.3-1). Suppose that there exist positive constants $0 < k_i < K_i$, $i = 1,2$ such that:

$$a + f_1(K_1,0) < 0, \quad b + f_2(0,K_2) < 0$$

$$a + f_1(k_1,K_2) > 0, \quad b + f_2(K_1,k_2) > 0$$

Then for boundary values $k_i < q_i(x) < K_i$, $i = 1,2$, the boundary value problem (2.3-4) has an equilibrium solution $(\hat{u}_1(x),\hat{u}_2(x))$, $k_i < \hat{u}_i < K_i$, $i = 1,2$ as it is proved in example (i). Let $r_1 = \max|(\partial f_1/\partial u_2)(u_1,u_2)|(\min|\partial f_1/\partial u_1|)^{-1}$, $R_1 = \min|\partial f_2/\partial u_2|(\max|\partial f_2/\partial u_1|)^{-1}$, $r_2 = \max|\partial f_2/\partial u_1|(\min|\partial f_2/\partial u_2|)^{-1}$, and $R_2 = \min|\partial f_1/\partial u_1|(\max|\partial f_1/\partial u_2|)^{-1}$; here max. and min. are taken over the rectangle $\{(u_1,u_2): k_i < u_i < K_i, i = 1,2\}$. If one has $K_1K_2r_i < k_1k_2R_i$, $i = 1,2$, then Theorem 2.3-1 implies that $(\hat{u}_1(x),\hat{u}_2(x))$ is asymptotically stable. (Roughly speaking, the condition that $|\partial f_i/\partial u_j|$, $i \neq j$, $1 < i, j < 2$ is "small" compared with $|\partial f_k/\partial u_k|$, $k = 1,2$ in the region of values of the solution will imply asymptotic stability.)

2.4 Competing Species with Homogeneous Boundary Conditions

Continuing with the assumptions in the last section (2.3-1) to (2.3-3), we consider boundary value problems when one or both competing species u_i are held at zero identically on $\delta\Omega$. It turns out that the relative sizes of the growth rates a,b in relation to the size of Ω and diffusion rates σ_1,σ_2 are important in determining the stable asymptotic states. Let $\lambda = \lambda_1 > 0$ be the principal eigenvalue of the eigenvalue problem $\Delta u + \lambda u = 0$ in Ω, $u = 0$ on $\delta\Omega$. In this section, we will assume when neccessary that there exists a positive C so that

$$(2.4\text{-}1) \qquad b + f_2(0,C) < 0$$

<u>Theorem</u> 2.4-1. <u>Assume</u> $a < \sigma_1\lambda_1$, <u>hypothesis</u> (2.4-1) <u>and</u> $u_2^*(x) > 0$ <u>in</u> $\bar{\Omega}$ <u>is a solution of the boundary value problem</u>:

$$(2.4\text{-}2) \qquad \begin{aligned} \sigma_2\Delta u + u[b + f_2(0,u)] &= 0 \qquad \text{in } \Omega \\ u = g(x) &> 0 \qquad \text{on } \delta\Omega \end{aligned}$$

<u>where</u> g <u>has an extension</u> $\hat{g} \in H^{2+\ell}(\bar{\Omega})$, (<u>such solution can be readily shown to exist by using</u> (2.4-1)). <u>Let</u> $(u_1(x,t), u_2(x,t))$ <u>with</u> $u_i \in H^{2+\ell/1+\ell/2}(\bar{\Omega}x[0,T])$, <u>each</u> $T > 0$, $i = 1,2$, <u>be a solution of the system</u> (2.3-1) <u>with initial boundary conditions</u>.

$$u_1(x,0) = \theta_1(x) \geq 0, \not\equiv 0, \; u_2(x,0) = \theta_2(x) \geq 0 \text{ for } x \in \bar{\Omega}, \text{ and}$$
$$u_1(x,t) = 0, \; u_2(x,t) = g(x) > 0 \text{ for } (x,t) \in \partial\Omega x[0,\infty),$$

<u>where</u> θ_1, θ_2, g <u>satisfy the compatibility conditions of order</u> 1 <u>at</u> $t = 0$ <u>as described in section</u> 1.3. <u>Then</u> $(u_1(x,t), u_2(x,t)) \to (0,u_2^*(x))$, <u>as</u> $t \to \infty$, $x \in \Omega$.

Remark. An immediate consequence is that a positive solution of (2.4-2) is unique. Analogous result is true when $b < \sigma_2\lambda_1$, and $u_1^* > 0$ in $\overline{\Omega}$ is a solution of $\sigma_1\Delta u + u[a + f_1(u,0)] = 0$ in Ω, $u = g(x) > 0$ in $\delta\Omega$. That is, we have $(u_1(x,t), u_2(x,t) \to (u_1^*(x),0)$, as $t \to \infty$, $x \in \overline{\Omega}$.

The above theorem is proved in [135]. We omits its proof here because it is similar and simpler than the following Theorem 2-4-2, which considers homogeneous boundary conditions for both species simultaneously. However, we have to additionally assume $b > \sigma_2\lambda_1$, so that nontrivial steady state of the second species can exist under homogeneous boundary condition.

Theorem 2.4-2. *Assume* $a < \sigma_1\lambda_1$, $b > \sigma_2\lambda_1$, *hypothesis* (2.4-1) *and* $u_2^*(x) \in H^{2+\ell}(\overline{\Omega})$ *is a solution of*

$$\sigma_2\Delta u + u[b + f_2(0,u)] = 0 \text{ in } \Omega,\ u = 0 \text{ on } \delta\Omega\ ,$$

with $u_2^*(x) > 0$ *for* $x \in \Omega$. *Let* $(u_1(x,t), u_2(x,t))$ *with* $u_i \in H^{2+\ell,1+\ell/2}(\overline{\Omega}x[0,T])$, *each* $T > 0$, $i = 1,2$, *be a solution of system* (2.3-1) *with initial-boundary conditions*

$$u_i(x,0) = \theta_i(x) > 0,\ x \in \overline{\Omega},$$

$$u_i(x,t) = 0, \qquad x \in \delta\Omega,\ t > 0,$$

$i = 1,2$, *where* $\theta_i \in H^{2+\ell}(\overline{\Omega})$, $i = 1,2$ *satisfy the compatibility conditions of order* 1 at $t = 0$ *as described in section* 1.3. *Then* $(u_1(x,t), u_2(x,t)) \to (0,u_2^*(x))$, *as* $t \to \infty$, *uniformly for* $x \in \overline{\Omega}$, *provided that* θ_1, θ_2 *and all their first partial derivatives are close enough to* 0, u_2^* *respectively and their corresponding first partial derivatives.*

Proof. The fact that $u_2^*(x)$ exists can be readily seen as follows. Let $\omega(x)$ be the principal eigenfunction for the eigenvalue problem $\Delta u + \lambda u = 0$ in Ω, $u = 0$ on $\delta\Omega$, with $\lambda = \lambda_1$ as the principal eigenvalue (thus $\omega(x) > 0$ in Ω). For $\delta > 0$ sufficiently small, we have $\sigma_2\Delta(\delta\omega) + (\delta\omega)[b + f_2(0,\delta\omega)] = \delta\omega[-\sigma_2\lambda_1 + b + f_2(0,\delta\omega$ > 0. While for $K > 0$ sufficiently large (2.4-1) implies that $\sigma_2\Delta K + K[b + f_2(0,K$ < 0. Thus by sections 1.4 and 5.1, $u_2^*(x)$ exists in $H^{2+\ell}(\bar{\Omega})$ with $\delta\omega(x) < u_2^*(x) <$ $x \in \bar{\Omega}$. We now proceed to apply Theorem 1.2-7 by constructing appropriate v_i, w_i, $i = 1,2$. Let $v_1(x,t) \equiv 0$ for $(x,t) \in \bar{\Omega}\times[0,\infty)$. If $\theta_1(x) \equiv 0$, define $w_1(x,t) \equiv 0$; otherwise, define $w_1(x,t)$ as the solution of the initial boundary value problem: $\sigma_1\Delta w_1 + w_1[a + f_1(w_1,0)] - \partial w_1/\partial t = 0$ for $(x,t) \in \Omega\times(0,\infty)$, $w_1(x,0) = \theta_1(x)$ for $x \in \Omega$, $w_1(x,t) = 0$ for $(x,t) \in \delta\Omega\times[0,\infty)$.

Existence of w_1 is by Theorem 1.3-1. Further, by the maximum principle for parabolic equations (Theorem 1.1-8 and its following remark) we have $w_1 > 0$ in $\Omega\times(0,\infty)$. We now show that $w_1 \to 0$ as $t \to \infty$. Let $a/\sigma_1 < \lambda_1' < \lambda_1$, Ω' be a domain containing $\bar{\Omega}$, and $\psi(x)$ be a function satisfying $\Delta\psi + \lambda_1'\psi = 0$ in Ω', $\psi|_{\delta\Omega'} = 0$, $\psi(x) > 0$ in $\bar{\Omega}$, $\sup_{x\in\Omega'}|\psi(x)| = 1$. Define $z(x,t)$ by $w_1(x,t) = z(x,t)\psi(x)e^{-\alpha_1 t}$, where $\alpha_1 > 0$ is chosen to satisfy $a - \sigma_1\lambda_1' + \alpha_1 < 0$. We have

$$\sigma_1\Delta z + 2\sigma_1 \frac{\nabla\psi}{\psi} \cdot \nabla z - \frac{\partial z}{\partial t} = -z[a - \sigma_1\lambda_1' + \alpha_1 + f_1(w_1,0)] > 0$$

in $\Omega\times(0,\infty)$, $z = 0$ on $\delta\Omega\times[0,\infty)$, $z(x,0) \geq 0$, $\not\equiv 0$, $x \in \Omega$. The maximum principle implies that $0 \leq z(x,t) \leq \sup\{\theta_1(x)/\psi(x): x\in\Omega\}$, and hence $0 \leq w_1(x,t) \leq Ke^{-\alpha_1 t}$ for $(x,t) \in \bar{\Omega}\times[0,\infty)$, and some constant K.

We let $v_2(x,t) = [1 - k(x)e^{-\bar{\ell}t}]u_2^*(x)$, $(x,t) \in \bar{\Omega}\times[0,\infty)$, where $k(x) = -\varepsilon u_2^*(x) + C$, $0 < C < 1$, and $\bar{\ell}$, ε are positive small constants to be chosen. We have

$$\begin{aligned}
&\sigma_2\Delta v_2 + v_2[b + f_2(w_1,v_2)] - \frac{\partial v_2}{\partial t} \\
&\quad = v_2[f_2(w_1,v_2) - f_2(0,v_2) + f_2(0,v_2) - f_2(0,u_2^*)] \\
&\quad -e^{-\bar{\ell}t}[\sigma_2(\Delta k)u_2^* + 2\sum_{i=1}^{n} k_{x_i}u_{2x_i}^*\sigma_2 + k\bar{\ell}u_2^*] \qquad (2.4\text{-}3)\\
&\quad > u_2^*[-qKe^{-\alpha_1 t} + (1-C)\min_{(1-C)<s<1}\left|\frac{\partial f_2}{\partial u_2}(0,su_2^*)\right|k(x)e^{-\bar{\ell}t}u_2^* \\
&\quad -e^{-\bar{\ell}t}(\sigma_2\Delta k + k\bar{\ell})] - 2\sigma_2 e^{-\bar{\ell}t}\sum_{i=1}^{n} k_{x_i}u_{2x_i}^*.
\end{aligned}$$

Here $-q = \min_{x\in\bar{\Omega}}$ in $\underset{1-C<s_2<1}{0<s_1<K}$ $(\partial f_2/\partial u_1)(s_1,s_2u_2^*(x)) < 0$, and ε is small enough such that $k(x) > 0$ in $\bar{\Omega}$. Choose $\bar{\ell} = \varepsilon\sigma_2$, thus $|u_2^*e^{-\bar{\ell}t}(\sigma_2\Delta k + k\bar{\ell})| = |u_2^*e^{-\bar{\ell}t}(-\sigma_2\varepsilon\Delta u_2^* + \varepsilon\sigma_2(-\varepsilon u_2^* + C))| < \varepsilon\sigma_2u_2^*e^{-\bar{\ell}t}R$, for some constant $R > 0$ independent of ε, for all $(x,t)\in\bar{\Omega}\times[0,\infty)$. In a neighborhood $\mathfrak{F}$ of $\delta\Omega$ in $\bar{\Omega}$, we have $-2\sigma_2e^{-\bar{\ell}t}\sum_{i=1}^n k_{x_i}u_{2x_i}^* - \varepsilon\sigma_2u_2^*e^{-\bar{\ell}t}R = 2\sigma_2e^{-\bar{\ell}t}\,\varepsilon\sum_{i=1}^n u_{2x_i}^{*2} - \varepsilon\sigma_2u_2^*e^{-\bar{\ell}t}R > \sigma_2\varepsilon e^{-\bar{\ell}t}P$ for some positive constant P independent of ε (because $u_2^*(x) > \delta\omega(x)$, thus $\sum_{i=1}^n u_{2x_i}^{*2} \neq 0$ on $\delta\Omega$). Further, the first term $-u_2^*qKe^{-\alpha_1 t}$ in the last line of (2.4-3) will in absolute value $<(1/2)\sigma_2\varepsilon e^{-\bar{\ell}t}P$, provided that ε is small enough so that $\bar{\ell} = \varepsilon\sigma_2 < \alpha_1$, and t is large enough; and the second term in the same line is always > 0 for $(x,t)\in\bar{\Omega}\times[0,\infty)$. Consequently, $\sigma_2\Delta v_2 + v_2[b + f_2(w_1,v_2)] - \partial v_2/\partial t > 0$ for $x\in\mathfrak{F}$, t large enough. In the complement of $\mathfrak{F}$ in Ω, the second term in the last line of (2.4-3) is bounded below by $Qe^{-\bar{\ell}t}$ for some positive constant Q which can remain unchanged if $\varepsilon > 0$ is reduced. The remaining terms will have absolute value $<(Q/2)e^{-\bar{\ell}t}$ for small enough $\varepsilon > 0$ and t large enough, as before. Therefore the expression in (2.4-3) is > 0, for $x\in\Omega\setminus\mathfrak{F}$, t large enough; and $v_2(x,t)$ is a lower solution for $t > \bar{T}$, for some large $\bar{T}$.

Next, we let $w_2(x,t) = [1 + \hat{k}(x)e^{-\hat{\ell}t}]\, u_2^*(x)$, $(x,t) \in \bar{\Omega}\times[0,\infty)$, where $\hat{k}(x) = \hat{C} - \hat{\varepsilon}u_2^*(x)$, and $\hat{C}$, $\hat{\ell}$, $\hat{\varepsilon}$ are positive constants to be chosen ($\hat{\varepsilon}$ at least small enough so that $\hat{k}(x) > 0$ in $\bar{\Omega}$). We have

$$\sigma_2\Delta w_2 + w_2[b + f_2(0,w_2)] - \frac{\partial w_2}{\partial t}$$

$$(2.4\text{-}4) \qquad = w_2[f_2(0,w_2) - f_2(0,u_2^*)] + e^{-\hat{\ell}t}[\sigma_2(\Delta\hat{k})u_2^* + 2\sigma_2 \sum_{i=1}^{n} k_{x_i}^* u_{2x_i}^* + \hat{k}\hat{\ell}u_2^*]$$

$$< u_2^*[\min_{0<s<1+\hat{C}} |\frac{\partial f_2}{\partial u_2}(0,s)|(-\hat{k}e^{-\hat{\ell}t}u_2^*) + e^{-\hat{\ell}t}(\sigma_2\Delta\hat{k} + \hat{k}\hat{\ell})]$$

$$+ 2\sigma_2 e^{-\hat{\ell}t} \sum_{i=1}^{n} \hat{k}_{x_i} u_{2x_i}^*.$$

Choose $\hat{\ell} = \hat{\varepsilon}\sigma_2$, thus $|u_2^* e^{-\hat{\ell}t}(\sigma_2\Delta\hat{k} + \hat{k}\hat{\ell})\,| < \hat{\varepsilon}\sigma_2 u_2^* e^{-\hat{\ell}t}\hat{R}$, for some $\hat{R} > 0$ independent of $\hat{\varepsilon}$, for all $(x,t) \in \bar{\Omega}\times[0,\infty)$. In a neighborhood $\hat{\mathcal{F}}$ of $\delta\Omega$ in $\bar{\Omega}$, we have

$$2\sigma_2 e^{-\hat{\ell}t} \sum_{i=1}^{n} \hat{k}_{x_i} u_{2x_i}^* + \hat{\varepsilon}\sigma_2 u_2^* e^{-\hat{\ell}t}\hat{R} = -2\sigma_2 e^{-\hat{\ell}t}\hat{\varepsilon} \sum_{i=1}^{n} u_{2x_i}^{*2} + \hat{\varepsilon}\sigma_2 u_2^* e^{-\hat{\ell}t}\hat{R} < -\sigma_2\hat{\varepsilon}e^{-\hat{\ell}t}\hat{P} < 0$$

for some positive constant $\hat{P}$ independent of $\hat{\varepsilon}$. Consequently $\sigma_2\Delta w_2 + w_2[b + f_2(0,w_2)] - \partial w_2/\partial t < 0$ for $x \in \hat{\mathcal{F}}$, $t > 0$. In the complement of $\hat{\mathcal{F}}$ in Ω, the first term of the last line of (2.4-4) is bounded above by $-\hat{D}e^{-\hat{\ell}t}$ for some positive constant $\hat{D}$, and the remaining terms will have absolute value $<(\hat{D}/2)e^{-\hat{\ell}t}$ for small enough $\hat{\varepsilon}$. The expression in (2.4-4) is therefore < 0 for $x \in \Omega\setminus\hat{\mathcal{F}}$, $t > 0$. $w_2(x,t)$ is an upper solution.

Since $u_2^*(x) > \delta\omega(x)$, $\delta > 0$, and $u_2^* \in H^{2+\ell}(\bar{\Omega})$, the outward unit normal derivative of u_2^* is bounded above by a negative constant. Thus, by choosing $\theta_1(x)$, $\theta_2(x)$ and their first partial derivatives to be sufficiently close to that of 0 and $u_2^*(x)$ and their first partial derivatives respectively, we have $v_i(x,T) < u_i(x,T)$ $w_i(x,\bar{T})$ for $x \in \bar{\Omega}$, $i = 1,2$. By Theorem 1.2-7, we conclude that $v_i(x,t) < u_i(x,t)$

$< w_i(x,t)$, for $(x,t) \in \bar{\Omega}x[T,\infty)$, $i = 1,2$. By the choice of v_i, w_i, we see that $(u_1(x,t),u_2(x,t)) \to (0,u_2^*(x))$ as $t \to \infty$, uniformly for $x \in \bar{\Omega}$. This completes the proof.

When the growth rate of both competing species are "small" in the sense of the following Theorem, then (0,0) is a "globally" stable steady state for the homogeneous Dirichlet problem.

Theorem 2.4-3. Let $a < \sigma_1\lambda_1$, $b < \sigma_2\lambda_1$ and $(u_1(x,t), u_2(x,t))$ with $u_i \in H^{2+\ell,1+\ell/2}(\bar{\Omega}x[0,T])$, each $T > 0$, $i = 1,2$ be a solution of system (2.3-1) with initial boundary conditions:

$$
\begin{aligned}
&u_i(x,0) = \theta_i(x) > 0, \quad i = 1,2,\ x \in \bar{\Omega} \\
(2.4\text{-}5) \qquad & \\
&u_i(x,t) = 0 \qquad\qquad , \quad i = 1,2,\ (x,t) \in \partial\Omega x[0,\infty)
\end{aligned}
$$

where θ_i, $i = 1,2$ satisfy the compatibility condition of order 1 at $t = 0$ as described in section 1.3. Then $(u_1(x,t), u_2(x,t)) \to (0,0)$ uniformly for $x \in \bar{\Omega}$, as $t \to \infty$. (Note (2.4-1) is not assumed.)

Proof. Let $v_1(x,t) = v_2(x,t) \equiv 0$ for $(x,t) \in \bar{\Omega}x[0,\infty)$. Define $w_1(x,t)$ exactly as it is in Theorem 2.4-2. Let $w_2(x,t)$ be the solution of the initial value problem: $\sigma_2\Delta w_2 + w_2[b + f_2(0,w_2)] - \frac{\partial w_2}{\partial t} = 0$ for $(x,t) \in \Omega x(0,\infty)$, $w_2(x,0) = \theta_2(x)$ for $x \in \Omega$, $w_2(x,t) = 0$ for $(x,t) \in \partial\Omega x[0,\infty)$. As for w_1, we proved in the same way as in the first part of the proof in Theorem 2.4-2 that $w_2(x,t) \to 0$ uniformly too, for $x \in \bar{\Omega}$, as $t \to \infty$. By Theorem 1.2-7, the proof is complete.

Remark. An immediate consequence is that, under the conditions of the last theorem (0,0) is the only possible nontrivial nonnegative steady state.

The most interesting situation arise when the growth rates of both competing species are large enough to sustain coexistence under <u>zero</u> Dirichlet boundary conditions. Suppose that

(2.4-6) $$a > \sigma_1\lambda_1 \; , \; b > \sigma_2\lambda_1$$

and there are positive constants $\bar{k}_1$, $\bar{k}_2$ such that

$$a - \sigma_1\lambda_1 + f_1(0,\bar{k}_2) > 0$$

(2.4-7) $$b + f_2(0,\bar{k}_2) < 0$$

$$b - \sigma_2\lambda_1 + f_2(\bar{k}_1,0) > 0$$

$$a + f_1(\bar{k}_1,0) < 0$$

The following Theorem illustrates the existence of positive coexistence. (Note that (2.4-7) can happen when $|\partial f_i/\partial u_j|$, $i \neq j$ is small compared with $|\partial f_i/\partial u_i|$, $1 \leq i, j \leq 2$)

<u>Theorem</u> 2.4-3. <u>Consider the boundary value problem</u>

(2.4-8)
$$\begin{aligned} &\sigma_1\Delta u_1 + u_1[a + f_1(u_1,u_2)] = 0 \\ &\sigma_2\Delta u_2 + u_2[b + f_2(u_1,u_2)] = 0 \end{aligned} \qquad x \in \Omega$$

$$u_i(x) = 0, \; x \in \delta\Omega, \; i = 1,2$$

<u>where</u> a, b, σ_1, σ_2, f_1, f_2 <u>satisfy conditions in section</u> 2.3 <u>through</u> (2.3-2) <u>to</u> (2.3-3). <u>Suppose further that</u> (2.4-6) <u>and</u> (2.4-7) <u>are valid, then</u> (2.4-8) <u>has a solution</u> $(u_1^0(x), u_2^0(x))$ <u>with</u> $u_i^0(x) \in H^{2+\ell}(\bar{\Omega})$, $u_i^0(x) > 0$, <u>for</u> $x \in \Omega$, $i = 1,2$.

Proof. Let $\omega(x)$ be the principal eigenfunction for the eigenvalue problem $\Delta u + \lambda u = 0$ in Ω, $u = 0$ on $\delta\Omega$, with $\lambda = \lambda_1$ as the principal eigenvalue (thus $\omega(x) > 0$ in Ω). For $k_1 > 0$ small enough, we have $\sigma_1\Delta(k_1\omega) + k_1\omega[a + f_1(k_1\omega,u_2)] = k_1\omega[a - \sigma_1\lambda_1 + f_1(k_1\omega,u_2)] > 0$, and $\sigma_1\Delta\bar{k}_1 + \bar{k}_1[a + f_1(\bar{k}_1,u_2)] < 0$ for $0 \leq u_2 \leq \bar{k}_2$. Also for $k_2 > 0$ small enough, we have $\sigma_2\Delta(k_2\omega) + k_2\omega[b + f_2(u_1,k_2\omega)] = k_2\omega[b - \sigma_2\lambda_1 + f_2(u_1,k_2\omega)] > 0$, and $\sigma_2\Delta\bar{k}_2 + \bar{k}_2[b + f_2(u_1,\bar{k}_2)] < 0$ for $0 \leq u_1 \leq \bar{k}_1$. Thus by Theorem 1.4-2 there exists a solution $(u_1^0(x), u_2^0(x))$ to (2.4-8) with $k_i\omega(x) \leq u_i^0(x) \leq \bar{k}_i$, $u_i^0(x) \in H^{2+\ell}(\bar{\Omega})$, $i = 1,2$. Since $\omega(x) > 0$ in Ω, the theorem is proved.

The following stability theorem can be applicable to the steady state solution of Theorem 2.4-3.

Theorem 2.4-4. *Consider system (2.3-1) with boundary conditions*

$$u_i(x,t) \equiv 0,\ i = 1,2 \text{ for } t > 0,\ x \in \partial\Omega,$$

and a, b, σ_1, σ_2, f_1, f_2 *satisfying conditions in section* 2.3 *through* (2.3-2) *to* (2.3-3). *Here*, (2.3-3) *is assumed to hold in the first closed quadrant. Let* $(\bar{u}_1(x),\bar{u}_2(x))$, *with* $\bar{u}_i(x) \in H^{2+\ell}(\bar{\Omega})$ $i = 1,2$ *be a steady state solution* (i.e., *solution of* (2.4-8)), *which satisfies* $\bar{u}_i(x) > 0$ *in* Ω, $\partial\bar{u}_i/\partial\eta < 0$ *on* $\delta\Omega$, $i = 1,2$, *and*

$$\sup_{x\in\Omega}\left|\frac{\bar{u}_i(x)}{\bar{u}_j(x)}\,\frac{(\partial f_j/\partial u_i)(\bar{u}_1(x),\bar{u}_2(x))}{(\partial f_j/\partial u_j)(\bar{u}_1(x),\bar{u}_2(x))}\right| < \inf_{x\in\Omega}\left|\frac{\bar{u}_i(x)}{\bar{u}_j(x)}\,\frac{(\partial f_i/\partial u_i)(\bar{u}_1(x),\bar{u}_2(x))}{(\partial f_i/\partial u_j)(\bar{u}_1(x),\bar{u}_2(x))}\right| < \infty$$

for each $1 \leq i,\ j \leq 2$, $i \neq j$.

(*Here* $\partial\eta$ *means outward normal derivative*). *Then* $(\bar{u}_1(x), \bar{u}_2(x))$ *is asymptotically stable*. (Here, asymptotic stability means that for any solution $(u_1(x,t), u_2(x,t))$ of the initial boundary value problem (2.3-1) (2.4-5) with $\theta_i(x)$ and its first partial derivatives close enough to that of $\bar{u}_i(x)$ for all $x \in \bar{\Omega}$, $i = 1, 2$, one has $u_i(x,t) \to \bar{u}_i(x)$ uniformly as $t \to \infty$, $i = 1,2$.)

Remark. Since we assume the existence of steady state $(\bar{u}_1(x),\bar{u}_2(x))$ here, the hypotheses (2.4-6) and (2.4-7) in Theorem 2.4-3 can be omitted.

The proof of Theorem 2.4-4 is similar to that of Theorem 2.3-1, combining with some of the methods in Theorem 2.4-2 in the analysis near the boundary. Details and an example can be found in [136]; they are hence omitted here. For further recent developments in the topics related to this section, see the note at the end of this chapter.

2.5. Related Basic Existence, Uniqueness Theory and A-Priori Estimates

In this section we describe some fundamental results similar to those in section 1.3. The theory is presented in slightly different function spaces. This will lead to an existence theorem concerning an initial-boundary value problem for a partial differential equation coupled with an ordinary differential equation. The subject is closely related to topics in this chapter in studying the interaction of two biological species or chemical reactants when one is affected by diffusion while the other is not. More related and general results will be further discussed in the next section. We assume that Ω is a bounded domain in R^n.

Let $u(x,t)$ be defined on $\Omega x(0,T]$, and α be any number $0 < \alpha < 1$, we define the following norms:

(2.5-1) $$|u|_0 = \sup\{|u(x,t)| : (x,t) \in \Omega x(0,T]\}$$

(2.5-2) $$|u|_\alpha = |u|_0 + \sup\left\{\frac{|u(x,t)-u(x',t')|}{(|x-x'|^2+|t-t'|)^{\alpha/2}} : \begin{array}{l}(x,t),\ (x't') \in \Omega x(0,T],\\ (x,t) \neq (x',t')\end{array}\right\}$$

(2.5-3) $$|u|_{1+\alpha} = |u|_\alpha + \sum_{i=1}^{n} \left|\frac{\partial u}{\partial x_i}\right|_\alpha$$

(2.5-4) $$|u|_{2+\alpha} = |u|_{1+\alpha} + \sum_{i=1}^{n} \left|\frac{\partial u}{\partial x_i}\right|_{1+\alpha} + \left|\frac{\partial u}{\partial t}\right|_\alpha$$

when the corresponding derivatives exist. For $q = 0, \alpha, 1+\alpha, 2+\alpha$, if $|u|_q < \infty$, we say $u \in C^q$. Such norms will also be used later in section 5.6. Further, we define

(2.5-5) $$|u|_1 = |u|_0 + \sup \left\{ \frac{|u(x,t)-u(x',t')|}{|x-x'| + |t-t'|} : (x,t),(x',t') \in \Omega \times (0,T] \atop (x,t) \neq (x',t') \right\}$$

If $|u|_1 < \infty$, we say $u \in \tilde{C}^1$; and if $\partial u/\partial x_i \in \tilde{C}^1$ for each i, then we say $u \in \tilde{C}^2$.

The cylindrical surface $\partial\Omega \times (0,T]$ is said to be of class $\tilde{C}^2$ or $C^{2+\alpha}$ if it can be locally represented as $x_i = X(x_1, \ldots x_{i-1}, x_{i+1}, \ldots x_n, t)$ for some i, with the function X in the class $\tilde{C}^2$ or $C^{2+\alpha}$ respectively. The surface $\partial\Omega \times (0,T]$ is covered by a finite number of neighborhoods, in each of which there is a fixed representation of the form described. When the functions $u(x,t)$ is only considered in $\partial\Omega \times (0,T]$, it can be written in each cover neighborhood as a function of $(x_1,\ldots x_{i-1},x_{i+1}\ldots,x_n,t)$, and $|u|_1^{\partial\Omega \times (0,T]}$ is taken as the maximum of such $|u|_1$ over all these neighborhoods.

We denote the parabolic operator L on $\bar{\Omega}_T \overset{(def)}{\equiv} \bar{\Omega} \times [0,T]$ by:

(2.5-6) $$L \equiv \sum_{i,j=1}^{n} a_{ij}(x,t) \frac{\partial^2}{\partial x_i \partial x_j} + \sum_{i=1}^{n} b_i(x,t) \frac{\partial}{\partial x_i} + c(x,t) - \frac{\partial}{\partial t},$$ satisfying

there exists a positive member μ such that

(2.5-7) $$\sum_{i,j=1}^{n} a_{ij}\, \xi_i\, \xi_j \geq \mu \sum_{i=1}^{n} \xi_i^2$$

for all ξ_i, $(x,t) \in \bar{\Omega}_T$, and

(2.5-8) a_{ij}, b_i, c are continuous in $\bar{\Omega}_T$ with all their $|\ |_\alpha < \infty$, and further a_{ij} are in $\tilde{C}^1$.

These assumptions will always be made for the operator L in this section.

The following two theorems by A. Friedman are fundamental in analyzing the coupled system in Theorem 2.5-3.

Theorem 2.5-1. *Suppose that* $\partial\Omega x(0,T]$ *is in the classes* $\tilde{C}^2$ *and* $C^{2+\alpha}$; *and* $f(x,t)$ *is a continuous function on* $\overline{\Omega}_T$, *with* $f(x,0) = 0$ *for* $x \in \delta\Omega$. *Let* $u(x,t)$ *be a solution of the problem:*

$$L[u] = f(x,t) \quad , \quad (x,t) \in \Omega x(0,T]$$

$$u = 0 \quad , \quad (x,t) \in (\partial\Omega x(0,T]) \bigcup (\overline{\Omega} x\{0\}),$$

Then for any $0 < \delta < 1$, *there exist a constant* K *depending only on* δ, T, Ω, μ, $\sum_{i,j=1}^{n} |a_{ij}|_\alpha$, $\sum_{i=1}^{n} |b_i|_0$, $|c|_0$ *and* $\sum_{i,j=1}^{n} |a_{ij}|_1^{\partial\Omega x(0,T]}$ *such that*

$$|u|_{1+\delta} \leq K|f|_0$$

(Note that if the boundary surface is not cylindrical, a $C^{2+\alpha}$ surface is not necessarily $\tilde{C}^2$).

Theorem 2.5-2. *Suppose that* $\partial\Omega x(0,T]$ *is in class* $\tilde{C}^2$ *and* $C^{2+\alpha}$, $f(x,t,u)$ *is Hölder continuous in bounded subsets of* (x,t,u)*-space* $(-\infty < u < \infty)$ *with some exponent in* $(0,1)$, *and that there exists constants* K_1, K_2 *such that:*

$$|f(x,t,u)| \leq K_1 + K_2|u| \quad \text{for all } (x,t,u).$$

Then for any function $\hat{\phi}(x,t)$ *in the class* $C^{2+\alpha}$ *in* $\overline{\Omega}_T$ *there exists a solution to the problem*

$$L[u] = f(x,t,u) \quad \text{for } (x,t) \in \Omega x(0,T]$$

$$u(x,t) = \phi(x,t) \quad \text{for } (x,t) \in (\partial\Omega x(0,T]) \bigcup (\overline{\Omega} x\{0\})$$

(*Here* ϕ *is the restriction of* $\hat{\phi}$ *to the initial and lateral boundary surface*). *Furthermore, the solution is of the class* $C^{1+\beta}$ *in* $\Omega x(0,T]$ *for any* $0 < \beta < 1$ *and of the class* $C^{2+\gamma}$ *in* $\overline{\Omega}_T$ *for some* $\gamma > 0$.

The proof of Theorems 2.5-1 and 2.5-2 can be found in [82]. Theorem 2.5-2 is proved by using Theorem 2.5-1 and Schauder's fixd point theorem 1.3-5 (cf [60]).

Theorems 2.5-1, 2.5-2 and differential inequalities as in section 1.2 lead to the following theorem in [224] through an iterative argument.

Theorem 2.5-3. *Assume that:*

(i) $\Omega\times(0,T]$ *belongs to* $\tilde{C}^2$ *and* $C^{2+\alpha}$,

(ii) $f(x,t,u,v)$ *and* $g(x,t,u,v)$ *are of class* $\tilde{C}^1$ *in* $\bar{\Omega}_T$, *and both satisfy the uniform Lipschitz condition:*

(2.5-9) $$|h(x,t,u_1,v_1) - h(x,t,u_2,v_2)| \leq M(|u_1-u_2| + |v_1-v_2|) \text{ for all } (u_1,v_1),\ (u_2,v_2),\ (x,t) \in \bar{\Omega}_T,$$

(iii) *There exists a function* ϕ *of class* $C^{2+\alpha}$ *in* $\bar{\Omega}_T$ *which coincides with the boundary values* $u_0(x,t)$ *of* u *on* $\partial\Omega\times(0,T]$ *and* $\bar{\Omega}\times\{0\}$,

(iv) *the initial values* $v_0(x,0)$ *is of class* $\tilde{C}^1$ *on* $\bar{\Omega}$,

(v) *there exist functions* $\underline{u}$ *and* $\bar{u}$, $\underline{v}$ *and* $\bar{v}$ *continuous in* $\bar{\Omega}_T$ *and having continuous bounded derivatives in* $\Omega\times(0,T]$ *such that for all* v *satisfying* $\underline{v} \leq v \leq \bar{v}$,

$$L(\underline{u}) - f(x,t,\underline{u},v) \geq 0 \geq L(\bar{u}) - f(x,t,\bar{u},v) \text{ in } \bar{\Omega}_T,$$

$$\underline{u} \leq u_0 \leq \bar{u} \quad \text{on } (\partial\Omega\times(0,T]) \cup (\bar{\Omega}\times\{0\}),$$

and for all u *satisfying* $\underline{u} \leq u \leq \bar{u}$,

$$\frac{\partial\bar{v}}{\partial t} - g(x,t,u,\bar{v}) \geq 0 \geq \frac{\partial\underline{v}}{\partial t} - g(x,t,u,\underline{v}) \text{ in } \bar{\Omega}_T$$

$$\underline{v} \leq v_0 \leq \bar{v} \quad \text{on } \Omega\times\{0\}.$$

Then there exists a solution of the system

$$
(2.5\text{-}10)\qquad
\begin{aligned}
&L[u] = f(x,t,u,v) \\
&\frac{\partial v}{\partial t} = g(x,t,u,v)
\end{aligned}
\quad \text{for } (x,t) \in \Omega\times(0,T]
$$

$$u = u_0(x,t) \quad \text{on} \quad (\partial\Omega\times(0,T]) \quad (\bar{\Omega}\times\{0\})$$

$$v = v_0(x,0) \quad \text{on} \quad \bar{\Omega}\times\{0\}$$

such that u *is of class* $C^{1+\beta}$ *in* $\Omega\times(0,T]$ *for any* $0 < \beta < 1$ *and of class* $C^{2+\gamma}$ *in* $\bar{\Omega}_T$ *for some* $\gamma > 0$, v *is of class* $\tilde{C}^1$ *in* $\bar{\Omega}_T$ *and* $\underline{u} \le u \le \bar{u}$, $\underline{v} \le v \le \bar{v}$.

The proof of Theorem 2.5-3 is similar to Theorem 2.6-1 in the next section, and will be omitted here. The following uniqueness theorem applicable to problem (2.5-10) is also included in [224].

Theorem 2.5-4. *Let* f *and* g *satisfy the uniform Lipschitz condition* (2.5-9). *Suppose* (u_1,v_1), (u_2,v_2) *are two solutions of problem* (2.5-10) *with the property that* u_1, v_1, u_2, v_2 *are all continuous in* $\bar{\Omega}_T$ *and their second-order* x_i *derivatives and first-order* t-*derivatives exist and uniformly bounded in* $\Omega\times(0,T]$. *Then* $u_1 \equiv u_2$, $v_1 \equiv v_2$.

The proof of the above Theorem uses maximum principle and differential inequalities as in section 1.2. For details, see [224].

We finally state another theorem concerning a-priori gradient estimates for solutions of nonlinear equations with a different type of boundary condition which may also be nonlinear. The theorem is similar to Theorem 2.5-1, and will be used in the next section and chapter for proving existence of solutions. We consider the problem

$$\frac{\partial u}{\partial t} = \sigma\Delta u + f(x,t,u) \quad \text{for} \quad (x,t) \in \Omega x(0,T)$$

(2.5-11)

$$\frac{\partial u}{\partial n} = h(x,t,u) \quad \text{for} \quad (x,t) \in \partial\Omega x[0,T].$$

Here n is the outward unit normal at the boundary $\partial\Omega$, and $\sigma > 0$. The function h is differentiable with respect to u, x, t and $\frac{\partial^2 h}{\partial u^2}$, $\frac{\partial^2 h}{\partial x_i \partial u}$, $\frac{\partial^2 h}{\partial t \partial u}$ exist in the region

(2.5-12) $\quad (x,t) \in \bar{\Omega}_T \ , \ |u| \leq M \ .$

In the region (2.5-12), h satisfies

$$|h| \leq \mu, \ |\frac{\partial h}{\partial u}| \leq \mu \ , \ |\frac{\partial h}{\partial t}| \leq \mu \ , \ |[\sum_{i=1}^{n} (\frac{\partial h}{\partial x_i})^2]^{1/2}| \leq \mu$$

(2.5-13)

$$|\frac{\partial^2 h}{\partial u^2}| \leq \mu, \ |\frac{\partial^2 h}{\partial t \partial u}| \leq \mu \ , \ |[\sum_{i=1}^{n} (\frac{\partial^2 h}{\partial x_i \partial u})^2]^{1/2}| \leq \mu \ .$$

The function f is defined and differentiable with respect to u and t in region (2.5-12); and it satisfies in there

(2.5-14) $\quad |f| \leq \mu, \ |\frac{\partial f}{\partial u}| \leq \mu, \ |\frac{\partial f}{\partial t}| \leq \mu$

Theorem 2.5-5. *Suppose that the functions* f(x,t,u) *and* h(x,t,u) *have the smoothness properties described above, and satisfy* (2.5-13) *and* (2.5-14) *in region* (2.5-12). *Let* u(x,t) *be a function with continuous derivative in* t *and continuous first and second derivatives with respect to* x *in* $\bar{\Omega}_T$, *and* $\max_{\bar{\Omega}_T} |u(x,t)| \leq M$. *If* $\partial\Omega \in H^{2+\alpha}$ *for some* $0 < \alpha < 1$, *and* u(x,t) *satisfies* (2.5-11), *then it must have the properties:*

(2.5-15) $\quad \max_{\bar{\Omega}_T} |[\sum_{i=1}^{n} (\frac{\partial u}{\partial x_i})^2]^{1/2}| \leq M_1 \ , \ |u|_{\Omega_T}^{(1+\delta)} \leq c \ .$

Here, the constants M_1, c *and* $0 < \delta < 1$ *depend only on* σ, M *and* μ *from* (2.5-11) *to* (2.5-13), *the maximum in* $\bar{\Omega}$ *for* $|u(x,0)|, |\frac{\partial u}{\partial x_i}(x,0)|$, $|\frac{\partial^2 u}{\partial x_i \partial x_j}(x,0)|$, $1 \le i$, $j \le n$ *and the boundary* $\partial\Omega$. (See section 1.3 for the definition of $|\cdot|_{\Omega_T}^{(1+\delta)}$).

This theorem (in a even more general form) can be found in section 7 of Chapter V in [126].

2.6. P.D.E. Coupled with a System of O.D.E. Several Species Competing for One Prey.

In the study of resource managements and biochemical reactions, one encounters components which are affected by diffusion and some others which are not. This leads to partial differential equations with boundary conditions coupled with an initial-value problem for ordinary differential equations. When one possesses more specific knowledge of the properties of the interactions, one can consider various steady states and stabilities as in sections 2.2 to 2.4. Such considerations are more elaborate study of problems of the type in Theorem 2.5-3. In this section, we follow the results of A. Leung and B. Bendjilali in [142]. Although the systems considered are not very general, however, the assumptions and methods used are typical of such problems. They are also related to sections 2.3 and 2.4. We consider n capital investments $K_1, \ldots, K_n$ exploiting a renewable resource u growing in a domain Ω. The intrinsic growth rate a(x) of u is spatially dependent; and u has diffusion rate σ, with no flux at the boundary. The mixed initial-boundary value problem is:

$$\frac{\partial u}{\partial t} = \sigma\Delta u + u[a(x) + f(u, K_1, \ldots, K_n)]$$

(2.6-1) for $x \in \Omega$, $t > 0$

$$\frac{\partial K_i}{\partial t} = K_i[-b_i + c_i u - \sum_{k=1}^{n} d_{ik}K_k] , \quad i = 1, \ldots, n$$

$$u(x,0) = u_0(x) \quad , \quad x \in \bar{\Omega}$$
$$(2.6\text{-}2) \qquad K_i(x,0) = g_i(x) \quad , \quad x \in \bar{\Omega} \; , \; i = 1, \ldots, n$$
$$\frac{\partial u}{\partial n} = 0 \qquad , \quad (x,t) \in \partial\Omega \times [0,T].$$

where $(x,t) = (x_1, \ldots, x_m, t) \in \bar{\Omega}\times[0,T]$, Ω is a bounded domain in R^m and n is the unit outward normal at the boundary $\partial\Omega$. The parameters σ, b_i, c_i and d_{ii} are positive constants for $i = 1, \ldots, n$, and $d_{ik} > 0$ for all i,k with $i \neq k$. The n capitals K_i are competing against each other (with Volterra-Lotka type competition) while the resource u acts as the prey. The following hypotheses will be made on the function $f(u, K_1, \ldots, K_n)$ and $a(x)$ in (2.6-1):

(i) f is differentiable everywhere with respect to u and K_i, $i = 1, \ldots, n$. $\partial f/\partial u$, $\partial f/\partial K_i$ satisfy the Lipschitz condition in compact sets, i.e.
$|h(u, K_1, \ldots, K_n) - h(\tilde{u}, \tilde{K}_1, \ldots, \tilde{K}_n)| < C(|u-\tilde{u}| + \sum_{i=1}^{n} |K_i - \tilde{K}_i|)$
for some $C > 0$ depending on the compact set.

(ii) $\partial f/\partial u < 0$ and $\partial f/\partial K_i < 0$ for $i = 1, \ldots, n$ in the set $\{(u, K_1, \ldots, K_n) | u > 0, K_i > 0, i = 1, \ldots, n\}$.

(iii) The function $a(x)$ is in $H^{\alpha}(\bar{\Omega})$, and there exists a constant $U^* > 0$ such that $a(x) + f(U^*, 0, \ldots, 0) < 0$ for all $x \in \bar{\Omega}$.

The model describes commercial fishing with u as the fish population, and the conditions in (ii) reflects the crowding effect of overpopulation and the exploitive effect of capital investment.

To state results specifically, we need the following notations. Let $\Omega_T = \Omega\times(0,T)$ and $\bar{\Omega}_T$ be its closure. $\mathcal{C}^1(\bar{\Omega})$ denotes the class of real functions u in $\bar{\Omega}$ with $\sup_{x\in\Omega} |u(x)| + \sup\{\frac{|u(x)-u(x')|}{|x-x'|} : x,x' \in \Omega, x \neq x'\}$ finite; and $\mathcal{C}^1(\bar{\Omega}_T)$ denotes the real functions u in $\bar{\Omega}_T$ with

$$\sup_{(x,t)\in\bar{\Omega}_T} |u(x,t)| + \sup\{\frac{|u(x,t)-u(x',t')|}{|x-x'| + |t-t'|} : (x,t), (x',t') \in \bar{\Omega}_T, (x,t) \neq (x',t')\}$$

finite. $C^{2,1}(\bar{\Omega}_T)$ denotes the class of real function in $\bar{\Omega}_T$ with continuous second derivatives in x and continuous first derivative in t in $\bar{\Omega}_T$.

The following existence theorem for problem (2.6-1), (2.6-2) is similar to Theorem 2.5-3. However, the equation is now x-dependent, the boundary condition is of different type, the solution function space is different, and the ordinary differential equations form a system with more specific property. Consequently, a complete proof is given below, and it includes the essential ideas in Theorem 2.5-3. Bounds for the solution are also included.

Theorem 2.6-1. Let $u_0(x)$ and $g_i(x)$, $i = 1, \ldots, n$ be nonnegative functions defined on $\bar{\Omega}$, with none of them identically zero. Suppose u_0 is in $H^{2+\alpha}(\bar{\Omega})$ and g_i is in $\tilde{C}^1(\bar{\Omega})$, $i = 1, \ldots, n$, where $(\partial u_0/\partial n)(x) = 0$ for all $x \in \delta\Omega$. Then for any $T > 0$, equations (2.6-1) for $(x,t) \in \bar{\Omega}_T$, (under hypotheses (i) to (iii)), with initial boundary conditions (2.6-2), has a solution $(\hat{u}(x,t), \hat{K}_1(x,t), \ldots, \hat{K}_n(x\ t))$ with $\hat{u}(x,t) \in H^{2+\alpha,1+\alpha/2}(\bar{\Omega}_T)$ and $\hat{K}_i(x,t) \in \tilde{C}^1(\bar{\Omega}_T)$, $i = 1, \ldots, n$. Furthermore, $0 \leqq \hat{u}(x,t) \leqq M_0$ and $0 \leqq \hat{K}_i(x,t) \leqq M_i$ for $(x,t) \in \bar{\Omega}_T$, $i = 1, \ldots, n$. Here $M_0 = \max\{\sup_{\bar{\Omega}} u_0(x), U^*\} + 1$, and $M_i = \max\{\sup_{\bar{\Omega}} g_i(x) + 1, (1/d_{ii})(c_iM_0 - b_i + 1\}$, $i = 1, \ldots, n$.

Proof. We will construct sequences of functions $u_j(x,t)$ and $K_{ij}(x,t)$, $j = 1, 2,$ from which subsequences will be extracted to converge respectively to $\hat{u}(x,t)$ and $\hat{K}_i(x,t)$ in $\bar{\Omega}_T$. Define $K_{i1}(x,t)$ in $\bar{\Omega}_T$ to be the solution of

$$\frac{\partial K_{i1}}{\partial t}(x,t) = K_{i1}(x,t)[-b_i + c_iM_0 - \sum_{k=1}^{n} d_{ik}K_{k1}(x,t)] \text{ for } (x,t) \in \bar{\Omega}_T,$$

$$K_{i1}(x,)) = g_i(x) \geqq 0 \quad \text{for } x \in \bar{\Omega}$$

$i = 1, \ldots, n$. The local existence theorem for ordinary differential equations uniquely defines $K_{i1}(x,t)$ for small t. Moreover, if $K_{i1}(\hat{x},\hat{t}) = 0$ for some $(\hat{x},\hat{t})$, then $K_{i1}(\hat{x},t) \equiv 0$ for all t in the interval of existence of K_{i1}. Consequently $K_{i1}(x,t) \geqq 0$ as long as they exist. On the other hand $-b_i + c_iM_0 - \sum_{k=1}^{n} d_{ik}K_{k1}\ (x,t) \leqq -b_i + c_iM_0 - d_{ii}K_{i1}(x,t)$ which is <0 when $K_{i1}(x,t) = M_i$; therefore we have $0 \leqq K_{i1}(x,t) \leqq M_i$ for $i = 1, \ldots, n$, and they exist for $(x,t) \in \bar{\Omega}_T$. Since $g_i(x)$ is in $\tilde{C}^1(\bar{\Omega})$, we conclude by a variant of Theorem 7.4 of [49, Chap. 1] that $K_{i1}(x,t)$ is in $\tilde{C}^1(\bar{\Omega}_T)$, (here, x is viewed as a parameter). The details here involve finding a bound for $|K_{i1}(x,t) - K_{i1}(\tilde{x},\tilde{t})|/[|x-\tilde{x}| + |t-\tilde{t}|]$ for all (x,t), $(\tilde{x},\tilde{t})$ in $\bar{\Omega}_T$. Let $h_i(t) = K_{i1}(x,t) - K_{i1}(\tilde{x},t)$, $i = 1, \ldots, n$, for $t \in [0,T]$, where x, $\tilde{x}$ are arbitrary points in $\bar{\Omega}$. From the differential equations for K_{i1}, we deduce that

$$\left|\frac{dh_i}{dt}(t)\right| < A_1 \sum_{k=1}^{n} |h_k(t)| \ , \text{ for } t \in [0,T]$$

where A_1 depends on b_i, c_i, d_{ik}, M_i, $i = 1, \ldots, n$ and M_0. The function

$$H(t) \overset{(def)}{\equiv} \sum_{k=1}^{n} |\, h_k(t)|, \ t \in [0,T]$$

thus satisfies:

$$H(t) = \sum_{i=1}^{n} \left| \int_0^t \frac{dh_i}{dt}(s)\, ds + (g_i(x) - g_i(\tilde{x}))\right| \leqq nA_1 \int_0^t H(s)\, ds + A_2|x-\tilde{x}|$$

for some positive constant A_2. The above inequality gives, by means of Gronwall's inequality (see e.g. [98]):

$$|K_{i1}(x,t) - K_{i1}(\tilde{x},t)| \leqq A_3|x-\tilde{x}|$$

where A_3 is a positive constant, for $t \in [0,T]$, x, $\tilde{x}$ arbitrary points in $\bar{\Omega}$. Similarly, we deduce that

$$|K_{i1}(x,t) - K_{i1}(x,\tilde{t})| \leqq A_4|t-\tilde{t}|$$

for some positive constant A_4, $t, \tilde{t} \in [0,T]$, x arbitrary in $\bar{\Omega}$. The last two inequalities imply that $K_{i1}(x,t)$ is in $\tilde{C}^1(\bar{\Omega}_T)$. (Even more details of similar nature are given later in deducing inequalities (2.6-8) t (2.6-11)).

We now inductively define, for $j = 1, 2, \ldots$, $u_j(x,t)$ and $K_{ij+1}(x,t)$ to be solutions of:

$$\frac{\partial u_j}{\partial t} = \sigma\Delta u_j + u_j[a(x) + f(u_j, K_{1j}(x,t), \ldots, K_{nj}(x,t)] \text{ for } (x,t) \in \bar{\Omega}_T,$$

$$(2.6\text{-}3) \qquad u_j(x,0) = u_0(x), \quad x \in \bar{\Omega},$$

$$\frac{\partial u_j}{\partial \eta}(x,t) = 0, \quad (x,t) \in \delta\Omega x[0,T].$$

$$(2.6\text{-}4) \qquad \frac{\partial K_{ij+1}}{\partial t}(x,t) = K_{ij+1}[-b_i + c_i u_j - \sum_{k=1}^{n} d_{ik}K_{kj+1}], \quad (xt) \in \bar{\Omega}_T,$$

$$K_{ij+1}(x,0) = g_i(x) \geqq 0, \quad x \in \bar{\Omega}$$

for $i = 1, \ldots, n$. We will inductively prove the assertion that: $u_j(x,t)$ is in $H^{2+\alpha\,1+\alpha/2}(\bar{\Omega}_T)$ and $0 \leqq u_j(x,t) \leqq M_0$ in $\bar{\Omega}_T$; while K_{ij+1} is in $\tilde{C}^1(\bar{\Omega}_T)$ and $0 \leqq K_{ij+1}(x,t) \leqq M_i$ in $\bar{\Omega}_T$ for $i = 1, \ldots, n$. For this purpose, define $u_0(x,t) \equiv u_0(x$ in $\bar{\Omega}_T$, so by assumption and the last paragraph the induction hypothesis is true for $j = 0$.

We first define a few symbols and terms so that Theorems A1.4 and A1.5 can be readily applied. Let $\bar{f}(u, K_1, \ldots, K_n) = f(h(u), K_1, \ldots, K_n)$ everywhere, where

$$h(u) = \begin{cases} u & \text{if } |u| \leqq M_0 \\ 2M_0 - M_0 e^{(M_0-u)/M_0} & \text{if } u \geqq M_0 \\ -h(-u) & \text{if } u \leqq -M_0. \end{cases}$$

Note that $h(u)$ is a continuously differentiable function for $-\infty < u < \infty$ and $\lim_{u\to\pm\infty} h(u) = \pm 2M_0$. Let

$$b(x,t,v) = -(v + u_0(x))[a(x) + \overline{f}(v + u_0(x), K_{1j}(x,t), \ldots, K_{nj}(x,t)] - \sigma\Delta u_0(x)$$

for $(x,t) \in \overline{\Omega}_T$, $-\infty < v < \infty$. We consider the family of problems:

$$(2.6\text{-}5)\qquad \begin{aligned} &\frac{\partial v}{\partial t} = \sigma\Delta v - rb(x,t,v), \quad (x,t) \in \overline{\Omega}_T, \\ &\sigma\frac{\partial v}{\partial n} + \frac{(1-r)\sigma^2}{(1-r)\sigma+r} v = 0, \quad (x,t) \in \delta\Omega \times [0,T], \\ &v(x,0) = 0, \quad x \in \overline{\Omega} \end{aligned}$$

for $0 \leqq r \leqq 1$. We apply Theorem A1.4 with u, $a_{ij}(x,t)$, $f(x,t,u)$ and $\psi(x,t,u)$ respectively replaced by v, σ, $rb(x,t,v)$ and $((1-r)\sigma^2/((1-r)\sigma + r))v$. Since $|\overline{f}(v + u_0(x), K_{1j}(x,t), \ldots, K_{nj}(x,t))| \leqq \max\{|f(u, K_1, \ldots, K_n)| : |u| \leqq 2M_0, 0 \leqq K_i \leqq M_i, i = 1, \ldots, n\}$ provided $0 \leqq K_{ij}(x,t) \leqq M_i$, for $(x,t) \in \overline{\Omega}_T$, $i = 1, \ldots, n$, the second condition in (A1.8) is readily satisfied. From Theorem A1.4 we obtain a uniform bound:

$$\max_{\overline{\Omega}_T} |v(x,t)| < \tilde{M}$$

for any $C^{2,1}(\overline{\Omega}_T)$ solution of (2.6-5) all $r \in [0,1]$, $(x,t) \in \overline{\Omega}_T$. We next apply Theorem A1.5 with $a_{ij}(x,t)$, $f(x,t,u)$, $\psi(x,t,u)$ respectively replaced by σ, $b(x,t,v)$, 0. Here we assume inductively that $K_{ij}(x,t)$ are in $\tilde{C}^1(\overline{\Omega}_T)$, $i = 1, \ldots, n$, for the definition of $b(x,t,v)$. Condition (d) is satisfied with u, M respectively replaced by v, $\tilde{M}$. (Note that $b(x,t,v)$ is Hölder continuous in x with exponent α). Consequently, we conclude the existence of a unique solution for (2.6-5) in $H^{2+\alpha,1+\alpha/2}(\overline{\Omega}_T)$, for each $r \in [0,1]$. For $r = 1$, we define $v \equiv v_j(x,t)$, and $u_j(x,t) = v_j(x,t) + u_0(x)$ for $(x,t) \in \overline{\Omega}_T$. The function $u_j(x,t)$ satisfies:

$$\frac{\partial u_j}{\partial t} = \sigma\Delta u_j + u_j[a(x) + \overline{f}(u_j, K_{1j}(x,t), \ldots, K_{nj}(x,t))] \text{ in } \overline{\Omega}_T,$$

(2.6-6)
$$u_j(x,0) = u_0(x), \quad x \in \overline{\Omega},$$

$$\frac{\partial u_j}{\partial n} = 0, \quad (x,t) \in \delta\Omega x[0,T]$$

by (2.6-5) when $r = 1$. Moreover $u \equiv 0$ is a lower solution for (2.6-6) with u_j replaced by u; and $u \equiv M_0$ is a corresponding upper solution, see remark 1.2-3 (because

$$M_0[a(x) + \overline{f}(M_0, K_{1j}(x,t), \ldots, K_{nj}(x,t))]$$
$$= M_0[a(x) + f(M_0, K_{1j}(x,t), \ldots)] \leqq M_0[a(x) + f(M_0, 0, \ldots, 0)] < 0$$

by (iii)). By a comparison principle (see Theorem 1.2-4 and remark 1.2-1, or [28]) we conclude $0 \leqq u_j(x,t) \leqq M_0$ in $\overline{\Omega}_T$. We thus proved the first part of the induction assertion, because $\overline{f}(u, K_1, \ldots, K_n)$ is the same as f when $|u| \leqq M_0$, and $u_j(x,t)$ actually satisfies (2.6-3) as well as (2.6-6).

Having obtained $u_j(x,t)$ in $H^{2+\alpha,1+\alpha/2}(\overline{\Omega}_T)$, with $0 \leqq u_j \leqq M_0$, we now can assert the existence of K_{ij+1} for $i = 1, \ldots, n$, satisfying problem (2.6-4). The proof is exactly the same as that in the first paragraph of this proof for K_{i1}. Note that now $-b_i + c_i u_j(x,t) - \sum_{k=1}^{n} d_{ik} K_{kj+1} \leqq -b_i + c_i M_0 - d_{ii}K_{ij+1}(x,t)$ which is <0 when $K_{ij+1}(x,t) = M_i$. As before, we therefore have $K_{ij+1}(x,t)$ in $\mathcal{C}^1(\overline{\Omega}_T)$ with $0 \leqq K_{ij+1}(x,t) \leqq M_i$ for $i = 1, \ldots, n$.

We next deduce some uniform bound for $|u_j(x,t)|_{\Omega_T}^{(1+\delta)}$ for some $\delta > 0$, and for $|K_{ij}(x,t) - K_{ij}(\tilde{x},\tilde{t})/[|x-\tilde{x}| + |t-\tilde{t}|]$, all (x,t), $(\tilde{x},\tilde{t})$ in $\overline{\Omega}_T$, $i = 1, \ldots, n$, uniformly for all $j = 1, 2, \ldots$. For the functions u_j, we apply Theorem 2.5-5, since u_j satisfies (2.6-3), the functions $f(x,t,u)$, $h(x,t,u)$ correspond to $u_j[a(x) + f(u_j, K_{1j}(x,t), \ldots, K_{nj}(x,t))]$ and 0 respectively. Since we have uniform bounds for $|u_j(x,t)|$ and $|K_{ij}(x,t)|$ for all $j = 1, 2, \ldots$, and $\partial K_{ij}/\partial t$ can be bounded by using the right side of (2.6-4), (this is necessary for the condition on $\partial f/\partial t$ in (2.5-14)), we conclude that (2.5-13) and (2.5-14) are all satisfied uniformly for all j. Consequently, we have

(2.6-7) $$|u_j|_{\Omega_T}^{(1+\delta)} \leqq c$$

for some c and $\delta > 0$, all $j = 1, 2, \ldots$. To estimate the difference of K_{ij} at two different points uniformly in j, we let

$$h_{ij}(t) \equiv K_{ij}(x,t) - K_{ij}(\tilde{x},t), \text{ for } t \in [0,T], \ i = 1, \ldots, n,$$

where x, $\tilde{x}$ are two arbitrary points in $\bar{\Omega}$. By means of (2.6-4), $j \geqq 0$, we readily deduce that for $j = 1, 2, \ldots$:

(2.6-8) $$\left|\frac{dh_{ij}(t)}{dt}\right| \leqq B_1 \sum_{k=1}^{n} |h_{kj}(t)| + B_2|u_j(x,t) - u_j(\tilde{x},t)|$$

$i = 1, \ldots, n$, $t \in [0,T]$, where B_1 and B_2 depend on the uniform bounds M_i, $i = 0, \ldots, n$ and the constants b_i, c_i, d_{ik}. Letting $H_j(t) \equiv \sum_{i=1}^{n} |h_{ij}(t)|$, $t \in [0,T]$, we have the inequality:

(2.6-9) $$H_j(t) = \sum_{i=1}^{n} \left| \int_0^t \frac{dh_{ij}}{ds}(s)ds + [g_i(x)-g_i(\tilde{x})]\right| \leqq nB_1 \int_0^t H_j(s)ds + [nB_2cT + B_3]|x-\tilde{x}|$$

by means of (2.6-7) and (2.6-8). Here B_3 is a constant, and $B_3|x - \tilde{x}|$ arises from estimating $\sum_{i=1}^{n}|g_i(x) - g_i(\tilde{x})|$. By the Gronwall's inequality, (2.6-9) gives for $t \in [0,T]$:

(2.6-10) $$|K_{ij}(x,t) - K_{ij}(\tilde{x},t)| < B_4| x-\tilde{x}|$$

uniformly for $j = 1,2, \ldots$, where B_4 is a constant, and x, $\tilde{x}$ are arbitrary points in $\bar{\Omega}$. To estimate $|K_{ij}(x,t) - K_{ij}(x,\tilde{t})|$ for an arbitrary fixed $x \in \bar{\Omega}$, we use (2.6-4) to deduce

$$|K_{ij}(x,t) - K_{ij}(x,\tilde{t})| \leq \int_{\tilde{t}}^{t} |K_{ij}(x,s)\,[-b_i + c_i u_{j-1}(x,s) - \sum_{k=1}^{n} d_{ik}K_{kj}(x,s)]|ds.$$

Hence from the uniform bound for u_j and K_{ij} etc., we obtain for $t,\tilde{t} \in \{0,T]$, $i = 1, \ldots, n$:

(2.6-11) $$|K_{ij}(x,t) - K_{ij}(x,\tilde{t})| \leq B_5|t-\tilde{t}|$$

uniformly for $j = 1, 2, \ldots$, where B_5 is a constant, and x arbitrary in $\bar{\Omega}$. (Note that (2.6-10) and (2.6-11) also inductively imply that $K_{ij}(x,t)$ are in $\tilde{C}^1(\bar{\Omega}_T)$). Inequalities (2.6-10) and (2.6-11) give uniform bounde for

$$|K_{ij}(x,t) - K_{ij}(\tilde{x},\tilde{t})|/[|x-\tilde{x}| + |t-\tilde{t}|], \quad \text{for all } (x,t), (\tilde{x},\tilde{t}) \text{ in } \bar{\Omega}_T.$$

Inequalities (2.6-7), (2.6-10) and (2.6-11) together with Ascoli's lemma imply that we can choose subsequences $\{u_{j_k}\}$ and $\{K_{ij_k}\}$, $i = 1, \ldots, n$, $k = 1, 2, \ldots$ convergent uniformly in $\bar{\Omega}_T$ respectively to say $\tilde{u}(x,t)$ and $\tilde{K}_i(x,t)$, $i = 1, \ldots, n$. Moreover, since $0 < \alpha < 1$, conditions (2.6-7), (2.6-10) and (2.6-11) imply that we can assert $u_{j_k}(x,t) \to \tilde{u}(x,t)$ and $K_{ij_k}(x,t) \to \tilde{K}_i(x,t)$ in the space $H^{\alpha,\alpha/2}(\bar{\Omega}_T)$ as $k \to \infty$, for $i = 1, \ldots, n$ (for analogous arguments, see e.g. [83, Thm. 1, p. 188].

To obtain a solution of (2.6-1) and (2.6-2), we finally consider

$$\frac{\partial u}{\partial t} = \sigma\Delta u + \tilde{u}(x,t)[a(x) + f(\tilde{u}(x,t), \tilde{K}_1(x,t), \ldots, \tilde{K}_n(x,t)] \text{ in } \bar{\Omega}_T.$$

(2.6-12) $$u(x,0) = u_0(x), \qquad x \in \bar{\Omega},$$

$$\frac{\partial u}{\partial \eta}(x,t) = 0, \qquad (x,t) \in \delta\Omega\times[0,T].$$

$$\frac{\partial K_i}{\partial t} = \tilde{K}_i[-b_i + c_i\tilde{u} - \sum_{k=1}^{n} d_{ik}\tilde{K}_i] \quad \text{in } \bar{\Omega},$$

(2.6-13)

$$K_i(x,0) = g_i(x), \quad x \in \bar{\Omega},\ i = 1, \ldots, n.$$

Since $\tilde{u}$ and $\tilde{K}_i$, $i = 1, \ldots, n$ are in $H^{\alpha,\alpha/2}(\bar{\Omega}_T)$, the inhomogeneous term of the first linear equation in (2.6-12) is in $H^{\alpha,\alpha/2}(\bar{\Omega}_T)$. By Theorem 1.3-2, (2.6-12) has a unique solution $u = \bar{u}(x,t)$ in $H^{2+\alpha,1+\alpha/2}(\bar{\Omega}_T)$. We will next show that $\bar{u} \equiv \tilde{u}$ in $\bar{\Omega}_T$. Letting $w_k = \bar{u} - u_{j_k}$, it satisfies:

$$\frac{\partial w_k}{\partial t} = \sigma\Delta w_k + \phi_k(x,t) \text{ in } \bar{\Omega}_T,$$

(2.6-14)

$$w_k(x,0) = 0, \quad x \in \bar{\Omega},$$

$$\frac{\partial w_k}{\partial \eta} = 0, \quad (x,t) \in \delta\Omega x[0,T]$$

where $\phi_k(x,t) = \tilde{u}[a + f(\tilde{u}, \tilde{K}_1, \ldots, \tilde{K}_n)] - u_{j_k}[a + f(u_{j_k}, K_{1j_k}, \ldots, K_{nj_k})]$.

From the theorem just mentioned, we have $|w_k|_{\Omega_T}^{(2+\alpha)} \to 0$ provided that $|\phi_k|_{\Omega_T}^{(\alpha)} \to 0$ as $k \to \infty$. For convenience, we let $\omega(x, u, K_1, \ldots, K_n) = u[a(x) + f(u, K_1, \ldots, K_n)]$ for $x \in \bar{\Omega}$, $|u| < \infty$, $|K_i| < \infty$, $i = 1, \ldots, n$. The main difficulty to see that $|\phi_k(x,t)|_{\Omega_T}^{(\alpha)} \to 0$ as $k \to \infty$ is to show that expressions of the form $|\phi_k(x,t) - \phi_k(\hat{x},t)|/|x-\hat{x}|^{\alpha} \to 0$ as $k \to \infty$ where $x, \hat{x} \in \bar{\Omega}$. To show this, we first write:

$$\begin{aligned}
&\phi_k(x,t) - \phi_k(\hat{x},t) \\
&= \omega(x,\tilde{u}(x,t),\tilde{K}_1(x,t), \ldots, \tilde{K}_n(x,t)) - \omega(x,u_{j_k}(x,t), K_{1j_k}(x,t) \ldots) \\
&\quad - \omega(\hat{x},\tilde{u}(\hat{x},t), \tilde{K}_1(\hat{x},t), \ldots) + \omega(\hat{x},u_{j_k}(\hat{x},t), K_{1j_k}(\hat{x},t)\ldots) \\
&= [\int_0^1 \frac{\partial\omega}{\partial u} (x,s\tilde{u}(x,t) + (1-s)u_{j_k}(x,t),s\tilde{K}_1(x,t) + (1-s)K_{1j_k}(x,t),\ldots)ds] \\
&\qquad \cdot[\tilde{u}(x,t) - u_{j_k}(x,t)]
\end{aligned}$$

$$+\sum_{i=1}^{n}\left[\int_0^1 \frac{\partial\omega}{\partial K_i}(x,s\tilde{u}(x,t)+(1-s)u_{j_k}(x,t),s\tilde{K}_1(x,t)+(1-s)K_{1j_k}(x,t),\ldots)ds\right]$$

$$\cdot[\tilde{K}_i(x,t)-K_{ij_k}(x,t)]$$

$$-\left[\int_0^1 \frac{\partial\omega}{\partial u}(\hat{x},s\tilde{u}(\hat{x},t)+(1-s)u_{j_k}(\hat{x},t),s\tilde{K}_1(\hat{x},t)+(1-s)K_{1j_k}(\hat{x},t),\ldots)ds\right]$$

$$\cdot[\tilde{u}(\hat{x},t)-u_{j_k}(\hat{x},t)] \tag{2.6-15}$$

$$-\sum_{i=1}^{n}\left[\int_0^1 \frac{\partial\omega}{\partial K_i}(\hat{x},s\tilde{u}(\hat{x},t)+(1-s)u_{j_k}(\hat{x},t),\ldots)ds\right][\tilde{K}_i(\hat{x},t)-K_{ij_k}(\hat{x},t)]]$$

$$=\left[\int_0^1 \frac{\partial\omega}{\partial u}(x,s\tilde{u}(x,t)+(1-s)u_{j_k}(x,t),s\tilde{K}_1(x,t)+(1-s)K_{1j_k}(x,t),\ldots)ds\right]$$

$$\cdot\{[\tilde{u}(x,t)-u_{j_k}(x,t)]-[\tilde{u}(\hat{x},t)-u_{j_k}(\hat{x},t)]\}$$

$$-\left[\int_0^1 \frac{\partial\omega}{\partial u}(\hat{x},s\tilde{u}(\hat{x},t)+(1-s)u_{j_k}(\hat{x},t),\ldots)\right.$$

$$\left.\frac{\partial\omega}{\partial u}(x,s\tilde{u}(x,t)+(1-s)u_{j_k}(x,t),\ldots)ds\right][\tilde{u}(\hat{x},t)-u_{j_k}(\hat{x},t)]$$

$$+\sum_{i=1}^{n}\left[\int_0^1 \frac{\partial\omega}{\partial K_i}(x,s\tilde{u}(x,t)+(1-s)u_{j_k}(x,t),\ldots)ds\right]$$

$$\cdot\{[\tilde{K}_i(x,t)-K_{ij_k}(x,t)]-[\tilde{K}_i(\hat{x},t)-K_{ij_k}(\hat{x},t)]\}$$

$$-\sum_{i=1}^{n}\left[\int_0^1 \frac{\partial\omega}{\partial K_i}(\hat{x},s\tilde{u}(\hat{x},t)+(1-s)u_{j_k}(\hat{x},t),\ldots)\right.$$

$$\left.-\frac{\partial\omega}{\partial K_i}(x,s\tilde{u}(x,t)+(1-s)u_{j_k}(x,t),\ldots)ds\right][\tilde{K}_i(\hat{x},t)-K_{ij_k}(\hat{x},t)].$$

Then we use the fact $|[\tilde{u}(x,t)-u_{j_k}(x,t)]-[\tilde{u}(\hat{x},t)-u_{j_k}(\hat{x},t)]|/|x-\hat{x}|^\alpha \to 0$ as $k\to\infty$, and that terms of the form

$$|\frac{\partial\omega}{\partial u}(\hat{x},s\tilde{u}(\hat{x},t)+(1-s)u_{j_k}(\hat{x},t),s\tilde{K}_1(\hat{x},t)+(1-s)K_{1j_k}(\hat{x},t),\ldots)$$

$$-\frac{\partial\omega}{\partial u}(x,s\tilde{u}(x,t)+(1-s)u_{j_k}(x,t),s\tilde{K}_1(x,t)+(1-s)K_{1j_k}(x,t)\ldots)|$$

can be estimated by means of (i) and the uniform bound for u_{j_k}, K_{j_k} to be less than:

$$C_1[s|\tilde{u}(\hat{x},t)-\tilde{u}(x,t)|+(1-s)|u_{j_k}(\hat{x},t)-u_{j_k}(x,t)|$$

$$+\sum_{i=1}^{n}\{s|\tilde{K}_i(\hat{x},t)-\tilde{K}_i(x,t)|+(1-s)|K_{ij_k}(\hat{x},t)-K_{ij_k}(x,t)|\}]\leq C_2|x-\hat{x}|^{\alpha}$$

where C_1, C_2 are constants (for the last inequality above, we use (2.6-7), (2.6-10) and (2.6-11). Terms on the right of (2.6-15) involving $\tilde{K}_i$, K_{j_k} and $\partial\omega/\partial K_i$ in place of $\tilde{u}$, u_{j_k} and $\partial\omega/\partial u$ are estimated in the analogous way. We can subsequently conclude from (2.6-15) that $|\phi_k(x,t)-\phi_k(\hat{x},t)|/|x-\hat{x}|^{\alpha}\to 0$ as $k\to\infty$, and further that $|\phi_k(x,t)|_{\Omega_T}^{(\alpha)}\to 0$ as $k\to\infty$. Going back to (2.6-14) we deduce that $|w_k|_{\Omega_T}^{(2+\alpha)}\to 0$ as $k\to\infty$; that is, $u_{j_k}\to\bar{u}$ in Ω_T. We therefore must have $\bar{u}(x,t)\equiv\tilde{u}(x,t)$ for $(x,t)\in\bar{\Omega}_T$.

For equation (2.6-13) it must have a solution $K_i=\bar{K}_i(x,t)$ in $\bar{\Omega}_T$, $i=1$, $\ldots$, n (with argument similar to the first paragraph of the proof). Considering the ordinary differential equations for the functions $[\bar{K}_i(x,t)-K_{ij_k}(x,t)]$, together with their initial conditions, we can prove as in the previous paragraph that $K_{ij_k}(x,t)\to\bar{K}_i(x,t)$ uniformly in $\bar{\Omega}_T$ as $k\to\infty$, $i=1,\ldots,n$. Hence $\bar{K}_i(x,t)\equiv\tilde{K}_i(x,t)$ in $\bar{\Omega}_T$, $i=1,\ldots,n$.

Finally, let $(\hat{u}(x,t),\hat{K}_1(x,t),\ldots,\hat{K}_n(x,t))=(\tilde{u}(x,t),\tilde{K}_1(x,t),\ldots,\tilde{K}_n(x,t))$ in $\bar{\Omega}_T$. From (2.6-12), (2.6-13) and the last two paragraphs, it is the solution described in the statement of the theorem. Since $0\leq u_{j_k}\leq M_0$ and

$0 \leqq K_{ij_k} \leqq M_i$, $i = 1, \ldots, n$, the corresponding bounds must be satisfied by $\hat{u}$, $\hat{K}_i$ in Ω_T. The function $\hat{u} \equiv \tilde{u} \equiv \bar{u}$ is in $H^{2+\alpha,1+\alpha/2}(\bar{\Omega}_T)$; and (2.6-10), (2.6-11) imply that $\hat{K}_i \equiv \tilde{K}_i$ is in $\tilde{C}^1(\bar{\Omega}_T)$, $i = 1, \ldots, n$.

Theorem 2.6-1 extends to more general situations when the subsystem of ordinary differential equations is replaced by $\partial K_i/\partial t = K_i q_i(u_1, K_1, \ldots, K_n)$, $i = 1, \ldots, n$. With appropriate assumptions on q_i, one has an invariant region for the ordinary differential equations in the form $\Sigma = \{(K_1, \ldots, K_n): 0 < K_i < M_i\}$ for all $0 < u < M_0$. With further smoothness condition of q_i, the arguments of the proof of Theorem 2.6-1 can be completed in the same way. Consideration of uniqueness will be omitted in order not to prolong the present discussion excessively.

In the remaining part of this section, we investigate the existence of nonnegative steady-state solutions for (2.6-1), (2.6-2). We also deduce a sufficient condition for its asymptotic stability. To avoid excessive complexity, we only restrict to the case $d_{ik} = 0$ when $i \neq k$ (and consequently abbreviate d_{ii} as d_i for convenience). In [142], d_i represents the rate of depreciation of K_i; the quantity d_{ik}, $i \neq k$, is appropriately assumed to be zero because the depreciation of K_i should not influence the rate of change of K_k, $k \neq i$. More precisely, we consider:

(2.6-16)
$$\sigma\Delta u + u[a(x) + f(u, K_1, \ldots, K_n)] = 0, \quad \text{for } x \in \bar{\Omega},$$
$$K_i[-b_i + c_i u - d_i K_i] = 0, \; i = 1, \ldots, n, \quad \text{for } x \in \bar{\Omega}$$

(2.6-17)
$$\frac{\partial u}{\partial \eta} = 0, \text{ for } x \in \partial\Omega$$

where σ, b_i, c_i, d_i, $i = 1, \ldots, n$, are positive constants. The functions $a(x)$ and f satisfy all the assumptions (i) to (iii) described at the beginning of this section.

Theorem 2.6-2. Let $P = \max\{(b_i/c_i) | i = 1, \ldots, n\}$. Suppose that

(2.6-18) $$a(x) + f(P, R_1, \ldots, R_n) \geqq 0 \quad \text{for all } x \in \bar{\Omega},$$

where $R_i = d_i^{-1}(c_i P - b_i)$, $i = 1, \ldots, n$. Then the boundary value problem (2.6-16), (2.6-17) has a nonnegative equilibrium solution $(\bar{u}(x), \bar{K}_1(x), \ldots, \bar{K}_n(x))$ with $\bar{u}(x) > 0$, $\bar{K}_i(x) \geqq 0$ in $\bar{\Omega}$, $i = 1, \ldots, n$. Each component of the solution is in $H^{2+\alpha}(\bar{\Omega})$. (Note that $R_i \geqq 0$, $i = 1, \ldots, n$, and $R_{i*} = 0$ if $P = b_{i*}/c_{i*}$).

Proof. Expressing each K_i in terms of u by means of the second equation in (2.6-16) and substituting into the first equation, we obtain the problem:

(2.6-19)
$$\sigma\Delta u + u[a(x) + f(u, G_1(u), \ldots, G_n(u))] = 0 \quad \text{in } \bar{\Omega}$$
$$\frac{\partial u}{\partial \eta} = 0 \quad , x \in \partial\Omega$$

where $G_i(u) = d_i^{-1}[c_i u - b_i]$. Let $v(x) \equiv P$ in $\bar{\Omega}$. We have $\sigma\Delta v + v[a(x) + f(v, G_1(v), \ldots, G_n(v))] = P[a(x) + f(P, R_1, \ldots, R_n)] \geq 0$ by (2.6-18); and $\frac{\partial v}{\partial \eta} \leq 0$. That is, v is a lower solution for (2.6-19). Condition (2.6-18) and hypothesis (ii), (iii) imply that $P < U^*$. Otherwise, suppose $U^* \leq P$, then $0 \leq a(x) + f(P, R_1, \ldots, R_n) \leq a(x) + f(P, 0, \ldots, 0) \leq a(x) + f(U^*, 0, \ldots, 0) < 0$ which is a contradiction. Let $w(x) \equiv U^*$ for $x \in \bar{\Omega}$. We have $\sigma\Delta w + w[a(x) + f(w, G_1(w), \ldots, G_n(w))] < U^*[a(x) + f(U^*, 0, \ldots, 0)] < 0$ in $\bar{\Omega}$, because $G_i(U^*) > G_i(P) = R_i \geq 0$; and $0 \leq \frac{\partial w}{\partial \eta}$. Therefore w is an upper solution of (2.6-19). By the remark following Theorem 1.4-2, or section 5.1 and [71] there is a solution to the boundary value problem (2.6-19), $u = \bar{u}(x)$, with $0 < P \equiv v \leq \bar{u}(x) \leq w \equiv U^*$, for all $x \in \bar{\Omega}$. Define $\bar{K}_i(x) = G_i(\bar{u}(x))$, $x \in \bar{\Omega}$, $i = 1, \ldots, n$. Clearly $(\bar{u}(x), \bar{K}_1(x), \ldots, \bar{K}_n(x))$ satisfies (2.6-16) and (2.6-17). Further $K_i(x) \geq d_i^{-1}[c_i P - b_i] = R_i \geq 0$ for all $x \in \bar{\Omega}$, $i = 1, \ldots, n$.

Remarks. (i) Suppose that in Theorem 2.6-2 condition (2.6-18) is modified to

(2.6-20) $$a(x) + f(P, R_1, \ldots, R_n) > 0 \text{ for all } x \in \bar{\Omega},$$

then the equilibrium solution $(\bar{u}(x), \bar{K}_1(x), \ldots, \bar{K}_n(x))$ described would have the additional property that $\bar{K}_i(x) > 0$ in $\bar{\Omega}$, $i = 1, \ldots, n$ (as well as $\bar{u}(x) > 0$). To see this, we let $R_i^\varepsilon = d_i^{-1}[c_i(P+\varepsilon)-b_i]$; one has $a(x) + f(P+\varepsilon, R_1^\varepsilon, \ldots, R_n^\varepsilon) > 0$ for $x \in \bar{\Omega}$ if $\varepsilon > 0$ is small enough. Reduce $\varepsilon > 0$ if necessary so that $P + \varepsilon < U^*$. We have $P+\varepsilon$ and U^* as constant lower and upper solutions respectively for (2.6-19). Consequently (2.6-19) has a solution $\bar{u}(x)$, with $P+\varepsilon \le \bar{u}(x) \le U^*$. Thus $\bar{K}_i(x) = d_i^{-1}[c_i\bar{u}(x)-b_i] \ge d_i^{-1}[c_i(P+\varepsilon)-b_i] > 0$ for $x \in \bar{\Omega}$.

(ii) If $a(x) \equiv$ constant, $n = 1$ and $f(u,K)$ is linear of the form $f(u,K) = -\lambda u-\gamma K$ (with $\lambda > 0$, $\gamma > 0$). The existence of a critical point with both species positive for the vector field $(u[a + f(u,K)], K[-b+cu-dK])$ is equivalent to $a/\lambda > b/c$. On the other hand, condition (2.6-20) becomes $a-\lambda(b/c) - \gamma R_1 = a - (\lambda b/c) > 0$, (with $P = b/c$, $R = R_1 = 0$). These conditions are thus equivalent.

Finally, we will study the stability of the equilibrium solution described above, as a solution of the parabolic problem:

(2.6-21)
$$\frac{\partial u}{\partial t} = \sigma\Delta u + u[a(x) + f(u, K_1, \ldots, K_n)]$$
$$\frac{\partial K_i}{\partial t} = K_i[-b_i + c_i u - d_i K_i] \qquad (x,t) \in \bar{\Omega}_T$$

$i = 1, \ldots, n$, with initial and boundary conditions (2.6-2) for each $T > 0$.

The following lemma is needed in proving the stability Theorem 2.6-3.

Lemma 2.6-1. For $(x,t) \in \bar{\Omega}\times[0,\infty)$, let $v(x,t)$, $w(x,t)$ be functions in $H^{2+\alpha,1+\alpha/2}($ and let $\tilde{v}_i(x,t)$, $\tilde{w}_i(x,t)$, $i = 1, \ldots, n$ be functions in $\mathcal{C}^1(\bar{\Omega}_T)$, each $T > 0$.

Suppose that in $\bar{\Omega}x[0,\infty)$:

$$0 < v < w \text{ and } 0 < \tilde{v}_i < \tilde{w}_i, \; i = 1, \dots, n .$$

(2.6-22)

$$\sigma\Delta v + v[a(x) + f(v, \tilde{w}_1, \dots, \tilde{w}_n)] - \frac{\partial v}{\partial t} > 0$$

$$\sigma\Delta w + w[a(x) + f(w, \tilde{v}_1, \dots, \tilde{v}_n)] - \frac{\partial w}{\partial t} < 0$$

(2.6-23)

$$\tilde{v}_i[-b_i + c_i v - d_i\tilde{v}_i] - \frac{\partial \tilde{v}_i}{\partial t} > 0 \qquad i = 1, \dots, n$$

$$\tilde{w}_i[-b_i + c_i w - d_i\tilde{w}_i] - \frac{\partial \tilde{w}_i}{\partial t} < 0$$

Let $(u(x,t), K_1(x,t), \dots, K_n(x,t))$ *be a solution of* (2.6-21), u *in* $H^{2+\alpha,1+\alpha/2}(\bar{\Omega}_T)$ *and* K_i *in* $\mathcal{C}^1(\bar{\Omega}_T)$ *for each* $T > 0$, $i = 1, \dots, n$, *with initial boundary conditions such that*:

(2.6-24)

$$v(x,0) < u(x,0) < w(x,0) \; , \; x \in \bar{\Omega}$$

$$\frac{\partial v}{\partial \eta} < \frac{\partial u}{\partial \eta} < \frac{\partial w}{\partial \eta} \quad \text{for} \quad (x,t) \in \partial\Omega x[0,\infty)$$

(2.6-25)

$$\tilde{v}_i(x,0) < K_i(x,0) < \tilde{w}_i(x,0), \; x \in \bar{\Omega}, \; i = 1, \dots, n.$$

Then $(u(x,t), K_1(x,t), \dots, K_n(x,t))$ *will satisfy*:

(2.6-26)

$$v(x,t) < u(x,t) < w(x,t), \; (x,t) \in \bar{\Omega}x[0,\infty)$$

$$\tilde{v}_i(x,t) < K_i(x,t) < \tilde{w}_i(x,t) \; , \; (x,t) \in \bar{\Omega}x[0,\infty), \; i = 1, \dots, n$$

The proof of Lemma 2.6-1 is similar to those in Section 1.2, and will be omitted here. For more details, see [142].

The next theorem gives a sufficient condition for the asymptotic stability of an equilibrium (steady-state) solution.

Theorem 2.6-3. Let $(\bar{u}(x), \bar{K}_1(x), \ldots, \bar{K}_n(x))$ be an equilibrium solution to (2.6-16), (2.6-17), with each component function strictly positive in $\bar{\Omega}$, and in $H^{2+\alpha}(\bar{\Omega})$, as described in the remark following Theorem 2.6-2. Suppose further that:

(2.6-27)
$$\min_{1\le i\le n}\{d_i \min \bar{K}_i(x)/\max c_i\bar{u}(x)\}\ \frac{[\min \bar{u}(x)][-\max\frac{\partial f}{\partial u}(\bar{u}(x),\bar{K}_1(x),\ldots,\bar{K}_n(x))]}{n \max\limits_{1\le i\le n}\{\max \bar{K}_i(x)\cdot|\min \frac{\partial f}{\partial K_i}(\bar{u}(x),\bar{K}_1(x),\ldots,\bar{K}_n(x))|\}} > 1$$

Then $(\bar{u}(x), \bar{K}_1(x), \ldots, \bar{K}_n(x))$ is asymptotically stable. (In (2.6-27) all max and min are taken over $x \in \bar{\Omega}$).

(Here asymptotic stability is interpreted to mean that for any solution $(u(x,t), K_1(x,t), \ldots, K_n(x,t))$ with $u \in H^{2+\alpha,1+\alpha/2}(\bar{\Omega}_T)$, $K_i \in \tilde{C}^1(\bar{\Omega}_T)$, $i = 1, \ldots, n$, each $T > 0$, of the equation (2.6-21) with boundary condition $\frac{\partial u}{\partial n} = 0$, $(x,t) \in \delta\Omega\times[0,\infty)$ and initial conditions $(u_0(x), g_1(x), \ldots, g_n(x))$ described in Theorem 2.6-1 close enough to $(\bar{u}(x), \bar{K}_1(x), \ldots, \bar{K}_n(x))$ for all $x \in \bar{\Omega}$, one has $u(x,t) \to \bar{u}(x)$ and $K_i(x,t) \to \bar{K}_i(x)$, $i = 1, \ldots, n$ uniformly as $t \to +\infty$).

Proof. Inequality (2.6-27) remains valid if the max and min for $\frac{\partial f}{\partial u}$ and $\frac{\partial f}{\partial K_i}$, $i = 1, \ldots, n$ are taken over a small neighborhood $\mathcal{D}$ of the graph of $(\bar{u}, \bar{K}_1, \ldots,$ in the $(x, u, K_1, \ldots, K_n)$ space. In the remaining part of the proof, all such max and min will be taken over such a neighborhood, (of course max and min for $\bar{u}, \bar{K}_1, \ldots, \bar{K}_n$ are taken over $x \in \bar{\Omega}$), unless otherwise stated. For $M > 0$, sufficiently large, one clearly has

(2.6-28) $$\left(\frac{N-1}{N}\right)^3 Q^2 > 1$$

where Q denotes the expression on the left of (2.6-27).

We will construct v, $\tilde{v}_i$ and w, $\tilde{w}_i$ (lower and upper solutions) as described in Lemma 2.6-1, with the additional property of tending to $\bar{u}$, $\bar{K}_i$ as $t \to \infty$. Let $R = \max_{1 \le i \le n} \{\max_\Omega \bar{K}_i(x) \cdot | \min \frac{\partial f}{\partial K_i} |\}$, and $\delta_2 > 1$ close enough to 1 so that

(2.6-29) $$\frac{(\delta_2 - 1)}{nR} (\min \bar{u})(-\max \frac{\partial f}{\partial u}) < 1$$

(Note that for small enough around $(\bar{u}, \bar{K}_1, \dots, \bar{K}_n)$, $\max \frac{\partial f}{\partial u}$ is negative). Let

$$\alpha_1 = 1 - \left(\frac{N-1}{N}\right) \frac{(\delta_2 - 1)}{nR} (\min \bar{u})(-\max \frac{\partial f}{\partial u})$$

(2.6-30) $$\delta_1 = 1 - (1-\alpha_1) \left(\frac{N-1}{N}\right) \min_{1 \le i \le n} \left\{\frac{d_i \min \bar{K}_i}{d_i \max \bar{K}_i + b_i}\right\}$$

$$\alpha_2 = 1 + (1-\delta_1)\left(\frac{N-1}{N}\right)\left(\frac{1}{nR}\right) (\min \bar{u})(-\max \frac{\partial f}{\partial u})$$

By (2.6-29), we have $0 < \alpha_1 < 1$, and $0 < \delta_1 < 1$. Define $w(x,t) = \hat{\delta}_2(t)\bar{u}(x)$, $v(x,t) = \hat{\delta}_1(t)\bar{u}(x)$ and $\tilde{w}_i(x,t) = \hat{\alpha}_2(t)\bar{K}_i(x)$, $\tilde{v}_i(x,t) = \hat{\alpha}_1(t)\bar{K}_i(x)$, $i = 1, \dots, n$, where $\hat{\delta}_2(t) = 1+(\delta_2-1)e^{-\varepsilon t}$, $\hat{\delta}_1(t) = 1-(1-\delta_1)e^{-\varepsilon t}$, $\hat{\alpha}_2(t) = 1+(\alpha_2-1)e^{-\varepsilon t}$, $\hat{\alpha}_1(t) = 1 - (1-\alpha_1)e^{-\varepsilon t}$. Here ε is a sufficiently small positive constant to be determined. Reduce $\delta_2 > 1$, if necessary, so that for each $(x,t) \in \bar{\Omega} \times [0,\infty)$, the rectangle $\{(u, K_1, \dots, K_n): v(x,t) \le u \le w(x,t), \tilde{v}_i(x,t) \le K_i \le \tilde{w}_i(x,t), i = 1, \dots, n\}$ is contained in the cross section $\{(u, K_1, \dots, K_n): (x, u, K_1, \dots, K_n) \in \}$. We now check the inequalities (2.6-22) and (2.6-23). We have:

$$\sigma \Delta w + w[a(x) + f(w, \tilde{v}_1, \dots, \tilde{v}_n)] - \frac{\partial w}{\partial t}$$

$$= \bar{u}(x)[\hat{\delta}_2(t) \{f(w, \tilde{v}_1, \dots, \tilde{v}_n) - f(w, \bar{K}_1, \dots, \bar{K}_n) + f(w, \bar{K}_1, \dots, \bar{K}_n)$$

$$\begin{aligned}
(2.6\text{-}31) \qquad & -f(\bar{u}, \bar{K}_1, \ldots, \bar{K}_n)\} + \varepsilon(\delta_2 - 1)e^{-\varepsilon t}] \\
& < \bar{u}(x)e^{-\varepsilon t}[\hat{\delta}_2(t)\{\sum_{i=1}^{n} (-\min \frac{\partial f}{\partial K_i})(1-\alpha_1)\max \bar{K}_i(x) + (\delta_2 - 1)\min \bar{u} \max \frac{\partial f}{\partial u}\} + \varepsilon(\delta_2 - 1)] \\
& < \bar{u}e^{-\varepsilon t}[\hat{\delta}_2(t)\{(\frac{N-1}{N}) \frac{(\delta_2 - 1)}{nR} (\min \bar{u})(-\max \frac{\partial f}{\partial u})\, nR + (\delta_2 - 1)(\min \bar{u})(\max \frac{\partial f}{\partial u}) + \varepsilon(\delta_2 - 1)] \\
& < 0
\end{aligned}$$

for all $(x,t) \in \bar{\Omega}x[0,\infty)$, provided ε satisfies $0 < \varepsilon < \frac{1}{N}(\min \bar{u})(-\max \frac{\partial f}{\partial u})$. For $\tilde{v}_i$, we have for $i = 1, \ldots, n$:

$$\begin{aligned}
& \tilde{v}_i[-b_i + c_i v - d_i \tilde{v}_i] - \frac{\partial \tilde{v}_i}{\partial t} \\
& = \bar{K}_i(x)[\hat{\alpha}_1(t)\{-b_i + \hat{\delta}_1(t)(b_i + d_i \bar{K}_i(x)) - d_i \hat{\alpha}_1(t)\bar{K}_i(x)\} - \varepsilon(1-\alpha_1)e^{-\varepsilon t}] \\
& = \bar{K}_i(x)e^{-\varepsilon t}[\hat{\alpha}_1(t)\{-b_i(1-\delta_1) + d_i \bar{K}_i(x)(\delta_1 - \alpha_1)\} - \varepsilon(1-\alpha_1)] \\
(2.6\text{-}32) \quad & = \bar{K}_i e^{-\varepsilon t}[\hat{\alpha}_1(t)\{\delta_1(b_i + d_i \bar{K}_i(x)) - (\alpha_1 d_i \bar{K}_i(x) + b_i)\} - \varepsilon(1-\alpha_1)] \\
& = \bar{K}_i e^{-\varepsilon t}[\hat{\alpha}_1(1-\alpha_1)\{d_i \bar{K}_i(x) - (\frac{N-1}{N}) \min_{1 \le i \le n} \{\frac{d_i \min \bar{K}_i}{d_i \max \bar{K}_i + b_i}\} (b_i + d_i \bar{K}_i(x)) - \varepsilon\}] \\
& > \bar{K}_i e^{-\varepsilon t}[\hat{\alpha}_1(1-\alpha_1)\{d_i \bar{K}_i(x) - (\frac{N-1}{N})\, d_i \min \bar{K}_i - \varepsilon\}] > 0
\end{aligned}$$

for all $(x,t) \in \bar{\Omega}x[0,\infty)$, provided ε satisfies $0 < \varepsilon < \frac{1}{N} \min_{\bar{\Omega}} \bar{K}_i(x)$, $i = 1, \ldots, n$. For v, we can prove as in (2.6-31) that

$$\sigma \Delta v + v[a(x) + f(v, \tilde{w}_1, \ldots, \tilde{w}_n)] - \frac{\partial v}{\partial t} > 0$$

for all $(x,t) \in \bar{\Omega}x[0,\infty)$, provided ε satisfies $0 < \varepsilon < \frac{\delta_1}{N} (\min \bar{u})(-\max \frac{\partial f}{\partial u})$. Finally, for $\tilde{w}_i$, we can show by using (2.6-30) and hypothesis (2.6-27) that for $\varepsilon > 0$ sufficiently small that

$$\tilde{w}_i[-b_i + c_i w - d_i\tilde{w}_i] - \frac{\partial \tilde{w}_i}{\partial t} < 0,$$

if $(x,t) \in \bar{\Omega}\times[0,\infty)$. More details for deducing these inequalities can be found in [142].

It remains to verify that the initial-boundary inequalities (2.6-24) and (2.6-25) are satisifed so that Lemma 2.6-1 can be applied. They are all readily satisfied by construction, and we conclude that (2.6-26) is true. Clearly, $w(x,t)$ and $v(x,t) \to \bar{u}(x)$, $\tilde{w}_i(x,t)$ and $\tilde{v}_i(x,t) \to \bar{K}_i(x)$, $i = 1, \ldots, n$ as $t \to \infty$ uniformly for $x \in \bar{\Omega}$, by our construction. This proves the theorem.

As an example for application of the theorems in this section, we first consider the Volterra-Lotka type equations:

$$\frac{\partial u}{\partial t} = \sigma\Delta u + u[a(x) - 20u - 10K_1 - 30K_2 - 7K_3], \ \sigma > 0 \text{ arbitrary},$$

$$\frac{\partial K_1}{\partial t} = K_1[-1 + u - K_1]$$

$$(2.6\text{-}33) \qquad \frac{\partial K_2}{\partial t} = K_2[-3/2 + u - K_2]$$

$$\frac{\partial K_3}{\partial t} = K_3[-2/5 + 3/5u - 2/5K_3]$$

for $(x,t) \in \bar{\Omega}\times[0,\infty)$, with initial boundary conditions (2.6-2) satisfying conditions described in Theorem 2.6-1. We assume $a(x) \in H^{\alpha}(\bar{\Omega})$, $0 < \alpha < 1$, and $45 \le a(x) \le 50$, $x \in \bar{\Omega}$. Theorem 2.6-1 readily gives existence of nonnegative solutions to the initial-boundary value problem (2.6-33), (2.6-2) with appropriate upper bounds to the components.

For steady-state solutions to (2.6-33), (2.6-17), corresponding to $\frac{\partial K_i}{\partial t} \equiv 0$, $i = 1, 2, 3$, we apply Theorem 2.6-2. Note that $a(x) + f(u, K_1, \ldots, K_n) \equiv a(x) - 20u - 10K_1 - 30K_2 - 7K_3$; $P = 3/2$, $R_1 = 1/2$, $R_2 = 0$ and $R_3 = 5/4$.

Consequently, $a(x) + f(P, R_1, R_2, R_3) > 45 - 30 - 5 - 0 - 35/4 > 0$ and condition (2.6-20) is satisfied. Referring to the symbols in the remark following Theorem 2.6-2 and letting $\varepsilon = 0.01$, we find that $P+\varepsilon = 1.51$, $R_1^\varepsilon = 0.51$, $R_2^\varepsilon = 0.01$, $R_3^\varepsilon =$ and $a(x) + f(P+\varepsilon, R_1^\varepsilon, R_2^\varepsilon, R_3^\varepsilon) > 45-44.455 = 0.545 > 0$. Referring to hypothesis (iii), we find that we can choose $U^* = 2.6$. By the remark following Theorem 2.6-2, we consequently have a steady state solution $(\overline{u}(x), \overline{K}_1(x), \overline{K}_2(x), \overline{K}_3(x))$ for (2.6-33), (2.6-17) with $1.51 = P+\varepsilon < \overline{u}(x) < U^* = 2.6$, $0.51 = R_1^\varepsilon < \overline{K}_1(x) < \frac{1}{d_1}[c_1U^*-b_1] = 1.6$, and similarly $0.01 < \overline{K}_2(x) < 1.1$, $1.265 < \overline{K}_3(x) < 2.9$, $x \in \overline{\Omega}$.

The stability of the above steady state is not so readily obtainable by using Theorem 2.6-3. For a simple application of Theorem 2.6-3 we consider equations (2.6-33) with only its first equation changed to

$$(2.6\text{-}34) \qquad \frac{\partial u}{\partial t} = \sigma\Delta u + u[a(x) - 20u - K_1 - K_2 - 0.8K_3]$$

and the rest unchanged. (Recall that $45 < a(x) < 50$). As in the above paragraph we find that $P = 3/2$, $R_1 = 1/2$, $R_2 = 0$, $R_3 = 5/4$ and $a(x) + f(P, R_1, R_2, R_3) > 0$. However, letting $\varepsilon = 0.5$, then we have the new $P+\varepsilon = 2$, $R_1^\varepsilon = 1$, $R_2^\varepsilon = 0.5$ and $R_3^\varepsilon = 2$. Moreover $a(x) + f(P+\varepsilon, R_1^\varepsilon, R_2^\varepsilon, R_3^\varepsilon) > 0$, and we have the existence of steady state solution $(\overline{u}(x), \overline{K}_1(x), \overline{K}_2(x), \overline{K}_3(x))$ with boundary condition (2.6-17), with $2 < \overline{u}(x) < 2.6$, $1 < \overline{K}_1(x) < 1.6$, $0.5 < \overline{K}_2(x) < 1.1$ and $2 < \overline{K}_3(x) < 2.9$ for $x \in \overline{\Omega}$. (Here, we again apply the remark following theorem 2.6-2). Furthermore, we can now verify condition (2.6-27) in Theorem 2.6-3, we have $\min_{1<i<3} \{d_i \min\overline{K}_i(x)/\max c_i\overline{u}(x)\} > 0.192$, $[-\max\frac{\partial f}{\partial u}(\overline{u}(x), \overline{K}_1(x), \overline{K}_2(x), \overline{K}_3(x))] = 2$ $\max_{1<i<3} \{\max\overline{K}_i(x) \cdot |\min\frac{\partial f}{\partial K_i}|\} < \max\{(1.6)(1), (1.1)(1), (2.9)(0.8)\} = 2.32$. The expression on the right of (2.6-27) is $> (0.192)\frac{(2)(20)}{(3)(2.32)} \cong 1.103 > 1$. We can therefore apply Theorem 2.6-3 to conclude that the steady state solution $(\overline{u}(x), \overline{K}_1(x), \overline{K}_2(x), \overline{K}_3(x))$ is asymptotically stable.

Notes

Theorem 2.1-1 is a modification of a theorem in William and Chow [229] using methods given in Ladyženskaja, Solonnikov and Ural'ceva [127]. The materials in Section 2.2 and 2.3 are obtained from Leung and Clark [145] and Leung [135], and those in Section 2.4 are from Leung [135], [136]. Further work on competing species concerning existence and uniqueness of positive coexistence steady states with Dirichlet boundary conditions can be found in Cosner and Lazer [57], Pao [178], Dancer [61], and more recently in McKenna and Walter [154], Cantrell and Cosner [39], Li [151], Korman and Leung [121]. Other work on related problems can be found in e.g. Conway, Gardner and Smoller [55], Hadeler et al. [92], [93], de Mottoni [63], Fife [74], Rothe [197], Mimura et al. [158], [159], Blat and Brown [25], [26], Matano [154], [155], Gardner [86], Hernandez [100], Schiaffiao and Tesei [205]. Section 2.5 gathers some results from Friedman [82], Ladyženskaja, Solonnikov and Ural'ceva [127], and Wake [224]. Section 2.6 is adapted from Leung and Bendjilali [142]. Related results concerning order preserving flow by Matano [155] and Hirsch [104] will be presented in Section 4.6. Numerous work concerned primarily with the time-dynamical aspect of interacting species had been done recently, as described by Freedman [78] and Britton [31].

CHAPTER III

Other Boundary Conditions, Nonlinear Diffusion, Asymptotics

3.1 Introduction

In this Chapter, we consider various extensions of the theories in the last chapter in order to include more realistic and general problems. In section 3.2, we study the situation where the boundary condition is coupled, mixed and nonlinear. In prey-predator interaction for example, the predator may be under control at the boundary of the medium, while the prey cannot move across the boundary. Diffusion of predators at the boundary may be adjusted nonlinearly according to populations present, and there might also be some physical limitations to the process. In section 3.3, we analyze the problem when the diffusion rate is density dependent, and thus the Laplacian operator will be modified to become nonlinear and u-dependent. Moreover, the nonlinear nonhomogeneous terms become highly spatially dependent. In section 3.4, we consider the case when diffusion rate of some component is small. More thorough results concerning large-time behavior can be obtained by asympptotic methods. Estimates can be obtained by using an appropriate "reduced" problem.

In every section of this chapter, Ω represents an open bounded connected set in R^n, with $\bar{\Omega}$ denoting its closure. If $n \geq 2$, we always assume that the boundary $\delta\Omega$ is in $H^{2+\alpha}$, $0<\alpha<1$. The symbol η denotes outward unit normal at the boundary. We will not pose the theorems in the most general form, in order to avoid excessive technicalities. For example, we simply use the Laplacian operator rather than general elliptic operator. Many immediate generalizations of such nature is possible. For any nonintegral $s>0$, recall the definition of $H^{s,s/2}(\bar{\Omega}_T)$ and $| \ |^s_{\Omega_T}$ from section 1.3.

3.2 Nonlinear Monotone Boundary Conditions

We first prove an existence theorem for elliptic systems, which is a generalization of Theorem 1.4-2. Theorem 3.2-1 is a slight modification of a theorem in [138]. We consider the following boundary value problem:

$$(3.2\text{-}1)\qquad \begin{aligned} &\Delta u + f(x,u) = 0, && x \in \Omega,\\ &\frac{\partial u}{\partial \eta} + g(x,u) = \beta(x), && x \in \delta\Omega \end{aligned}$$

Here, $u = (u_1,\ldots,u_m)$, the Laplacian Δ and normal derivatives $\partial/\partial\eta$ are interpreted componentwise. The following assumptions will be made for Theorem 3.2-1:

(i) For bounded open sets $B \subset R^m$, $f \in H^{\alpha}(\bar{\Omega}xB,R^m)$, $g \in H^{1+\alpha}(\delta\Omega xB,R^m)$ with respect to $x \in \bar{\Omega}$ or $\delta\Omega$ respectively; and with respect to variables in B, f is Lipschitz uniformly in $\bar{\Omega}$, g is in $H^{1+\alpha}$. $\beta(x) \in H^{2+\alpha}(\delta\Omega,R^m)$.

(ii) We introduce a closed, bounded set $J \subset \bar{\Omega}xR^m$ such that for all $x \in \bar{\Omega}$, $J_x \equiv \{v \in R^m\colon (x,v) \in J\}$ is nonempty and convex. Define ϕ, F: $\bar{\Omega}xR^m \to R$, and G: $\delta\Omega xR^m \to R$ by

$\phi(x,\xi) = \max\,\{\xi\cdot v : v \in J_x\}$;

$F(x,\xi) = \max\,\{\xi\cdot f(x,v) : v \in I_{x,\xi}\}$,

where $I_{x,\xi} = \{v \in J_x\colon \xi\cdot v = \phi(x,\xi)\}$;

$G(x,\xi) = \min\,\{\xi\cdot[g(x,v) - \beta(x)] : v \in I_{x,\xi}\}$.

(iii) For each $\xi \in R^m$, $\phi(x,\xi)$ is in $C^2(\Omega)\cap C^1(\bar{\Omega})$ with respect to x.

(iv) There exists a function $\hat{\beta} \in H^{2+\alpha}(\bar{\Omega},R^m)$ such that $\hat{\beta}(x) \in J_x$ for all $x \in \bar{\Omega}$ and $\hat{\beta}(x) = \beta(x)$ for all $x \in \delta\Omega$.

Theorem 3.2-1. *Under assumptions (i) to (iv), suppose that*

$$(3.2\text{-}2)\qquad \Delta\phi(x,\xi) + F(x,\xi) \le 0 \qquad \text{for all } (x,\xi) \in \Omega xR^m, \text{ and}$$

$$(3.2\text{-}3)\qquad \frac{\partial\phi}{\partial\eta}(x,\xi) + G(x,\xi) \ge 0 \qquad \text{for all } (x,\xi) \in \delta\Omega xR^m,$$

then (3.2-1) *has a solution* u *in* $C^2(\Omega, R^m) \cap C^1(\bar{\Omega}, R^m)$, *and whose graph lies in* J.

(In (3.2-2) and (3.2-3), the inequalities are interpreted as to hold for each component.)

Outline of Proof. We will only consider the more restrictive case when both (3.2-2) and (3.2-3) are modified to satisfy strict inequalities. Such restrictions can be removed by taking appropriate limiting procedures. Moreover, for the time being, we modify assumption (iv) to (iv') where we further assume $\hat{\beta}(x) \in$ interior of J_x for all $x \in \bar{\Omega}$. We can thus find a constant k large enough so that

$$(3.2\text{-}4) \qquad \xi \cdot [f(x,q) + \Delta\hat{\beta}(x) + k(q - \hat{\beta}(x))] > 0$$

for all $\xi \in R^m - \{0\}$, $x \in \bar{\Omega}$, $q \in I_{x,\xi}$, and

$$(3.2\text{-}5) \qquad \xi \cdot [g(x,q) - \hat{\beta}(x) + \frac{\partial\hat{\beta}}{\partial\eta}(x) - k(q - \hat{\beta}(x))] < 0$$

for all $\xi \in R^m - \{0\}$, $x \in \delta\Omega$, $q \in I_{x,\xi}$. Define Π to be the set of $v \in C^1(\bar{\Omega}, R^m)$ whose graph lies in J, and Π_K to be $\Pi \cap \{v : \|v\|_1 \le K\}$ where K will be defined below as a sufficiently large constant (Here $\| \ \|_1$ is the usual norm in $C^1(\bar{\Omega}, R^m)$). For v restricted in $H^{1+\alpha}(\bar{\Omega}, R^m)$, we can define the map $w = T_k v$ such that

$$(3.2\text{-}6) \qquad \Delta w - kw + f(x,v) + kv = 0 \qquad \text{for } x \in \Omega, \text{ and}$$

$$(3.2\text{-}7) \qquad \frac{\partial w}{\partial \eta} + kw + g(x,v) - \beta(x) - kv = 0 \qquad \text{for } x \in \delta\Omega.$$

By extending the domain of the map T_k to $C^1(\bar{\Omega}, R^m)$ with range in the Sobolev space of those u whose weak derivatives up to order 2 are p integrable (p large enough), one can view T_k as a compact operator from Π_K to $C^1(\bar{\Omega}, R^m)$ (for details, see [201] or appendix A3). Note that Π_K is a closed, bounded and convex set in $C^1(\bar{\Omega},R^m)$. We set $H(v,\lambda) = (v - \hat{\beta}) + \lambda(T_k v - \hat{\beta})$ for $(x,\lambda) \in \Pi_K x\ [0,1]$. Clearly $H(v,0) \neq 0$ for $v \in \delta\Pi_K$, (the boundary of Π_K), as $\hat{\beta}$ is an interior point. To obtain a $v \in \Pi_K$ such that $H(v,1) = 0$, the usual degree theoretic argument implies that it remains to show that $H(v,\lambda) \neq 0$ for any $(v,\lambda) \in \delta\Pi_K x\ (0,1]$, if K is properly selected (cf. Theorems 1.3-7 and 1.3-8). Suppose that $(v,\lambda) \in \delta\Pi_K x\ (0,1]$ such that $H(v,\lambda) = 0$, then (3.2-6) implies that

$$(3.2\text{-}8) \qquad \Delta[v - (1-\lambda)\hat{\beta}] + k(1-\lambda)(\hat{\beta} - v) + \lambda f(x,v) = 0 \qquad \text{for } x \in \Omega.$$

By a more general version of Lemma 1.4-1 (see e.g. [126]), one can obtain from (3.2-8) an apriori bound of the form $\|v\|_1 \leq K'$. Taking $K > K'$, the condition $v \in \delta\Pi_K$ implies that for some $y \in \bar{\Omega}$, we have $v(y) \in \delta J_y$. If $y \in \Omega$, choose $\xi \in R^m$ such that $v(y) \in I_{y,\xi}$. We have

$$(3.2\text{-}9) \qquad \xi\cdot[\Delta v(y) + f(y,v)] \leq \Delta\phi(y,\xi) + F(y,\xi) < 0,$$

the last strict inequality being true due to the modified (3.2-2) hypothesis. On the other hand, (3.2-8) implies that $\Delta v(y) + f(y,v) = (1 - \lambda)[f(y,v) + \Delta\hat{\beta}(y) + k(v(y) - \hat{\beta}(y))]$; and together with (3.2-4), we have

$$(3.2\text{-}10) \qquad \xi\cdot[\Delta v(y) + f(y,v)] \geq 0,$$

contradicting (3.2-9). We must therefore have $y \in \delta\Omega$. Now, from the modified version of (3.2-3), we can show that for some $\xi \in R^m$ with $v(y) \in I_{y,\xi}$, we have

$$(3.2\text{-}11) \qquad \xi\cdot\ [\frac{\partial v}{\partial \eta}(y) + g(y,v) - \beta(y)] > 0.$$

But $H(v,\lambda) = 0$ means that at $y \in \delta\Omega$,

$$\frac{\partial}{\partial\eta}[v - (1+\lambda)\hat{\beta}] + k(1-\lambda)(v-\beta) + \lambda g(y,v) + \lambda\beta = 0.$$

Hence, $(\partial v/\partial\eta)(y) + g(y,v) - \beta(y) = (1-\lambda)[\partial\hat{\beta}/\partial\eta + k(\beta-v) + g(y,v) - \beta(y)]$, and together with (3.2-5), we have

$$\text{(3.2-12)} \qquad \xi\cdot[\frac{\partial v}{\partial\eta}(y) + g(y,v) - \beta(y)] \le 0,$$

contradicting (3.2-11). Consequently, for some $v \in \Pi_K$, we have $H(v,1) = 0$ and such v is a solution of (3.2-1) lying in $\Pi_K \cap C^2(\Omega, R^m)$.

To strengthen assumptions (iv') back to (iv), we define $J^\varepsilon = \{(x,v): \text{dist}(v, J_x) \le \varepsilon\}$, for $\varepsilon > 0$. J^ε is closed, bounded and has convex sections J_x^ε. Define $q(x,v)$ to be the (unique) nearest point in J_x to v, and $\tilde{f}(x,v) = f(x,q(x,v))$. Analogously, define $\phi^\varepsilon(x,\xi) = \max\{\xi\cdot v : v \in J_x^\varepsilon\}$, $F^\varepsilon(x,\xi) = \max\{\xi\cdot\tilde{f}(x,v) : v \in R^m,\ \xi\cdot v = \phi^\varepsilon(x,v)\}$. We have $\phi^\varepsilon = \phi + \varepsilon$, $F^\varepsilon = F$ and $\Delta\phi^\varepsilon + F^\varepsilon = \Delta\phi + F$. Now $\hat{\beta}(x) \in$ interior of J_x^ε, if $\hat{\beta}(x) \in J_x$. From the previous paragraphs, we can therefore find a solution u^ε to the problem: $\Delta u + \tilde{f}(x,u) = 0$ in Ω, $\partial u/\partial\eta + g(x,u) = \beta(x)$ on $\delta\Omega$, with the graph of u^ε in J^ε. Letting $\varepsilon \to 0^+$ and choosing a subsequence in $\{u^\varepsilon\}$, we obtain a limiting function u which is a solution as asserted by the theorem.

To illustrate the use of the previous theorem in ecological problems, we consider the reaction-diffusion equation for prey-predator interaction:

$$\text{(3.2-13)} \qquad \begin{aligned} \frac{\partial u_1}{\partial t} &= \sigma_1\Delta u_1 + u_1[a + f_1(u_1,u_2)] \\ \frac{\partial u_2}{\partial t} &= \sigma_2\Delta u_2 + u_2[-r + f_2(u_1,u_2)] \end{aligned} \qquad x \in \Omega,\ t > 0,$$

where a, r, σ_1, σ_2 are positive constants, with u_1, u_2 respectively representing the concentrations of prey and predator. Here we assume that f_i: $R^2 \to R$ have Hölder continuous partial derivatives up to second order in compact sets, i = 1,2;

$$\text{(3.2-14)} \qquad \begin{aligned} &f_1(0,0) = f_2(0,0) = 0, \\ &\frac{\partial f_1}{\partial u_1} < 0, \ \frac{\partial f_1}{\partial u_2} < 0, \ \frac{\partial f_2}{\partial u_1} > 0, \ \frac{\partial f_2}{\partial u_2} < 0 \end{aligned}$$

for (u_1,u_2) in the first open quadrant. (Note that the sign of $\partial f_2/\partial u_1$ indicates that the first species is the prey for the second one.) Further, we assume that for each $\hat{m} \geq 0$

$$\text{(3.2-15)} \qquad \lim_{u_2 \to +\infty} f_2(\hat{m},u_2) = -\infty, \quad \lim_{u_1 \to +\infty} f_2(u_1,\hat{m}) = +\infty, \quad \lim_{u_1 \to +\infty} f_1(u_1,0) = -\infty .$$

These hypotheses on f_i are satisfied by Volterra-Lotka type interaction with crowding effect.

In realistic ecological situations, one expects that boundary conditions to be coupled, mixed and nonlinear. For instance, diffusion of predators at the boundary may be adjusted nonlinearly according to prey and predator populations present, while the prey may have no flux at the boundary. We will not try to consider the most general boundary conditions. However, we will only study a simple prototype case:

$$\text{(3.2-16)} \qquad \frac{\partial u_1}{\partial \eta} = 0, \quad \frac{\partial u_2}{\partial \eta} - p(u_1) - q(u_2) = 0 \qquad \text{on } \partial\Omega.$$

where p and q have Hölder continuous first derivatives in compact subsets of $[0,\infty)$. Moreover, for $u \in [0,\infty)$ we assume either:

$$(3.2\text{-}17)\qquad \begin{aligned} &0 \le p(u) \le P < \infty \quad \text{and} \quad 0 < q(u) \le Q < \infty; \text{ or} \\ &0 < p(u) \le P < \infty \quad \text{and} \quad 0 \le q(u) \le Q < \infty. \end{aligned}$$

The following theorem gives sufficient conditions for time-independent steady-state solution for (3.2-13) under boundary condition (3.2-16) (assuming, of course, all the conditions stated above for the parameters and those for f_i, p, q).

Theorem 3.2-2. *Suppose that $|\partial f_2/\partial u_2|$ is uniformly bounded in the first open quadrant, and there is an $\hat{R} > 0$, such that for α, $\beta > \hat{R}$:*

$$(3.2\text{-}18)\qquad \left|\frac{\partial f_1}{\partial u_2}(0,\beta)\frac{\partial f_2}{\partial u_1}(\alpha,\beta)\right| \le \rho\left|\frac{\partial f_2}{\partial u_2}(\alpha,\beta)\frac{\partial f_1}{\partial u_1}(\alpha,0)\right| ,$$

for a constant $\rho < 1$. Then for $a > 0$ sufficiently large, the equation:

$$(3.2\text{-}19)\qquad \begin{aligned} &\sigma_1\Delta u_1 + u_1[a + f_1(u_1,u_2)] = 0, \\ &\sigma_2\Delta u_2 + u_2[-r + f_2(u_1,u_2)] = 0, \end{aligned} \qquad \text{for } x \in \Omega,$$

with boundary conditions (3.2-16) has a steady-state (equilibrium) solution $(\bar{u}_1(x),\bar{u}_2(x))$, such that $\bar{u}_i \in C^1(\bar{\Omega})\cap C^2(\Omega)$ and $\bar{u}_i(x) > 0$ for all $x \in \bar{\Omega}$.

Remark: a more specific condition on the parameter a will be given in the proof.

Proof. Let $\lambda = \lambda_1$ be the principal eigenvalue for the eigenvalue problem $\Delta u + \lambda u = 0$ in Ω, $u = 0$ on $\delta\Omega$; and let $\omega(x)$ be the corresponding eigenfunction (hence $\omega(x) > 0$ in Ω, $\partial\omega/\partial\eta < 0$ on $\delta\Omega$). We will construct a set J in $\bar{\Omega}\times R^2$ with its boundary surfaces satisfying appropriate differential inequalities, and apply Theorem 3.2-1 to assert the existence of a solution to (3.2-19), (3.2-16) with its graph contained in J.

Let k be any positive number satisfying:

$$(3.2\text{-}20) \qquad k > (P + Q)\left(\min_{\delta\Omega}\left|\frac{\partial\omega}{\partial\eta}\right|\right)^{-1},$$

We will show below that there exist positive constants d, C such that the following conditions (3.2-21) to (3.2-24):

$$(3.2\text{-}21) \qquad a + f_1(d,0) = 0,$$

$$(3.2\text{-}22) \qquad -k\omega(x) + C > 0 \qquad \text{for all } x \in \bar{\Omega},$$

$$(3.2\text{-}23) \qquad f_2(d,-k\omega(x)+C) \le [Cr - k\omega(x)(\lambda_1\sigma_2+r)][C - k\omega(x)]^{-1}$$

for all $x \in \bar{\Omega}$, and

$$(3.2\text{-}24) \qquad |f_1(0,C)| < a - \sigma_1\lambda_1$$

are true for large enough a. Define J to be the set $J = \{(x,u_1,u_2) : x \in \bar{\Omega},\ \delta\omega(x) \le u_1 \le d_1,\ 0 \le u_2 \le -k\omega(x) + C\}$, where $\delta > 0$ is small enough so that $\delta\max_{\bar{\Omega}}\,\omega(x) < d$ and $|f_1(\delta\max_{\bar{\Omega}}\,\omega(x),C)| < a - \sigma_1\lambda_1$. Let $f(u_1,u_2) = (\sigma_1^{-1}u_1[a + f_1(u_1,u_2)],\ \sigma_2^{-1}u_2[-r + f_2(u_1,u_2)])$; define ϕ, $F : \bar{\Omega}\text{x}R^2 \to R$ by $\phi(x,\xi) = \max\{\xi\cdot u : u = (u_1,u_2),\ (x,u_1,u_2) \in J\}$ and $F(x,\xi) = \max\{\xi\cdot f(u_1,u_2) : (x,u_1,u_2) \in J \text{ with } \xi\cdot u = \phi(x,\xi)\}$. We will see that:

$$(3.2\text{-}25) \qquad \Delta\phi(x,\xi) + F(x,\xi) \le 0, \qquad \text{for all } (x,\xi) \in \Omega\text{x}R^2.$$

Observe that for each $x \in \Omega$, the cross section $J_x = \{(u_1,u_2)\colon (x,u_1,u_2) \in J\}$ is a closed rectangle. It suffices to see that the above differential inequality is true for $\xi = (-1,0)$, $(0,-1)$, $(1,0)$ and $(0,1)$.

For $\xi = (-1,0)$, $\Delta\phi(x,\xi) + F(x,\xi) = -\Delta\delta\omega(x) + \max\,\{-\sigma_1^{-1}\delta\omega(x)[a +$

$f_1(\delta\omega(x),u_2)] : 0\le u_2\le -k\omega(x)+C \} \le \delta\omega(x)[\lambda_1 - \sigma_1^{-1}f_1(\delta\omega(x),C)] < 0$, for each $x \in \Omega$, because of the choice of δ. For $\xi = (0,-1)$, $\Delta\phi(x,\xi) + F(x,\xi) \equiv 0$; and for $\xi = (1,0)$, $\Delta\phi(x,\xi) + F(x,\xi) = \Delta d + \max \{\sigma_1^{-1}d[a + f_1(d,u_2)] : 0\le u_2\le -k\omega(x)+C \}= 0$. Finally for $\xi = (0,1)$, $\Delta\phi(x,\xi) + F(x,\xi) = \Delta(-k\omega(x)+C) + \max \{\sigma_2^{-1}(-k\omega(x)+C)[-r + f_2(u_1,-k\omega+C)] : \delta\omega(x)\le u_1\le d \} = k\lambda_1\omega(x) + r\sigma_2^{-1}k\omega(x) - \sigma_2^{-1}Cr + (-k\omega(x)+C)f_2(d,-k\omega+C)\sigma_2^{-1} \le 0$ for all $x \in \Omega$, by the choice of C satisfying (3.2-23). This shows (3.2-25).

To take the boundary conditions into consideration, we let $G(x,\xi) = \min \{\xi\cdot(0,-p(u_1)-q(u_2)) : (x,u_1,u_2) \in J \text{ with } \xi\cdot(u_1,u_2) = \phi(x,\xi)\}$, for $(x,\xi) \in \delta\Omega x R^2$. We will show that

$$\text{(3.2-26)} \qquad \frac{\partial\phi}{\partial\eta} + G(x,\xi) \ge 0,$$

for all $(x,\xi) \in \delta\Omega x R^2$, and then apply Theorem 3.2-1, which asserts that (3.2-25) and (3.2-26) will imply the existence of a solution to our problem with components in $C^1(\bar{\Omega})\cap C^2(\Omega)$, with graph in J. Again, it suffices to prove (3.2-26) with $\xi = (-1,0)$, $(0,-1)$, $(1,0)$ and $(0,1)$. For $\xi =(-1,0)$, $\partial\phi/\partial\eta(x,\xi) + G(x,\xi) = (\partial/\partial\eta)(-\delta\omega) > 0$, for $x \in \delta\Omega$. For $\xi = (0,-1)$, $\partial\phi/\partial\eta + G = \min \{p(u_1)+q(0) : 0\le u_1\le d \} > 0$, for $x \in \delta\Omega$. For $\xi = (1,0)$, $\partial\phi/\partial\eta + G = 0$; and for $\xi = (0,1)$, $\partial\phi/\partial\eta + G = (\partial/\partial\eta)(-k\omega+C) + \min \{-p(u_1)-q(C) : 0\le u_1\le d \} > -k\partial\omega/\partial\eta - P - Q > 0$ for $x \in \delta\Omega$, by the choice of k satisfying (3.2-20). This proves (3.2-26), and hence the existence of a solution $(\bar{u}_1(x),\bar{u}_2(x))$ with $\bar{u}_i \in C^1(\bar{\Omega})\cap C^2(\Omega)$ and with graph in J.

It remains to prove that $\bar{u}_1(x) > 0$ for all $x \in \bar{\Omega}$. The graph of $(\bar{u}_1,\bar{u}_2)$ lies in J implies that $\bar{u}_i\ge 0$, $i = 1,2$, for $x \in \bar{\Omega}$. Since $\partial\bar{u}_2/\partial\eta = p(\bar{u}_1) + q(\bar{u}_2) > 0$ for $\bar{u}_1\ge 0$, $\bar{u}_2\ge 0$, so we must have $\bar{u}_2(x) > 0$ for $x \in \delta\Omega$. Let $|f_2(u_1,u_2)| \le M_1$, for all $0\le u_1\le d$, $0\le u_2\le C$, and let $M_2 > (1/r)(M_1 + r)$. Let $v(x)$ be the solution of $\sigma_2\Delta v - M_2 rv = 0$ in Ω, $v = \bar{u}_2$ on $\delta\Omega$. Clearly the minimum of $v(x)$ in $\bar{\Omega}$ vannot be negative. To see that the minimum of v cannot be zero, we compare with the solution w(x) of $\sigma_2\Delta w - 2M_2 rw = 0$ in Ω, $w = \bar{u}_2$ on $\delta\Omega$. Then $w \geq 0$ in $\bar{\Omega}$, $v - w = 0$ on $\delta\Omega$ and $\sigma_2\Delta(v-w) - M_2 r(v-w) = -M_2 rw \leq 0$ in Ω, with strict inequality holding in some portion of Ω. Representation of v-w by use of Green's function for the operator $\sigma_2\Delta - M_2 r$ shows that $v-w > 0$ in Ω. In particular, if there is an $x_0 \in \Omega$ where $v(x_0) = 0$, then $w(x_0) < 0$, which is not possible. We therefore have $0 < v(x) < C$, for all $x \in \bar{\Omega}$. Finally, we compare $\bar{u}_2$ with v by observing:

$$\begin{aligned}&\sigma_2\Delta(\theta v) + (\theta v)[-r + f_2(\bar{u}_1,\theta v)]\\ &= (\theta v)[M_2 r - r + f_2(\bar{u}_1,\theta v)] \geq 0,\end{aligned}$$

$x \in \Omega$, for all $0 \leq \theta \leq 1$, by the choice of M_2. For such θ, we have $\theta v \leq \bar{u}_2$ on $\delta\Omega$. The sweeping principle, Theorem 1.4-3, implies that $\bar{u}_2(x) \geq 1v(x) > 0$ for $x \in \bar{\Omega}$.

Clearly $\bar{u}_1(x) > 0$ for $x \in \Omega$. Since $\bar{u}_1(x) \geq \delta\omega(x)$ in Ω, we must have $\partial\bar{u}_1/\partial\eta \leq \delta(\partial\omega/\delta\eta) < 0$ on $\delta\Omega$ if $\bar{u}_1(x) = \omega(x)$ on $\delta\Omega$. But this contradicts the boundary condition $\partial\bar{u}_1/\partial\eta = 0$ in (3.2-16). We therefore have $\bar{u}_1(x) > 0$ for $x \in \bar{\Omega}$. Finally, to justify that constants d and C satisfying (3.2-21) to (3.2-24) do exist, we let $M > |(\partial f_2/\partial u_2)(u_1,u_2)|\cdot k\cdot \max_{x\in\bar{\Omega}}\omega(x)$, for all u_1, $u_2 > 0$ and define $\hat{C}(d)$ and $\hat{d}(a)$ as functions which satisfy :

$$f_2(d,\hat{C}(d)) = -M \qquad \text{and} \qquad f_1(\hat{d}(a),0) = -a.$$

It is clear that $\hat{C}(d)$ and $\hat{d}(a)$ are uniquely defined, $\hat{C}'(d) = -(\partial f_2/\partial u_1)(d,\hat{C}(d))\cdot[(\partial f_2/\partial u_2)(d,\hat{C}(d))]^{-1} > 0$, $\hat{d}'(a) = -[(\partial f_1/\partial u_1)(\hat{d}(a),0)]^{-1} > 0$ and $\hat{C}(d) \to +\infty$ as $d \to +\infty$, while $\hat{d}(a) \to +\infty$ as $a \to +\infty$. The hypothesis of this theorem implies that $-1 < -\rho \leq (\partial H/\partial a)(a) < 0$ for all large a, where $H(a) = f_1(0,\hat{C}(\hat{d}(a)))$, and thus $|f_1(0,\hat{C}(\hat{d}(a)))| < a - \sigma_1\lambda_1$ for large enough a. Further, $f_2(\hat{d}(a),\hat{C}(\hat{d}(a))-k\omega(x)) < f_2(\hat{d}(a),\hat{C}(\hat{d}(a))) + M = 0$ and $[Cr - k\omega(x)(\lambda_1\sigma_2+r)][C - k\omega(x)]^{-1} \to r > 0$ as $C \to +\infty$, uniformly for $x \in \bar{\Omega}$. Thus (3.2-21) to (3.2-24) are all satisfied by letting $d = \hat{d}(a)$, $C = \hat{C}(\hat{d}(a))$ and a sufficiently large. This completes the proof of the Theorem.

Remark: From the proof of the above theorem, we notice that if we replace (3.2-17) by

(3.2-17a) $\qquad 0 \le p(u) \le P < \infty$ and $0 \le q(u) \le Q < \infty$

for $u \in [0,\infty)$, we can still prove the existence of solution $(\bar{u}_1(x),\bar{u}_2(x))$ to (3.2-19), (3.2-16) with $\bar{u}_i(x) \ge 0$ in $\bar{\Omega}$, $i = 1,2$. The more restrictive positivity assumption in (3.2-17) is only used for proving $\bar{u}_i(x) > 0$ in $\bar{\Omega}$, $i =1,2$.

We will next prove the existence of solution for the initial boundary value problem associated with the reaction diffusion equation (3.2-13). Then, we will eventually discuss the stability of the equilibrium found in Theorem 3.2-2 under slightly more restrictive conditions.

Theorem 3.2-3. *Assume that* $p(u)$ *and* $q(u)$ *in* (3.2-16) *can be extended to be twice continuously differentiable in any compact subset of* $(-\infty,\infty)$. *Let* g_1, g_2 *be positive functions in* $H^{2+\alpha}(\bar{\Omega})$; *and* $(\partial g_1/\partial\eta)(x) = 0$, $(\partial g_2/\partial\eta)(x) = p(g_1(x)) + q(g_2(x))$ *for all* $x \in \delta\Omega$. *Then for any* $T > 0$, *there exists a* $\hat{\rho} > 0$, *so that if* $-\hat{\rho} \le \partial f_1/\partial u_2 < 0$ *in the first open quadrant, equation* (3.2-13) *for* $(x,t) \in \bar{\Omega}_T$ *together with initial boundary conditions:*

(3.2-27) $\qquad u_i(x,0) = g_i(x)$ *for* $x \in \bar{\Omega}$, $i = 1,2$,

(3.2-28) $\qquad \frac{\partial u_1}{\partial \eta}(x,t) = 0, \quad \frac{\partial u_2}{\partial \eta}(x,t) - p(u_1(x,t)) - q(u_2(x,t)) = 0$

<u>for</u> $(x,t) \in \delta\Omega x[0,T]$, <u>has a solution</u> $(u_1^*(x,t), u_2^*(x,t))$ <u>with</u> $u_i^*(x,t) \in H^{2+\alpha,(2+\alpha)/2}(\bar{\Omega}_T)$, $i = 1,2$. (<u>Here,</u> $\hat{\rho}$ <u>depends on the bound of</u> $\partial f_1/\partial u_1$ <u>on a fixed compact set which is determined by</u> P, Q, a, r <u>and the</u> $\max_{\bar{\Omega}} g_i$, $i = 1,2$.) <u>Further,</u> $u_i^*(x,t) > 0$ <u>for</u> $x \in \bar{\Omega}$, $0 \le t \le T$, $i = 1,2$.

<u>Proof</u> Let $T > 0$ be fixed. We will apply Leray-Schauder criterion, Theorem 1.3-7, for the existence of fixed point. We now choose an appropriate bound for the solutions and an appropriate Hölder space. Let $m_i > \max\{g_i(x) : x \in \bar{\Omega}\}$, $i = 1.2$. Let λ_1 and $\omega(x)$ be principal eigenvalue and eigenfunctions exactly as described in the proof of Theorem 3.2-2. Choose any $\varepsilon > r$ so that $b \overset{(def)}{\equiv} \varepsilon - r + f_2(2m_1e^{aT}, 0) > 0$; and finally choose positive constants K, C such that

(3.2-29) $$|K\frac{\partial\omega}{\partial\eta}| > P + Q \qquad \text{for all } x \in \delta\Omega; \text{ and}$$

(3.2-30) $$C > \max\{K\omega(x),\ K\omega(x)[1+(\sigma_2\lambda_1)/\varepsilon] - m_2\}, \text{ for all } x \in \bar{\Omega}.$$

(Note that K and C are independent of f_i, $i = 1,2$). We will define a mapping on the set $S \overset{(def)}{\equiv} \{(\tilde{z}_1, \tilde{z}_2) \in H^{1+\bar{\alpha},(1+\bar{\alpha})/2}(\bar{\Omega}_T)xH^{1+\bar{\alpha},(1+\bar{\alpha})/2}(\bar{\Omega}_T)$: $\tilde{z}_i(x,0) \equiv 0$, $x \in \bar{\Omega}$; $-g_1(x) \le \tilde{z}_1(x,t) \le m_1e^{aT}$, $-g_2(x) \le \tilde{z}_2(x,t) \le [m_2+C-K\omega(x)]e^{bT}$, $(x,t) \in \bar{\Omega}_T$; and $|\tilde{z}_i|_{\Omega_T}^{(1+\bar{\alpha})} \le \bar{c}_i + 1$, $i=1,2\}$, with $\bar{\alpha}, \bar{c}_1, \bar{c}_2$ to be determined presently.

For convenience, let $K_1 = m_1(e^{aT}+1)$, $K_2 = (m_2+C)e^{bT} + m_2$. For $i = 1,2$, truncate the functions $f_i(u_1,u_2)$ to define $\bar{f}_i(u_1,u_2)$ as :

$$\bar{f}_i(u_1,u_2) = f_i(u_1,u_2) \qquad \text{if } |u_i| \le K_i + 1 \text{ for each } i = 1,2$$

$$\bar{f}_i(u_1,u_2) = \begin{cases} f_i(K_1+1,u_2) & \text{if } u_1 > K_1+1 \text{ and } |u_2| \le K_2+1 \\ f_i(-K_1-1,u_2) & \text{if } u_1 < -K_1-1 \text{ and } |u_2| \le K_2+1 \\ f_i(u_1,K_2+1) & \text{if } |u_1| \le K_1+1 \text{ and } u_2 > K_2+1 \\ f_i(u_1,-K_2-1) & \text{if } |u_1| \le K_1+1 \text{ and } u_2 < -K_2-1 \\ f_i(K_1+1,K_2+1) & \text{if } u_1 > K_1+1 \text{ and } u_2 > K_2+1 \\ f_i(-K_1-1,K_2+1) & \text{if } u_1 < -K_1-1 \text{ and } u_2 > K_2+1 \\ f_i(-K_1-1,-K_2-1) & \text{if } u_1 < -K_1-1 \text{ and } u_2 < -K_2-1 \\ f_i(K_1+1,-K_2-1) & \text{if } u_1 > K_1+1 \text{ and } u_2 < -K_2-1 \end{cases}$$

For any fixed given $w(x,t) \in C^{1,1}(\bar{\Omega}_T)$, i.e. functions which are continuously differentiable with respect to x_i and t in $\bar{\Omega}_T$, with $\max_{\bar{\Omega}_T}|w(x,t)| \le K_2-m_2$, $0 \le \lambda \le 1$, define

$$F_1^w(\lambda,v,x,t) = (v+\lambda g_1(x))[a + \bar{f}_1(v+g_1(x),w(x,t)+g_2(x))] + \lambda\sigma_1\Delta g_1(x).$$

Suppose that $v_1(x,t)$ is twice continuously differentiable with respect to x and continuously differentiable with respect to t in $\bar{\Omega}_T$ and satisfies:

$$\text{(3.2-31)} \quad \begin{aligned} &\frac{\partial v_1}{\partial t} = \sigma_1\Delta v_1 + F_1^w(\lambda,v_1,x,t) \qquad \text{in } \Omega\times(0,T], \\ &\frac{\partial v_1}{\partial \eta}(x,t) = -\lambda\frac{\partial g_1}{\partial \eta}(x) \quad \text{on } \delta\Omega\times[0,T],\ v_1(x,0) = 0 \text{ in } \bar{\Omega}. \end{aligned}$$

By Theorem A1.3, in the appendix, there exist $\delta_1>0$ and $c_1>0$ so that if $\max_\Omega |v_1| \le K_1-m_1$, then $|v_1|_\Omega^{(1+\delta_1)} \le c_1$. The constants and δ_1 and c_1 depend on the bounds of $|F_1^w|$, $|\frac{\partial F_1^w}{\partial t}(\lambda,v,x,t)|$ and $|\frac{\partial F_1^w}{\partial v}(\lambda,v,x,t)|$ for $(x,t) \in \bar{\Omega}_T$, $0 \le \lambda \le 1$, $|v| \le K_1-m_1$. If w and others are changed, the constants δ_1, c_1 are the same as long as the bounds for $|F_1^w|$, $|\partial F_1^w/\partial t|$ and $|\partial F_1^w/\partial v|$ are unchanged (cf. Theorem A1.3, or (7.6) and (7.15) on p. 476 and p. 479 in [127]). Analogously, for any fixed given $\hat{w}(x,t) \in C^{1,1}(\bar{\Omega}_T)$ with $\max_{\bar{\Omega}_T}|\hat{w}(x,t)| \le K_1-m_1$, define

$$F_2^{\hat{w}}(\lambda,v,x,t) = (v+\lambda g_2(x))[-r+\bar{f}_2(\hat{w}(x,t)+g_1(x),v+g_2(x))] + \lambda\sigma_2\Delta g_2(x).$$

Suppose that $v_2(x,t)$ is twice continuously differentiable with respect to x and continuously differentiable with respect to t in $\bar{\Omega}_T$, and satisfies

$$\frac{\partial v_2}{\partial t} = \sigma_2\Delta v_2 + F_2^{\hat{w}}(\lambda,v_2,x,t) \quad \text{in } \Omega\times(0,T],$$

$$(3.2\text{-}32) \qquad \frac{\partial v_2}{\partial \eta} = \lambda p(\hat{w}(x,t)+g_1(x)) + \lambda q(v_2+g_2(x)) - \lambda\frac{\partial g_2}{\partial \eta} \quad \text{on } \delta\Omega\times[0,T],$$

$$v_2(x,0) = 0 \text{ on } \bar{\Omega}.$$

(Here p and q are fixed twice continuously differentiable extension on $[-K_i,K_i]$, i =1,2 respectively). By Theorem A1.3 again, there exist $\delta_2 > 0$ and $c_2 > 0$ so that if $\max_{\Omega_T}|v_2| \le K_2-m_2$, then $|v_2|_{\Omega_T}^{(1+\delta_2)} \le c_2$. The constants δ_2, c_2 depend on quantities which are determined by the bounds on $\hat{w}$ and $\partial\hat{w}/\partial t$. Let ρ_i, i =1,2 be large positive constants so that $\partial f_i/\partial u_i > -\rho_i$ in $\{(u_1,u_2) : u_i \ge 0,\ |u_i| \le K_i+1,\ i = 1,2\}$. Choose a fixed $\bar{w} \in C^{1,1}(\bar{\Omega}_T)$ with $|\bar{w}(x,t)| \le K_2-m_2$ and $\bar{w}(x,t) + g_2(x) \ge 0$ for $(x,t) \in \bar{\Omega}_T$. Let $P_0 = K_1[a + (K_1+1)\rho_1 + (K_2+1)\rho_2] + \sigma_1|g_1|_{\Omega_T}^{(2+\alpha)}$, which is a bound for $|F_1^{\bar{w}}(\lambda,v,x,t)|$ if $\bar{w} + g_2 \ge 0$, $(x,t) \in \bar{\Omega}_T$, $0 \le \lambda \le 1$, $|v| \le K_1-m_1$; and let $P_1 \ge \max\{\ |\frac{\partial F_1^{\bar{w}}}{\partial t}(\lambda,v,x,t)| : 0 \le \lambda \le 1,\ |v| \le K_1-m_1,\ (x,t) \in \bar{\Omega}_T\ \}$. The quantity $|\frac{\partial F_1^{w}}{\partial v}|$ is bounded by a fixed constant P_2 for all $|w| \le K_2-m_2$, $|v| \le K_1-m_1$, $(x,t) \in \bar{\Omega}_T$, $0 \le \lambda \le 1$. The quantities K_1,K_2,P_0,P_1,P_2 determine constants $\bar{\delta}_1$, $\bar{c}_1$ by Theorem A1.3, corresponding to δ_1, c_1 described above. Next, for any $\hat{w} \in C^{1,1}(\bar{\Omega}_T)$ with $\max_{\bar{\Omega}_T}|\hat{w}(x,t)| \le K_1-m_1$, $\max_{\bar{\Omega}_T}|\partial\hat{w}/\partial x_j| \le \bar{c}_1$ each j, $\max_{\bar{\Omega}_T}|\partial\hat{w}/\partial t| \le \hat{C}$, the same theorem determines constants $\bar{\delta}_2$ and $\bar{c}_2$ corresponding to δ_2, c_2 described above. Here $\hat{C}$ will be some large constant determined by $\bar{c}_1$. Once $\hat{C}$ is chosen, $\bar{\delta}_2$ and $\bar{c}_2$ will be determined. Finally, choose $\bar{\alpha} = \min\{\ \bar{\delta}_1,\bar{\delta}_2,\alpha\ \}$ and define S as the set described in the last paragraph with $\bar{\alpha}$, $\bar{c}_1$ and $\bar{c}_2$ just chosen.

For $(\tilde{z}_1,\tilde{z}_2) \in H^{1+\bar{\alpha},(1+\bar{\alpha})/2}(\bar{\Omega}_T) \times H^{1+\bar{\alpha},(1+\bar{\alpha})/2}(\bar{\Omega}_T)$, we define the norm $\|(\tilde{z}_1,\tilde{z}_2)\| = |\tilde{z}_1|^{(1+\bar{\alpha})}_{\Omega_T} + |\tilde{z}_2|^{(1+\bar{\alpha})}_{\Omega_T}$. The set S is the closure of a connected bounded open neighborhood of zero in a Banach subspace B contained in $H^{1+\bar{\alpha},(1+\bar{\alpha})/2} \times H^{1+\bar{\alpha},(1+\bar{\alpha})/2}$. (Note that B are those functions with $\tilde{z}_i(x,0) \equiv 0$ in $\bar{\Omega}$, $i = 1,2$). Define the mapping $\mathbb{C}$ from the set S into this Banach space B by $\mathbb{C} : (\tilde{z}_1,\tilde{z}_2) \to (z_1,z_2)$ where (z_1,z_2) is the solution of the linear problem:

$$
\begin{aligned}
&\frac{\partial z_1}{\partial t} = \sigma_1\Delta z_1 + \{\sigma_1\Delta g_1 + (z_1+g_1)[a+\bar{f}_1(\tilde{z}_1+g_1,\tilde{z}_2+g_2)]\} \qquad \text{in } \Omega\times(0,T] \\
&\frac{\partial z_2}{\partial t} = \sigma_2\Delta z_2 + \{\sigma_2\Delta g_2 + (z_2+g_2)[-r+\bar{f}_2(\tilde{z}_1+g_1,\tilde{z}_2+g_2)]\} \\
&z_i(x,0) = 0, \quad x \in \bar{\Omega},\ i = 1,2, \\
&\frac{\partial z_1}{\partial \eta} = -\frac{\partial g_1}{\partial \eta}, \quad \frac{\partial z_2}{\partial \eta} = p(\tilde{z}_1+g_1) + q(\tilde{z}_2+g_2) - \frac{\partial g_2}{\partial \eta} \quad \text{on } \partial\Omega\times[0,T].
\end{aligned}
\tag{3.2-33}
$$

Note that from Theorem 1.3-2, we have $z_i \in H^{2+\bar{\alpha},(2+\bar{\alpha})/2}(\bar{\Omega}_T)$; and the initial conditions thus make $(z_1,z_2) \in B$. By means of the fact that the embedding of $H^{2+\bar{\alpha},(2+\bar{\alpha})/2}(\bar{\Omega}_T)$ into $H^{1+\bar{\alpha},(1+\bar{\alpha})/2}(\bar{\Omega}_T)$ is compact and continuous (cf. e.g. [83], analogous to the Ascoli's Lemma), we further conclude that $\mathbb{C}$ is a compact and continuous map of S into B. Define the transformation $G : S\times[0,1] \to B$ by

$$G(\xi,\lambda) = \xi - \lambda\mathbb{C}(\xi), \qquad \text{for } \xi \in S,\ \lambda \in [0,1].$$

It remains to show that $G(\xi,\lambda) \neq 0$ for all ξ on the boundary of S, all $\lambda \in [0,1]$, so that the Leray-Schauder criterior, Theorem 1.3-7, can be applied.

Suppose $\hat{\xi} = (\hat{z}_1, \hat{z}_2) \in S$, $\hat{\lambda} \in (0,1]$, so that $G(\hat{\xi}, \hat{\lambda}) = 0$. We have $\hat{\xi} = \hat{\lambda}\mathbb{C}(\hat{\xi})$; thus

$$
\begin{aligned}
\frac{\partial \hat{z}_1}{\partial t} &= \sigma_1 \Delta \hat{z}_1 + \sigma_1 \hat{\lambda} \Delta g_1 + (\hat{z}_1 + \hat{\lambda} g_1)[a + \bar{f}_1(\hat{z}_1 + g_1, \hat{z}_2 + g_2)] \\
&= \sigma_1 \Delta \hat{z}_1 + F_1^{\hat{z}_2}(\hat{\lambda}, \hat{z}_1, x, t) \qquad \text{in } \Omega \times (0,T], \\
\frac{\partial \hat{z}_1}{\partial \eta} &= -\hat{\lambda} \frac{\partial g_1}{\partial \eta} \qquad \text{on } \delta\Omega \times [0,T], \\
\hat{z}_1(x,0) &= 0 \qquad \text{for } x \in \bar{\Omega}.
\end{aligned}
\tag{3.2-34}
$$

For convenicence, let $M \geq \max_{R^2} \bar{f}_i + a$, each $i = 1,2$; and let $\hat{u}_i = \hat{z}_i + \hat{\lambda} g_i$, $Q_i(x,t) = -\hat{u}_i(x,t)e^{-Mt}$, for $(x,t) \in \bar{\Omega}_T$, $i = 1,2$. The function $Q_1(x,t)$ satisfies:

$$
\begin{aligned}
&\sigma_1 \Delta Q_1 - \frac{\partial Q_1}{\partial t} + [a + \bar{f}_1(\hat{z}_1 + g_1, \hat{z}_2 + g_2) - M]Q_1 = 0 \quad \text{in } \Omega \times (0,T], \\
&Q_1(x,0) = -\hat{\lambda} g_1(x) < 0, \qquad x \in \bar{\Omega}, \\
&\frac{\partial Q_1}{\partial \eta}(x,0) = 0 \qquad \text{on } \delta\Omega \times [0,T].
\end{aligned}
\tag{3.2-35}
$$

The maximum priciples imply that $Q_1(x,t) < 0$ in $\bar{\Omega}_T$, i.e. $\hat{z}_1 > -\hat{\lambda} g_1$ in $\bar{\Omega}_T$. On the other hand, $\hat{z}_2$ satisfies

$$
\begin{aligned}
\frac{\partial \hat{z}_2}{\partial t} &= \sigma_2 \Delta \hat{z}_2 + \sigma_2 \hat{\lambda} \Delta g_2 + (\hat{z}_2 + \hat{\lambda} g_2)[-r + \bar{f}_2(\hat{z}_1 + g_1, \hat{z}_2 + g_2)] \\
&= \sigma_2 \Delta \hat{z}_2 + F_2^{\hat{z}_1}(\hat{\lambda}, \hat{z}_2, x, t) \qquad \text{in } \Omega \times (0,T], \\
\frac{\partial \hat{z}_2}{\partial \eta} &= \hat{\lambda} p(\hat{z}_1 + g_1) + \hat{\lambda} q(\hat{z}_2 + g_2) - \hat{\lambda} \frac{\partial g_2}{\partial \eta} \qquad \text{on } \delta\Omega \times [0,T], \\
\hat{z}_2(x,0) &= 0 \qquad \text{for } x \in \bar{\Omega}.
\end{aligned}
\tag{3.2-36}
$$

The function $Q_2(x,t)$ satisfies $\sigma_2\Delta Q_2 - \partial Q_2/\partial t + [-r + \bar{f}_2(\hat{z}_1+g_1, \hat{z}_2+g_2) - M]Q_2 = 0$ in $\Omega x(0,T]$, $Q_2(x,0) < 0$ for $x \in \bar{\Omega}$ and $\partial Q_2/\partial\eta = -\hat{\lambda}p(\hat{z}_1+g_1) - \hat{\lambda}q(\hat{z}_2+g_2) \le 0$ on $\delta\Omega x[0,T]$. (The last inequality is true because $\hat{z}_i+g_i \ge 0$). Maximum principles again imply that $Q_2(x,t) \le \max_{\bar{\Omega}} -\hat{\lambda}g(x) < 0$, i.e. $\hat{z}_2(x,t) > -\hat{\lambda}g_2(x)$ in $\bar{\Omega}_T$. To obtain upper estimates for $\hat{z}_1$, let $R_1(x,t) = (\hat{u}_1 - m_1e^{at})\cdot e^{-at}$ for $(x,t) \in \bar{\Omega}_T$. From (3.2-34), it satisfies

$$\sigma_1\Delta R_1 - \frac{\partial R_1}{\partial t} = e^{-at}(-\hat{u}_1)[a + \bar{f}_1(\hat{z}_1+g_1, \hat{z}_2+g_2)] + e^{-at}\hat{u}_1 a \ge 0$$

in $\Omega x(0,T]$. The last inequality is true because $\hat{u}_1 > 0$, $\hat{z}_i+g_i > 0$ and the way $\bar{f}_1$ is defined. For the initial and boundary values, we have $R_1(x,0) = \hat{\lambda}g_1(x) - m_1 < 0$ in $\bar{\Omega}$ and $\partial R_1/\partial\eta = (\partial\hat{z}_1/\partial\eta + \hat{\lambda}\cdot\partial g_1/\partial\eta)e^{-at} = 0$ on $\delta\Omega x[0,T]$. The maximum principles imply that $R_1(x,t) < 0$ in $\bar{\Omega}$; therefore $\hat{z}_1(x,t) < \hat{z}_1(x,t)+\hat{\lambda}g_1(x) = \hat{u}_1(x,t) < m_1e^{at} \le m_1e^{aT}$ for $(x,t) \in \bar{\Omega}_T$. Finally, let $R_2(x,t) = (\hat{u}_2 - [m_2+C-K\omega(x)]e^{bt})e^{-bt+\varepsilon t}$, where b, ε, K and C are chosen in the first paragraph of the proof. It satisfies

$$\begin{aligned}\sigma_2\Delta R_2 - \frac{\partial R_2}{\partial t} &= (\sigma_2\Delta\hat{u}_2 + \sigma_2 K\Delta\omega\cdot e^{bt})e^{-bt+\varepsilon t} - (\partial\hat{u}/\partial t)e^{-bt+\varepsilon t} + \\ &\quad [m_2+C-K\omega]\varepsilon e^{\varepsilon t} - \hat{u}_2(-b+\varepsilon)e^{-bt+\varepsilon t} \\ &= e^{-bt+\varepsilon t}\{-\hat{u}_2[-r + \bar{f}_2(\hat{z}_1+g_1, \hat{z}_2+g_2) - b + \varepsilon] + \\ &\quad e^{bt}[-\lambda_1\omega(x)K\sigma_2 + \varepsilon(m_2+C-K\omega(x)]\}\end{aligned}$$

in $\Omega x(0,T]$. We have $-r + \bar{f}_2(\hat{z}_1+g_1, \hat{z}_2+g_2) - b + \varepsilon < -r + f_2(2m_1e^{aT}, 0) - b + \varepsilon = 0$ and $-\lambda_1\omega K\sigma_2 + \varepsilon(m_2+C-K\omega) = \varepsilon[-K\omega(1+(\sigma_2\lambda_1)/\varepsilon) + C + m_2] > 0$ by (3.2-30); therefore $\sigma_2\Delta R_2 - \partial R_2/\partial t > 0$ in $\Omega x(0,T]$. On $\delta\Omega x[0,T]$, we have $\partial R_2/\partial\eta = (\hat{\lambda}p(\hat{z}_1+g_1) + \hat{\lambda}q(\hat{z}_2+g_2) + K\cdot\partial\omega/\partial\eta\cdot e^{bt})e^{-bt+\varepsilon t} < 0$, by (3.2-29); and $R_2(x,0) = \hat{\lambda}g_2(x) - [m_2+C-K\omega(x)] < 0$ by (3.2-30). The maximum principles consequently imply that $R_2(x,t) < 0$ in $\bar{\Omega}_T$; and $\hat{z}_2 + g_2 = \hat{u}_2 +$

$(1-\hat{\lambda})g_2 < [m_2+C-K\omega(x)]e^{bT} + g_2 < 0$ in $\bar{\Omega}_T$. We have therefore showed the strict inequalities:

$$\text{(3.2-37)} \qquad \begin{aligned} -g_1(x) &< \hat{z}_1(x,t) < m_1 e^{aT} \\ -g_2(x) &< \hat{z}_2(x,t) < [m_2+C-K\omega(x)]e^{bT} \end{aligned}$$

for $(x,t) \in \bar{\Omega}_T$.

We next show that if $\hat{\rho}$ satisfies appropriate conditions, then $|\hat{z}_i|^{(1+\bar{\alpha})}_{\Omega_T} < \bar{c}_i+1$, $i = 1,2$. Recall that P_0, P_1, P_2 had been now fixed, which completely determine $\bar{\delta}_1, \bar{c}_1$ and subsequently $\hat{C}$, then $\bar{\delta}_2, \bar{c}_2$. If $|\hat{z}_i|^{(1+\bar{\alpha})}_{\Omega_T} \le \bar{c}_i+1$, $i = 1,2$, the term $F^{\hat{z}_1}(\hat{\lambda}, \hat{z}_2(x,t),x,t)$ in (3.2-36) as a function of (x,t) will be uniformly bounded in the norm $|\ |^{(\bar{\alpha})}_{\Omega_T}$ for $\hat{\lambda} \in [0,1]$. From Theorem 1.3-2 for linear theory, we obtain a uniform bound for $|\partial\hat{z}_2/\partial t|$ in terms of $\bar{c}_1$, $\bar{c}_2$ (say $|\partial\hat{z}_2/\partial t| \le M_1$ in $\bar{\Omega}_T$). Suppose that $\partial f_1/\partial u_2 \ge -\hat{\rho}$ in the first open quadrant so that $K_1\hat{\rho}M_1 < P_1$ and $0 < \hat{\rho} \le \rho_2$. Since $(\hat{z}_1, \hat{z}_2) \in S$, we have $|\hat{z}_2(x,t)| \le K_2-m_2$, $|F_1^{\hat{z}_2}(\lambda,v,x,t)| \le P_0$, $|\frac{\partial F_1^{\hat{z}_2}}{\partial t}(\lambda,v,x,t)| \le |K_1\frac{\partial f_1}{\partial u_2}(v+g_1(x), \hat{z}_2(x,t)+g_2(x))\frac{\partial \hat{z}_2}{\partial t}(x,t)| \le K_1\hat{\rho}M_1 < P_1$, $|\frac{\partial F_1^{\hat{z}_2}}{\partial v}(\lambda,v,x,t)| \le P_2$ for all $(x,t) \in \bar{\Omega}_T$, $|v| \le K_1-m_1$, $0 \le \lambda \le 1$. (Here, without loss of generality, we may assume that $\partial f_1/\partial u_2(u_1,u_2) \ge -\hat{\rho}$ even for $u_1 < 0$, since it will be evaluated only at $u_1 = \hat{z}_1 + g_1 > 0$). Consequently, from (3.2-34) and (3.2-31) we have $|\hat{z}_1|^{(1+\bar{\delta}_1)}_{\Omega_T} \le \bar{c}_1$. Now, consider the term $F_1^{\hat{z}_2}(\hat{\lambda}, \hat{z}_1, x, t)$ in (3.2-34). Its $|\ |^{(\bar{\alpha})}_{\Omega_T}$ norm is bounded by an expression of the form $B_1|\hat{z}_1|^{(\bar{\alpha})}_{\Omega_T} + B_2|\hat{z}_2|^{(\bar{\alpha})}_{\Omega_T} + B_3$, where B_i are constants, with B_2 being a bound for $|\partial f_1/\partial u_2|$. From equation (3.2-34) and Theorem 1.3-2 for linear theory, we deduce that $|\partial\hat{z}_1/\partial t| \le D_1[B_1|\hat{z}_1|^{(\bar{\alpha})}_{\Omega_T} + B_2|\hat{z}_2|^{(\bar{\alpha})}_{\Omega_T}$

$+ B_3] + D_2$ for some constants D_1, D_2. In the choice of $\hat{C}$ earlier, we may assume that $\hat{C} > D_1B_1(\bar{c}_1+1) + D_1B_3 + D_2$. Thus by reducing $\hat{\rho}$ if necessary (which subsequently reduce B_2), we have $|\partial\hat{z}_1/\partial t| \le \hat{C}$ for all $\hat{\lambda} \in [0,1]$, if $|\hat{z}_i|^{(1+\bar{\alpha})} \le \bar{c}_i+1$, $i = 1,2$. Consequently, from (3.2-36) and (3.2-32), we obtain $|\hat{z}_2|_{\Omega_T}^{(1+\bar{\delta}_2)} \le \bar{c}_2$. Hence, we may conclude that

$$(3.2\text{-}38) \qquad |\hat{z}_i|_{\Omega_T}^{(1+\bar{\alpha})} < \bar{c}_i + 1, \quad i = 1, 2.$$

From (3.2-37) and (3.2-38), we find that the element $\hat{\xi} = (\hat{z}_1, \hat{z}_2)$ cannot be on the boundary of S. Thus Theorem 1.3-7 implies that the equation $G(\xi,1) = 0$ must have a root in the interior of S. Let this root be denoted by $\xi = \xi^* = (z_1^*, z_2^*)$. We have $0 < z_i^* + g_i < K_i$, $i = 1,2$; and consequently $\bar{f}_i(z_1^*+g_1, z_2^*+g_2) = f_i(z_1^*+g_1, z_2^*+g_2)$ in $\bar{\Omega}_T$. Let $u_i^* = z_i^* + g_i$, $i = 1,2$ in $\bar{\Omega}_T$; we see from equations (3.2-34) and (3.2-36) with $\hat{\lambda} = 1$ and $\hat{z}_i$ replaced by z_i^*, that (u_1^*, u_2^*) is a solution fo (3.2-13) for $(x,t) \in \Omega\times(0,T]$ and initial boundary conditions (3.2-27) and (3.2-28), with $u_i^* \in H^{1+\bar{\alpha},(1+\bar{\alpha})/2}$, $i = 1,2$. By Theorem 1.3-2 and the smoothness assumptions on f_i, p, q, g_i, we assert that $u_i^* \in H^{2+\bar{\alpha},(2+\bar{\alpha})/2}$. Again, by the smoothness of f_i etc. and Theorem 1.3-2, we conclude that $u_i^* \in H^{2+\alpha,(2+\alpha)/2}(\bar{\Omega}_T)$, $i = 1,2$.

In order to establish the time stability of the equilibrium solution found in Theorem 3.2-2 for problem (3.2-13), (3.2-16), we impose additional hypotheses on the functions p(u) and q(u) which control the flux at the boundary. It is reasonable to assume that they depend monotonically on concentrations. Thus, in addition to (3.2-17), we assume in [138] that for u > 0:

$$(3.2\text{-}39) \qquad p'(u) > 0, \quad p''(u) < 0 \quad \text{and} \quad q'(u) > 0, \quad q''(u) < 0.$$

In [138], it is proved that in such situation, if the steady state solution $(\bar{u}_1(x),\bar{u}_2(x))$ of Theorem 3.2-2 satisfies conditions similar to that in Theorem 2.3-1, it will be asymptotically stable. That is , any solution $(u_1(x,t),u_2(x,t))$ with $u_i \in H^{2+\alpha,1+\alpha/2}(\bar{\Omega}_T)$, each $T > 0$, $i =1,2$, of (3.2-13) with boundary conditions (3.2-16) and initial conditions $u_i(x,0) = g_i(x)$ close enough to $\bar{u}_i(x)$ for all $x \in \bar{\Omega}$, $i = 1,2$, one has $u_i(x,t) \to \bar{u}_i(x)$ uniformly as $t \to +\infty$, $i = 1, 2$.

3.3. Nonlinear Density-Dependent Diffusion and Spatially Varying Growth

In many ecological diffusion-reaction studies, it was found that one should include the effect of density dependent diffusion rates, drift terms and spatially varying growth rates, in order to obtain more accurate results. On the other hand, many mathematical results on reaction-diffusion system as those on earlier sections do not include such general setting. For example, conditions for existence of positive coexistence steady states for competing species in Theorem 2.4-3 are of the nature that growth rates of the species are uniformly larger than certain positive constants related to the first eigenvalue. In the case of highly heterogeneous environment, such conditions are difficult to satisfy. This section considers the behavior of competing-species reaction-diffusion under a more general situation. We also determine the nature of nonlinear density-dependent diffusion, which would still allow results analogous to those in section 2.4. Results are obtained concerning coexistence, survival and extinction.

In our equations below, we assume that diffusivity depends on concentration, giving rise to the term $\text{div}(\sigma_i(u_i)\nabla u_i)$ with $\sigma_i(u_i)$ expressing the concentration dependence. The intrinsic growth rates $a_i(x)$ will be assumed to be functions of position $x = (x_1,x_2,..,x_n)$ in an open connected bounded set Ω in R^n, $n \geq 2$. We follow the development given in

[141], and consider the following initial Dirichlet boundary value problem for m competing-species reaction-diffusion:

$$\frac{\partial u_i}{\partial t} = \mathrm{div}(\sigma_i(u_i)\nabla u_i) + u_i[a_i(x) + f_i(u_1,\dots,u_m)]$$

(3.3-1) $\quad$ for $(x,t) \in \Omega\times(0,T]$, $T > 0$, $i = 1,\dots,m$;

$$u_i(x,0) = \phi_i(x),\ x \in \bar{\Omega};\ u_i(x,t) = \Phi_i(x),\ (x,t) \in \delta\Omega\times[0,T].$$

Here, $\nabla = ((\partial/\partial x_1),\dots,(\partial/\partial x_n))$ denotes the gradient, $\mathrm{div}\ v = \sum_{i=1}^{n} \partial v_i/\partial x_i$ denotes the divergence. The following hypotheses on f_i, a_i and σ_i will always be made in this section. The functions f_i: $R^m \to R$ have Hölder continuous partial derivatives up to second order in compact sets, $i = 1, \dots, m$. The partial derivatives also satisfy:

(3.3-2) $\quad \partial f_i/\partial u_j < 0$, $i,j = 1,\dots,m$ in $\{(u_1,\dots u_m)$: $u_i \geq 0$, $i = 1,\dots,m\}$,

$f_i(0,0) = 0$, and

(3.3-3) $\quad \sup_{s\geq 0}(\partial f_i/\partial u_i)(0,\dots,0,s,0,\dots,0) \overset{(\mathrm{def})}{=} r_i < 0$,

where $s \geq 0$ occurs at the ith component, $i = 1,\dots,m$. The intrinsic growth rate function of the ith species, $a_i(x)$, is in $H^{1+\alpha}(\bar{\Omega})$ and

(3.3-4) $\quad a_i(x) \geq 0$ in $\bar{\Omega}$, $i = 1,\dots,m$.

The diffusivity functions $\sigma_i(s)$ satisfy:

(3.3-5) $\quad \sigma_i(0) > 0$, $\sigma_i(s)$ in $H^{\alpha}(R)$, $\sigma_i'(s) \geq 0$ in $[0,\infty)$, $\sigma_i''(s)$ is continuous in $[0,\infty)$, $i = 1,\dots,m$.

These assumptions and equations are biologically plausible, while the smoothness and other hypotheses are made convenient enough so that excessive technicalities do not arise.

The initial-boundary value problem (3.3-1), under conditions (3.3-2) to (3.3-5) and appropriate smoothness conditions for the initial boundary function, would possess a solution in $H^{2+\alpha,1+\alpha/2}(\bar{\Omega}_T)$, each T >0. The detailed proof for existence will be considered at the end of this section, because we will need the method of establishing a-priori bound for solutions given in Theorem 3.3-2. The following Theorem 3.3-1 and 3.3-3 study criteria for survival and coexistence. The intrinsic growth rate, $a_k(x)$, of a particular k^{th} species is assumed to be locally high in a subdomain Ω' of Ω. We will see that one only needs locally high growth rates to sustain survival.

Theorem 3.3-1. *Let k be an integer,* $1 \le k \le m$. *Let* $u = (u_1,\ldots,u_m)$ *be a solution of* (3.3-1) *in the class* $H^{2+\alpha,1+\alpha/2}(\bar{\Omega}_T)$, T>0, *initially satisfying:*

$$0 \le u_i(x,0) \le b_i \;, \; x \in \bar{\Omega}, \; i = 1,\ldots,m, \tag{3.3-6}$$

where b_i *are positive numbers satisfying* $b_i \ge |r_i^{-1}|\cdot\max\{a_i(x): x \in \bar{\Omega}\}$. *Suppose that there exists a subdomain* $\Omega' \subseteq \Omega$ *(with principal eigenvalue* λ' *i.e.* $\lambda = \lambda' > 0$ *is the first eigenvalue for the problem* $\Delta\phi + \lambda\phi = 0$ *in* Ω', $\phi = 0$ *on* $\delta\Omega'$*) with the properties:*

$$0 < u_k(x,0), \; x \in \bar{\Omega}' \; ; \tag{3.3-7a}$$

$$a_k(x) - \sigma_k(0)\lambda' + f_k(b_1,\ldots,b_{k-1},0,b_{k+1},\ldots,b_m) > 0 \tag{3.3-7b}$$

for all $x \in \bar{\Omega}'$. *Then the solution* u *satisfies:*

$$0 < u_k(x,t) \quad \text{for} \quad (x,t) \in \Omega' \times [0,T]. \tag{3.3-8}$$

Moreover, $u_k(x,t) \geq \delta > 0$ *for all* x *in any compact set contained in* Ω', $0 \leq t \leq T$ (*where* δ *is some constant depending on the compact set, independent of* T); *and*

$$0 \leq u_i(x,t) \leq b_i \quad \text{for } (x,t) \in \bar{\Omega}_T,\ i = 1,\dots,m. \tag{3.3-9}$$

Proof. We shall construct lower and upper solutions v_i, w_i satisfying differential inequalities (3.3-14), (3.3-15) below, with v_i, w_i replacing α_i, β_i respectively. Then, we apply Theorem 3.3-2 below to conclude $u_k(x,t) \geq v_k(x,t)$ in $\bar{\Omega}_T$. The function v_k will be positive for x in the interior of Ω', thus implying the survival of the kth species. Let $\theta(x)$ be a positive eigenfunction in Ω', associated with the principal eigenvalue $\lambda = \lambda'$. Define $v_i(x,t) \equiv 0$ in $\bar{\Omega}_T$ for $i \neq k$, $1 \leq i \leq m$; and

$$v_k(x,t) = \begin{cases} \varepsilon\theta(x), & \text{if } x \in \Omega' \\ 0, & \text{if } x \in \bar{\Omega}\backslash\Omega' \end{cases} \tag{3.3-10}$$

in $\bar{\Omega}_T$. Here ε is a sufficiently small positive constant to be determined later. For $i = 1,\dots,m$, define $w_i(x,t) \equiv b_i$ in $\bar{\Omega}_T$. We have the following inequality, for $i = 1,\dots,m$:

$$\begin{aligned} &\operatorname{div}(\sigma_i(w_i)\nabla w_i) + w_i[a_i(x) + f_i(v_1,\dots,v_{i-1},w_i,v_{i+1},\dots,v_m)] - \frac{\partial w_i}{\partial t} \\ &\leq w_i[a_i(x) + f_i(0,\dots,0,w_i,0,\dots,0)] \\ &= w_i[a_i(x) + \int_0^{w_i} \frac{\partial f_i}{\partial u_i}(0,\dots,0,u_i,0,\dots,0)du_i] \\ &\leq w_i[a_i(x) + r_i w_i] \leq b_i[a_i(x) - \max\{a_i(x): x \in \bar{\Omega}\}] \leq 0 \end{aligned} \tag{3.3-11}$$

for $(x,t) \in \Omega\times[0,T]$. For $i \neq k$, clearly we have

$$(3.3\text{-}12)\quad \operatorname{div}(\sigma_i(v_i)\nabla v_i)+ v_i[a_i(x)+f_i(w_1,\dots,w_{i-1},v_i,w_{i+1},\dots,w_m)] - \frac{\partial v_i}{\partial t} = 0$$

for $(x,t) \in \Omega x[0,T]$. For $i = k$, (3.3-12) is clearly valid for $(x,t) \in (\Omega\backslash\Omega')x[0,T]$. If $(x,t) \in \Omega' x[0,T]$, we have

$$\operatorname{div}(\sigma_k(v_k)\nabla v_k)+ v_k[a_k(x)+f_k(w_1,\dots,w_{k-1},v_k,w_{k+1},\dots,w_m)] - \frac{\partial v_i}{\partial t}$$

$$(3.3\text{-}13)\quad = \sigma_k(v_k)\Delta v_k+ \sigma_k'(v_k)|\text{grad } v_k|^2+ \varepsilon\theta(x)[a_k(x)+ f_k(w_1,\dots,v_k,\dots,w_m)]$$

$$= \varepsilon\theta(x)[a_k(x)-\sigma_k(\varepsilon\theta(x))\lambda'+ f_k(b_1,\dots,b_{k-1},\varepsilon\theta,b_{k+1},\dots,b_m)]+$$

$$\sigma_k'(\varepsilon\theta)|\text{grad } v_k|^2.$$

Now, choose $\varepsilon > 0$ sufficiently small so that the expression in (3.3-13) is positive in $\Omega' x[0,T]$. (This is possible due to hypotheses (3.3-5) and (3.3-7b). Let $(u_1,\dots,u_m)$ be a solution of (3.3-1) satisfying (3.3-6) and (3.3-7) as stated. Reduce the choice of $\varepsilon > 0$, if necessary, so that $u_k(x,0) > v_k(x,0) = \varepsilon\theta(x)$ for $x \in \bar{\Omega}'$ (note that this will not affect the sign of the expression in (3.3-13)). Utilizing inequalities (3.3-11) to (3.3-13) and Theorem 3.3-2 below, we conclude that

$$0 \le v_i \le u_i(x,t) \le w_i = b_i,\ i = 1,\dots,m$$

for $(x,t) \in \bar{\Omega}x[0,T]$. From the definition of v_k in (3.3-10), we have (3.3-8) and the strict positivity of u_k in compact subsets of Ω' as stated in the theorem.

The following theorem is a comparison result similar to those found in section 1.2. However, we presently use nonsmooth comparison functions, and the differential operator has its coefficients $\sigma_i(u_i)$ dependent on u_i.

It is used in the proof of the last Theorem.

Theorem 3.3-2. *Let $\Omega' \subseteq \Omega$ be a subdomain, with λ' as its principal eigenvalue and $\theta(x)$ a positive eigenfunction in Ω'. Let j be an integer $1 \le j \le m$; $\alpha_i(x,t) \equiv 0$ in $\bar{\Omega}_T$ if $i \ne j$, $1 \le i \le m$, and*

$$\alpha_j(x,t) = \begin{cases} \delta\theta(x) & \text{if } (x,t) \in \Omega' x[0,T] \\ 0 & \text{if } (x,t) \in (\bar{\Omega}\backslash\Omega')x[0,T] \end{cases}$$

where $\delta > 0$ is a constant. Let $\beta_i(x,t)$ be nonnegative functions in $H^{2+\alpha,1+\alpha/2}(\bar{\Omega}_T)$ for $i = 1,..,m$. Suppose that α_i, β_i satisfy:

$$\alpha_i(x,t) \le \beta_i(x,t) \qquad \text{for } (x,t) \in \bar{\Omega}_T;$$

$$\text{(3.3-14)} \quad \operatorname{div}(\sigma_i(\alpha_i)\nabla\alpha_i) + \alpha_i[a_i(x) + f_i(\beta_1,..,\beta_{i-1},\alpha_i,\beta_{i+1},..,\beta_m)] - \frac{\partial\alpha_i}{\partial t} \ge 0$$

$$\text{(3.3-15)} \quad \operatorname{div}(\sigma_i(\beta_i)\nabla\beta_i) + \beta_i[a_i(x) + f_i(\alpha_1,..,\alpha_{i-1},\beta_i,\alpha_{i+1},..,\alpha_m)] - \frac{\partial\beta_i}{\partial t} \le 0$$

for $(x,t) \in \Omega x(0,T]$, $i = 1,\dots,m$, except for $i = j$ in (3.3-14) valid only for $(x,t) \in (\Omega\backslash\delta\Omega')x(0,T)$. Let $(u_1,\dots,u_m)$, $u_i \in H^{2+\alpha,1+\alpha/2}(\bar{\Omega}_T)$, be a solution of the differential equation in (3.3-1) with boundary conditions satisfying:

$$\text{(3.3-16)} \quad \begin{aligned} &\alpha_i(x,0) \le u_i(x,0) \le \beta_i(x,0), \quad x \in \bar{\Omega}, \\ &\alpha_i(x,t) \le u_i(x,t) \le \beta_i(x,t), \quad (x,t) \in \delta\Omega x[0,T] \end{aligned}$$

for $i = 1,\dots,m$. Then we have

$$\text{(3.3-17)} \quad \alpha_i(x,0) = \alpha_i(x,t) \le u_i(x,t) \le \beta_i(x,t)$$

for $(x,t) \in \bar{\Omega}x[0,T]$, $i = 1,\ldots,m$.

Proof. Since $u_i, \alpha_i, \beta_i \in H^{2+\alpha,\,1+\alpha/2}(\bar{\Omega})$, there are constants K and M such that $|\alpha_i| \le K$, $|\beta_i| \le K$, $|\Delta u_i| \le M$, $|\text{grad } u_i|^2 \le M$ for all $(x,t) \in \bar{\Omega}_T$, $i = 1,\ldots,m$. The assumptions on f_i, a_i and σ_i imply that there are constants R and B so that for each $i = 1,\ldots,m$, we have $|\sigma_i'(s)| \le R$, $|\sigma_i''(s)| \le R$ for $0 \le s \le 2K$, and $|a_i(x) + f_i(s_1,\ldots,s_m)| \le B$ for $x \in \bar{\Omega}$, $0 \le s_i \le 2K$, $i = 1, \ldots, m$.

Let $0 < \varepsilon < K[1 + 3(B + 2MR + KLm)T]^{-1}$, where $(1/2)L$ is a bound for the absolute values of all first partial derivatives of $f_i(s_1,\ldots,s_m)$, $0 \le s_i \le 2K$, $i = 1,\ldots,m$. Define, for $(x,t) \in \bar{\Omega}_T$, $i = 1,\ldots,m$,

$$
\text{(3.3-18)} \qquad
\begin{aligned}
u_i^+(x,t) &= u_i(x,t) + \varepsilon[1 + 3(B + 2MR + KLm)t] \\
u_i^-(x,t) &= u_i(x,t) - \varepsilon[1 + 3(B + 2MR + KLm)t].
\end{aligned}
$$

By hypothesis, we have

$$
\text{(3.3-19)} \qquad \alpha_i(x,t) < u_i^+(x,t) \quad \text{and} \quad u_i^-(x,t) < \beta_i(x,t)
$$

for $x \in \bar{\Omega}$, $t = 0$, $i = 1,\ldots,m$. Suppose one of these inequalities fails at some point in $\bar{\Omega}x(0,\tau_1)$, where $\tau_1 = \min\{T, 1/(3(B + 2MR + KLm))\}$; and (x_1,t_1) is a point in $\bar{\Omega}x(0,\tau_1)$ with minimal t_1 where (3.3-19) fails. At (x_1,t_1), $\sigma_i = u_i^+$ or $u_i^- = \beta_i$ for some i. Assume the former is the case; a similar proof holds for the latter case.

Suppose further that at (x_1,t_1), $\alpha_j = u_j^+$ (a simplier proof will work if $\alpha_i = u_i^+$ at (x_1,t_1) for $i \ne j$), we consider separately the situations for $x_1 \in (\Omega\backslash\Omega')$ or $x_1 \in \Omega'$. If $x_1 \in \Omega\backslash\Omega'$, we have $u_j^+(x,t_1) > 0$ for $t < t_1$, $x \in \bar{\Omega}$ and $u_j^+(x_1,t_1) = 0$. Observe that $x_1 \notin \delta\Omega$ because $u_j^+(x,t_1) > u_j(x,t_1) \ge 0$ for $x \in \delta\Omega$, by (3.3-16). However, for $(x,t) \in \Omega x(0,T]$:

$$\frac{\partial}{\partial t}(-u_j^+) = \frac{-\partial}{\partial t}(u_j) - \varepsilon 3(B + 2MR + KLm)$$

$$= -\sigma_j(u_j)\Delta u_j - \sigma'_j(u_j)|\mathrm{grad}\ u_j|^2 - u_j[a_j(x)+f_j(u_1,\ldots,u_m)]$$

$$(3.3\text{-}20) \qquad -\varepsilon 3(B + 2MR + KLm)$$

$$= -\sigma_j(u_j^+)\Delta u_j + [\sigma_j(u_j^+) - \sigma_j(u_j)]\Delta u_j$$

$$-\sigma'_j(u_j)|\mathrm{grad}\ u_j|^2 + [u_j^+ - u_j][a_j + f_j(u_1,\ldots,u_m)]$$

$$-u_j^+[a_j + f_j(u_1,\ldots,u_m)] - \varepsilon 3(B + 2MR + KLm).$$

Recalling that $\sigma_j(u_j^+) > 0$; and at (x_1,t_1) we have grad u_j = grad $u_j^+ = 0$, $\Delta u_j = \Delta u_j^+ \geq 0$, (3.3-20) implies that

$$\frac{\partial}{\partial t}(-u_j^+)|_{(x_1,t_1)} \leq R\varepsilon[1 + 3(B{+}2MR{+}KLm)t_1]M + \varepsilon[1{+}\ 3(B{+}2MR{+}KLm)t_1]B$$

$$(3.3\text{-}21) \qquad -\varepsilon 3(B{+}2MR{+}KLm)$$

$$\leq MR\varepsilon 2 + B\varepsilon 2 - \varepsilon 3[B{+}2MR{+}KLm] < 0$$

contradicting the definition of (x_1,t_1).

If $x_1 \in \Omega'$, we have $u_j^+(x,t) > \alpha_j(x,t)$ for $t < t_1$, $x \in \bar{\Omega}$; and $u_j^+(x_1,t_1) = \alpha_j(x_1,t_1) = \delta\theta(x_1)$. But for $(x,t) \in \Omega' x(0,T]$

$$\frac{\partial}{\partial t}(\alpha_j - u_j^+) \leq \mathrm{div}(\sigma_j(\alpha_j)\nabla\alpha_j) + \alpha_j[a_j+f_j(\beta_1\ldots,\beta_{j-1},\alpha_j,\beta_{j+1},\ldots,\beta_m)]$$

$$-\mathrm{div}(\sigma_j(u_j)\nabla u_j) - u_j[a_j+f_j(u_1,\ldots,u_m)] - \varepsilon 3(B{+}2MR{+}KLm)$$

$$(3.3\text{-}22)$$

$$= \sigma_j(\alpha_j)\Delta\alpha_j + \sigma'_j(\alpha_j)|\mathrm{grad}\ \alpha_j|^2 + \alpha_j[a_j+f_j(\beta_1,\ldots,\beta_{j-1},\alpha_j,$$

$$\beta_{j+1},\ldots,\beta_m)] - \sigma_j(u_j^+)\Delta u_j + [\sigma_j(u_j^+) - \sigma_j(u_j)]\Delta u_j$$

$$+[\sigma'_j(u_j^+) - \sigma'_j(u_j)]|\mathrm{grad}\ u_j|^2 - \sigma'_j(u_j^+)|\mathrm{grad}\ u_j|^2 + (u_j^+ - u_j)[a_j +$$

$$f_j(u_1,\ldots,u_m)] - u_j^+[a_j+f_j(u_1,\ldots,u_m)] - \varepsilon 3(B+2MR+KLm).$$

At (x_1,t_1), we have $u_j^+ = \alpha_j$, grad u_j = grad u_j^+ = grad α_j, $\Delta(\alpha_j-u_j) = \Delta(\alpha_j-u_j^+) \le 0$, thus (3.3-22) gives

$$\frac{\partial}{\partial t}(\alpha_j-u_j^+)|_{(x_1,t_1)} = \sigma_j(\alpha_j)\Delta(\alpha_j-u_j)+\alpha_j[f_j(\beta_1,\ldots,\beta_{j-1},\alpha_j,\ldots,\beta_{j+1},\ldots,\beta_m)$$

(3.3-23)

$$-f_j(u_1,\ldots,u_m)] + [\sigma_j(u_j^+)-\sigma_j(u_j)]\Delta u_j + [\sigma_j'(u_j^+)-\sigma_j'(u_j)]\cdot$$

$$|\text{grad } u_j|^2 + (u_j^+-u_j)[a_j+f_j(u_1,\ldots,u_m)] - \varepsilon 3(B+2MR+KLm).$$

Moreover, at (x_1,t_1) we have

$$f_j(\beta_1,\ldots,\beta_j,\alpha_j,\beta_{j+1},\ldots,\beta_m) - f_j(u_1,\ldots,u_m)$$

(3.3-24)

$$\le f_j(\tilde{u}_1^-,\ldots,\tilde{u}_{j-1}^-,u_j^+,\tilde{u}_{j+1}^-,\ldots,\tilde{u}_m^-) - f_j(u_1,\ldots,u_m)$$

$$\le L\varepsilon[1 + 3(B + 2MR + KLm)t_1]m$$

where $\tilde{u}_i^-(x_1,t_1) = \max\{u_i^-(x_1,t_1),\alpha_i(x_1,t_1)\}$, because $|u_i-\tilde{u}_i^-| \le |u_i^+-u_i^-|$. Consequently, (3.3-23) gives

$$\frac{\partial}{\partial t}(\alpha_j-u_j^+)|_{(x_1,t_1)} \le KLm\varepsilon[1+3(B+2MR+KLm)t_1] + 2MR\varepsilon[1+3(B+2MR+KLm)t_1]$$

(3.3-25)

$$+ B\varepsilon[1+3(B + 2MR + KLm)t_1] - \varepsilon 3(B + 2MR + KLm)$$

$$\le KLm\varepsilon 2 + 4MR\varepsilon + 2B\varepsilon - \varepsilon 3(B + 2MR + KLm) < 0$$

contradicting the definition of (x_1,t_1). From these contradictions, we conclude that $u_j^+(x,t) > \alpha_j(x,t)$ for $(x,t) \in \bar{\Omega}\times[0,\tau_1)$. Passing to the limit as $\varepsilon \to 0^+$, we obtain $u_j(x,t) \ge \alpha_j(x,t)$ in $\bar{\Omega}\times[0,\tau_1]$.

If at (x_1,t_1), we have $\alpha_r = u_r^+$ for $r \ne j$, then $u_r^+(x,t) > 0$ for $t < t_1$,

$x \in \bar{\Omega}$ and $u_r^+(x_1,t_1) = 0$, with $x_1 \notin \delta\Omega$. For $x_1 \in \Omega$, repeat the arguments in (3.3-20) to (3.3-21), with j replaced by r. (There is no need for arguments analogous to (3.3-22) to (3.3-25).) We obtain $u_r^+ > \alpha_r = 0$ for $(x,t) \in \bar{\Omega}x[0,\tau_1)$, and consequently $u_r \geq \alpha_r = 0$ for $(x,t) \in \bar{\Omega}x[o,\tau_1]$.

If at (x_1,t_1), $u_i^- = \beta_i$ for some i, we show that

$$\frac{\partial}{\partial t}(\beta_i - u_i^-)|_{(x_1,t_1)} > 0 \tag{3.3-26}$$

by means of (3.3-15), in a way similar to the arguments that led to (3.3-22) to (3.3-25), but with inequalities reversed. Passing to the limit as $\varepsilon \to 0^+$, we again obtain $u_i \leq \beta_i$ for $(x,t) \in \bar{\Omega}x[0,\tau_1]$.

If $\tau_1 < T$, we repeat the above arguments by starting to define u_i^+, u_i^- with (3.3-18), with t in the square brackets on the right side of the formulas, replaced by $(t-\tau_1)$. This leads to $\alpha_i \leq u_i \leq \beta_i$ for $x \in \bar{\Omega}$, $\tau_1 \leq t \leq \min\{T, 2/3(B+2MR+KLm)\}$ etc. Eventually, we obtain (3.3-17) in $\bar{\Omega}x[0,T]$.

<u>Remark.</u> The assumption $\sigma_i'(s) \geq 0$, $i = 1,\dots,m$, has never been used in the proof of Theorem 3.3-2. However, $\sigma_k' \geq 0$ is essential for establishing the positivity of expression (3.3-13), in the proof of Thoerem 3.3-1.

The following is an immediate consequence of Theorem 3.3-1. It gives a sufficient condition for the coexistence of r species, $0 < r \leq m$, in Ω.

<u>Theorem</u> 3.3-3. <u>Let</u> $b_i \geq |r_i^{-1}|\max\{a_i(x): x \in \bar{\Omega}\}$, $i = 1,\dots,m$. <u>Suppose there exist</u> r <u>subdomains</u> $\Omega_{k_1},\dots,\Omega_{k_r}$ $(0<r\leq m$, $k_1,\dots,k_r$ <u>are distinct positive integers</u> $\leq m)$ <u>in</u> Ω, <u>with the property that:</u>

$$a_{k_i}(x) - \sigma_{k_i}(0)\lambda_{k_i} + f_{k_i}(b_1,\dots,b_{k_i-1},0,b_{k_i+1},\dots,b_m) > 0 \tag{3.3-27}$$

<u>for</u> $x \in \Omega_{k_i}$, $i = 1,\dots,r$. (Here, $\lambda = \lambda_{k_i} > 0$ <u>is the first eigenvalue for the problem:</u> $\Delta\phi + \lambda\phi = 0$ <u>in</u> Ω_{k_i}, $\phi = 0$ <u>on</u> $\delta\Omega_{k_i}$). <u>Let</u> $(u_1,\dots,u_m)$ <u>be a solution</u>

<u>of</u> (3.3-1) <u>with each component in</u> $H^{2+\alpha,1+\alpha/2}(\bar{\Omega}_T)$, $T>0$; <u>and assume initially that</u>

(3.3-28)
$$0 \le u_i(x,0) \le b_i, \quad x \in \bar{\Omega}, \quad i=1,\dots,m,$$
$$0 < u_{k_i}(x,0), \quad x \in \bar{\Omega}_{k_i}, \quad i=1,\dots,r.$$

<u>Then the solution satisfies</u>

(3.3-29)
$$0 < u_{k_i}(x,t), \quad (x,t) \in \bar{\Omega}_{k_i} x[0,T], \quad i=1,\dots,r.$$

<u>Moreover</u> $u_{k_i}(x,t) \ge \delta > 0$ <u>for all</u> x <u>in any compact set contained in</u> Ω_{k_i}, $0 \le t \le T$ (<u>where</u> δ <u>is some constant depending on the compact set, independent of</u> T); <u>and</u>

(3.3-30)
$$0 \le u_i(x,t) \le b_i \qquad (x,t) \in \bar{\Omega}_T.$$

Note that the k_i^{th} species will have, for all time under consideration, its concentration bounded below by positive constants in compact subsets of Ω_{k_i}. The simpliest situation happens when $\Omega_{k_1} = \Omega_{k_2} \dots = \Omega_{k_r}$; otherwise, the different species will primarily survive at different subregions in Ω.

We next study the problem of extinction for an arbitrary k^{th} species (i.e. u_k tending to zero). We consider the initial-boundary problem (3.3-1) as before, with the additional assumptions:

(3.3-31)
$$\Phi_k(x) \equiv 0, \ x \in \delta\Omega$$

(homogeneous boundary condition for the kth component), and nonnegative initial conditions $\phi_i(x) \ge 0$, $x \in \bar{\Omega}$, $i = 1,\dots,m$. Moreover, we impose the compatibility conditions $\phi_i(x) = \Phi_i(x)$ if $x \in \delta\Omega$, $\phi_i \in H^{2+\alpha}(\bar{\Omega})$, and

$$\{\operatorname{div}(\sigma_i(\phi_i)\nabla\phi_i) + \phi_i[a_i(x) + f_i(\phi_1,\dots,\phi_m)]\}|_{x\in\delta\Omega} = 0 \text{ for } i = 1,\dots,m$$

The following theorem describes a sufficient condition for the decay of the kth component (species).

Theorem 3.3-4. *Suppose that*

$$a_k(x) < \sigma_k(0)\lambda^1 \quad \text{for all } x \in \bar{\Omega} \tag{3.3-32}$$

(where $\lambda = \lambda^1$ *is the first eigenvalue of* $\Delta w + \lambda w = 0$ *in* Ω, $w = 0$ on $\delta\Omega$). *Let* $C_i > 0$, $i = 1,\dots,m$ *be such that for* $x \in \bar{\Omega}$

$$a_i(x) + f_i(0,\dots,0,C_i,0,\dots,0) \le 0 \tag{3.3-33}$$

(here C_i *appears in the* i^{th} *component), and* $\bar{C}_i = \max\{C_i, \sup_{x\in\bar{\Omega}} \phi_i(x)\}$. *Then there is a constant* $q > 0$ *so that the property;*

$$\sigma_k'(s) \le q, \quad \text{for all } 0 \le s \le \bar{C}_k \tag{3.3-34}$$

implies that for any solution $(u_1,\dots,u_m)$ *of* (3.3-1) (*under condition* (3,3-31), $\phi_i \ge 0$ *and compatibility condition described above) with each component in* $H^{2+\alpha,1+\alpha/2}(\bar{\Omega}_T)$, $T > 0$ *must satisfy;*

$$0 \le u_k(x,t) \le Ke^{-\varepsilon t} \quad \text{in } \bar{\Omega}_T \tag{3.3-35}$$

where K, ε *are positive constants independent of* T. *Moreover, we have*

$$0 \le u_i(x,t) \le \bar{C}_i, \quad i = 1,\dots,m \quad \text{in } \Omega_T. \tag{3.3.-36}$$

Remarks C_i exist by hypothesis (3.3-3); the size of q can be expressed in terms of the principal eigenfunction of a domain $\hat{\Omega} \supset \bar{\Omega}$. The details of proof is similar to that of Theorem 2.4-2 and can be found in [141].

To complete this section we finally discuss the conditions sufficient for the existence of a solution to the initial boundary value problem (3.3-1). We first state a relevant result concerning existence of solution to mixed initial boundary value problem for quasilinear parabolic systems:

$$\frac{\partial u}{\partial t} - \sum_{i,j=1}^{n} a_{ij}(x,t,u)\frac{\partial^2 u}{\partial x_i \partial x_j} + \sum_{i=1}^{n} b_i(x,t,u,u_x)\frac{\partial u}{\partial x_i} + c(x,t,u,u_x) = 0 \tag{3.3-37}$$

$$\text{for } (x,t) = (x_1,\dots,x_n,t) \in \Omega\times(0,T),\ T>0\ ;$$

$$u(x,0) = \phi(x),\ x \in \bar{\Omega};\ u(x,t) = 0,\ (x,t) \in \delta\Omega\times[0,T].$$

Here $u = (u_1,\dots,u_m)$, $u_x = (\partial u_1/\partial x_1,\dots,\partial u_1/\partial x_n,\dots,\partial u_m/\partial x_1,\dots,\partial u_m/x_n)$, $c = (c_1,\dots,c_m)$, $\phi = (\phi_1,\dots,\phi_m)$. Suppose a classical solution exists, the quantity, $\max_{\Omega_T}|u(x,t)|$ can be estimated, if for $(x,t) \in \bar{\Omega}_T\backslash\{(\delta\Omega\times[0,T])\cup(\Omega\times\{0\})\}$ and arbitrary u:

$$\sum_{i,j=1}^{n} a_{ij}(x,t,u)\xi_i\xi_j \ge 0 \qquad \text{for all } \xi = (\xi_1,\dots,\xi_n), \text{ and} \tag{3.3-38}$$

$$c_i(x,t,u,0)u_i \ge -k_1|u|^2 - k_2, \tag{3.3-39}$$

where k_1, k_2 are positive constant, $i = 1,\dots,m$. (Here $|u|^2 = \sum_{i=1}^{m} u_i^2$) Such estimate is obtained by means of the maximum principle and is done in the same way as for one equation given in chapter 1 in [127]. Namely, we have

$$\max_{\Omega_T}|u(x,t)| \le \min_{\lambda\ge k_1} e^{\lambda T}[\max_{\Omega}|u(x,0)| + \sqrt{k_2/(\lambda-k_1)}] \overset{(\text{def})}{\equiv} M \tag{3.3-40}$$

In order to obtain estimate for $|u_x|$, used for proving existence, we assume that for $(x,t) \in \Omega_T$, $|u| \le M$, and arbitrary mn dimensional vector $p = (p_1,\ldots,p_{mn})$:

$$\hat{\alpha}(|u|)\sum_{i=1}^{n} \xi_i^2 \le \sum_{i,j=1}^{n} a_{ij}(x,t,u)\xi_i\xi_j \le \hat{\beta}(|u|)\sum_{i=1}^{n} \xi_i^2 \quad , \ \alpha > 0, \tag{3.3-41}$$

for arbitrary $\xi = (\xi_1,\ldots,\xi_n)$, where $0 < \alpha_0 \le \hat{\alpha}(|u|) \le \hat{\beta}(|u|) \le \beta_0$ for $|u| \le M$;

$$\begin{aligned} |b_i(x,t,u,p)| &\le \hat{\beta}(|u|)(1 + |p|), \\ |c(x,t,u,p)| &\le [\varepsilon(|u|) + Q(|p|,|u|)](1 + |p|)^2, \end{aligned} \tag{3.3-42}$$

where $|p|^2 = \sum_{i=1}^{mn} p_i^2$, $Q(|p|,|u|)$ is a continuous function on the closed first quadrant, nondecreasing in $|u|$ and $Q(|p|,|u|) \to 0$ as $|p| \to \infty$, while $\varepsilon(|u|)$ is continuous and nondecreasing, with $\varepsilon(M)$ sufficiently small determined by M, α_0 and β_0. The functions $a_{ij}(x,t,u)$ are differentiable with respect to x_k and u_l satisfying

$$\left|\frac{\partial a_{ij}}{\partial x_k}(x,t,u)\right| \le \hat{\beta}(|u|), \text{ and } \left|\frac{\partial a_{ij}}{\partial u_l}(x,t,u)\right| \le \hat{\beta}(|u|). \tag{3.3-43}$$

The following existence and uniqueness theorem can be proved by using the Leray-Schauder fixed point thoerem.

<u>Theorem</u> 3.3-5. <u>Suppose that the following conditions are all satisfied for the initial boundary value problem</u> (3.3-37):

(i) <u>For</u> $(x,t) \in \bar{\Omega}_T \setminus \{(\partial\Omega\times[0,T])\cup(\Omega\times\{0\})\}$ <u>and arbitrary</u> u, <u>the inequality</u> (3.3-38) <u>and</u> (3.3-39) <u>are valid.</u>

(ii) <u>Let</u> M <u>be defined as in</u> (3.3-40), <u>then for</u> $|u| \le M$, $(x,t) \in \Omega_T$, <u>and arbitrary</u> p, <u>the functions</u> $a_{ij}(x,t,u)$, $b_i(x,t,u,p)$, $c(x,t,u,p)$, $(\partial a_{ij}/\partial x_k)(x,t,u)$, $(\partial a_{ij}/\partial u_l)(x,t,u)$ <u>are continuous and satisfy</u>

inequalitaies (3.3-41), (3.3-42) and (3.3-43).

(iii) $\delta\Omega \in H^{2+\alpha}$, $0 < \alpha < 1$.

(iv) Each component of $\phi(x)$ is in $H^{2+\alpha}(\bar{\Omega})$, and satisfies the compatibility conditions

$$\phi(x)|_{x\in\delta\Omega} = 0$$

$$[-\sum_{i,j=1}^{n} a_{ij}(x,0,0)\frac{\partial^2\phi}{\partial x_i\partial x_j} + \sum_{i=1}^{n} b_i(x,0,0,\phi_x)\frac{\partial\phi}{\partial x_i} + c(x,0,0,\phi_x)]|_{x\in\delta\Omega} = 0.$$

(v) (The above conditions determines a constant M_1, so that the a-priori estimate $\max_{\Omega_T}|u_x| \le M_1$ is true). For $(x,t) \in \bar{\Omega}_T$, $|u| \le M$, $|p| \le M_1$, the first derivatives of the functions $a_{ij}(x,t,u)$, $b_i(x,t,u,p)$, $c_i(x,t,u,p)$ with respect to x,t,u,p and the derivatives $\partial^2 a_{ij}/\partial u_s\partial u_k$, $\partial^2 a_{ij}/\partial u_s\partial x_r$, $\partial^2 a_{ij}/\partial u_s\partial t$, $\partial^2 a_{ij}/\partial x_r\partial t$ are continuous.

Then in the class $H^{2+\alpha,1+\alpha/2}(\bar{\Omega}_T)$, there exists a unique solution $u(x,t)$ for the problem (3.3-37).

(See [127], p.597, for reference to the above theorem.)

The problem (3.3-1) does not satisfy all the conditions given for (3.3-37) in the above. We will now show that with the aid of estimates, we can adapt Theorem 3.3-5 to our problem (3.3-1). Consequently, Theorem 3.3-6 justifies that the solutions for (3.3-1) in $H^{2+\alpha,1+\alpha/2}(\bar{\Omega}_T)$, assumed in Theorem 3.3-1 to Theorem 3.3-4, do indeed exist.

Theorem 3.3-6. Let Ω, f_i, a_i, σ_i, $i = 1,\ldots,m$ satisfy all the conditions as described in the beginning of this section. Let the initial boundary functions ϕ_i, Φ_i satisfy: $\phi_i(x) = \Phi_i(x)$ for $x \in \delta\Omega$, $\phi_i(x) \ge 0$ in $\bar{\Omega}$, ϕ_i has all third partial derivatives continuous in $\bar{\Omega}$, and

$$\{\mathrm{div}(\sigma_i(\phi_i)\nabla\phi_i) + \phi_i[a_i(x) + f_i(\phi_1(x),\ldots,\phi_m(x))]\}|_{x\in\delta\Omega} = 0 \qquad (3.3\text{-}44)$$

for $i = 1,\ldots,m$. Then, for any $T > 0$, in the class of functions in

$H^{2+\alpha,1+\alpha/2}(\bar{\Omega}_T)$, there exists a unique solution for the initial boundary value problem (3.3-1).

Proof. Let d_i be positive numbers satisfying:

$$d_i \geq |r_i^{-1}| \max\{a_i(x): x \in \bar{\Omega}\}, \quad \text{and}$$
$$0 \leq \phi_i(x) \leq d_i, \qquad x \in \bar{\Omega}$$

for $i = 1,\ldots,m$. Define $c_i(x,u_1,\ldots,u_m)$, $i = 1,\ldots,m$, $(x,u_1,\ldots,u_m) \in \bar{\Omega}\times R^m$ by:

$$c_i(x,u_1,\ldots,u_m) = h_i(u_i)[a_i(x) + f_i(h_1(u_1),\ldots,h_m(u_m))]$$

$$\text{where } h_i(s) = \begin{cases} s & \text{if } |s| \leq d_i \\ \rho_i(s) & \text{if } |s| > d_i \end{cases}$$

with $\rho_i(s)$ a twice continuously differentiable function for $|s| \geq d_i$, and $|\rho_i(s)| \leq 2d_i$, $\rho_i(\pm d_i) = \pm d_i$, $\rho_i'(\pm d_i) = 1$, and $\rho_i''(d_i) = 0$. Extend $\tilde{\sigma}_i(s)$ positively to $(-\infty,0)$ by letting $\tilde{\sigma}_i(s) = \sigma_i(s)$ for $s \in [0,\infty)$, with $\tilde{\sigma}_i(s)$ twice continuously differnetiable for $s \in (-\infty,\infty)$, and $\tilde{\sigma}_i(s) \geq (\sigma_i(0)/2) > 0$ for $s \in (-\infty,0)$, $i = 1,\ldots,m$.

We consider the initial boundary value problem:

$$\begin{aligned} \frac{\partial z_i}{\partial t}(x,t) &= \tilde{\sigma}_i(h_i(z_i+\phi_i(x))\Delta z_i + \tilde{\sigma}_i'(h_i(z_i+\phi_i))\sum_{j=1}^{m}[(z_i)_{x_j}+2(\phi_i)_{x_j}](z_i)_{x_j} \\ &+ \tilde{\sigma}_i'(h_i(z_i+\phi_i))\sum_{j=1}^{m}(\phi_i)_{x_j}^2 + \tilde{\sigma}_i(h_i(z_i+\phi_i))\Delta\phi_i \\ &+ c_i(x,z_1+\phi_1,\ldots,z_m+\phi_m) \end{aligned} \tag{3.3-45}$$

for $(x,t) \in \Omega\times(0,T]$, $i = 1,\ldots,m$;

(3.3-46) $\qquad z_i(x,0) = 0$ in $\bar{\Omega}$ and $z_i(x,t) = 0$ for $(x,t) \in \partial\Omega\times[0,T]$.

(Note that if we let $u_i(x,t) = z_i(x,t) + \phi_i(x)$, and if $0 \le u_i(x,t) \le d_i$, $i = 1,\dots,m$, then $u_i(x,t)$ satisfies:

(3.3-47) $$\frac{\partial u_i}{\partial t} = \sigma_i(u_i)\Delta u_i + \sigma_i'(u_i)|\text{grad } u_i|^2 + u_i[a_i(x)+f_i(u_1,\dots,u_m)].$$

Moreover, u_i satisfies the initial boundary conditions of (3.3-1)). Apply Theorem 3.3-5. The positivity of $\tilde{\sigma}_i$ and the boundedness of the last three terms of (3.3-45) imply that the condition (i) in Theorem 3.3-5 is satisfied. (3.3-42) of (ii) in Theorem 3.3-5 is satisfied by letting $Q(|p|,|u|) = C(1+|p|)^2$ for some large constant C and $\varepsilon(|u|) \equiv 0$. Compatibility condition (3.3-44) gives (iv). The smoothness of ϕ_i, $\tilde{\sigma}_i$ and h_i ensure that (v) is satisfied. Consequently, Theorem 3.3-5 gives a unique solution $z = (z_1(x,t),\dots,z_m(x,t))$ to (3.3-45), (3.3-46) for $(x,t) \in \bar{\Omega}\times[0,T]$, in the class $H^{2+\alpha,1+\alpha/2}(\bar{\Omega}_T)$.

We next show that $0 \le z_i(x,t) + \phi_i(x) \le d_i$, $i = 1,\dots,m$. Let $\hat{\alpha}_i(x,t) \equiv 0$ and $\hat{\beta}_i(x,t) \equiv d_i$, $i = 1,\dots,m$. Each function $\hat{\alpha}_i$ satisfies (3.3-14) in $\Omega\times(0,T]$ (with $\hat{\alpha}_i$, $\hat{\beta}_i$ replacing α_i, β_i respectively). Each function $\hat{\beta}_i$ satisfies:

$$\begin{aligned}
&\text{div}(\sigma_i(\hat{\beta}_i)\nabla\hat{\beta}_i) + \hat{\beta}_i[a_i(x)+f_i(\hat{\alpha}_1,\dots,\hat{\alpha}_{i-1},\hat{\beta}_i,\hat{\alpha}_{i+1},\dots,\hat{\alpha}_m)] - \frac{\partial\hat{\beta}_i}{\partial t}\\
&\quad = d_i[a_i(x) + f_i(0,\dots,0,d_i,0,\dots,0)]\\
&\quad = d_i[a_i(x) + \int_0^{d_i} \frac{\partial f_i}{\partial s_i}(0,\dots,0,s_i,0,\dots,0)ds_i]\\
&\quad \le d_i[a_i(x) + r_i d_i] \le d_i[a_i(x) - \max\{a_i(x): x \in \bar{\Omega}\}] \le 0
\end{aligned}$$

in $\Omega\times(0,T]$. For $i = 1,\dots,m$, $(x,t) \in \bar{\Omega}\times[0,T]$, let

(3.3-48) $$u_i(x,t) = z_i(x,t) + \phi_i(x).$$

The function u_i satisfies (3.3-47) for $x \in \Omega$, $0 < t \le t_1 \le T$ as long as $\hat{\alpha}_i \le u_i(x,t) \le \hat{\beta}_i$ for $(x,t) \in \bar{\Omega}x[0,t_1]$. By arguments exactly as given in Theorem 3.3-2, we can show that $\hat{\alpha}_i \le u_i \le \hat{\beta}_i$ for all $(x,t) \in \bar{\Omega}x[0,T]$, $i = 1,\dots,m$. (Note that our present situation is even simplier because all $\hat{\alpha}_i \equiv 0$, and we need only those arguments from (3.3-18) to (3.3-21). Those arguments from (3.3-22) to (3.3-25) for the $u_j \ge \alpha_j$ case will not be necessary). The a-priori bound, $\hat{\alpha}_i \le u_i \le \hat{\beta}_i$ in $\bar{\Omega}x[0,T]$, consequently implies that $u(x,t)$ is the unique solution of the initial boundary value problem (3.3-1) in $H^{2+\alpha,1+\alpha/2}(\bar{\Omega}_T)$.

Remark The above theorem shows that the solution $u(x,t)$ exists in $\bar{\Omega}_T$ for all $T > 0$. Under the assumptions of Theorems 3.3-4 and 3.3-6, we therefore have $u_k(x,t) \to 0$ uniformly for $x \in \bar{\Omega}$, as $t \to +\infty$.

3.4. Asymptotic Approximations for Small Diffusion Case

In the previous chapters and sections, none of the diffusion constants for the various components are assumed to be extremely small. When certain diffusion constant is small, one can use asymptotic methods to simplify the study of the behavior of the system. One can use the solution of a smaller subsystem of equations (or a scalar equation) to approximate the solution of the original full system. To illustrate the method, we consider the simple two competing-species system:

(3.4-1)
$$\begin{aligned} \frac{\partial u_1}{\partial t} &= \Delta u_1 + u_1[a - bu_1 - cu_2], \\ \frac{\partial u_2}{\partial t} &= \varepsilon\Delta u_2 + u_2[e - fu_1 - gu_2], \end{aligned} \qquad (x,t) \in \Omega x[0,\infty),$$

where a, b, c, e, f, g and ε are positive constants. We assume here that ε is small. For large time, the usual formal asymptotic "singular perturbation" procedure is to set $\varepsilon = 0$ and $(\partial u_2/\partial t) = 0$, solve u_2 in terms of u_1 in the equation

$$u_2[e - fu_1 - gu_2] = 0, \tag{3.4-2}$$

and substitute back into the first equation in (3.4-1). One then analyze the resulting scalar equation for u_1 alone and finally uses (3.4-2) again to study the behavior of u_2. This procedure reduces the study of the full system (3.4-1) to that of a scalar equation, and is therefore of significant simplification for numerical as well as analytical investigation.

One difficulty for this present problem is that (3.4-2) describes two natural solutions of u_2 in terms of u_1, namely:

$$u_2 = 0, \quad \text{or} \tag{3.4-3}$$

$$u_2 = g^{-1}(e - fu_1). \tag{3.4-4}$$

There is, therefore, a choice between them when we express u_2 in terms of u_1 in the first equation in (3.4-1). It turns out that the appropriate procedure is to switch between the two choices as u_1 crosses the value ef^{-1}. More precisely, we will use the following solution of (3.4-2):

$$u_2 = h(u_1) = \begin{cases} g^{-1}(e - fu_1) & \text{if } u_1 \le ef^{-1} \\ 0 & \text{if } u_1 \ge ef^{-1} \end{cases} \tag{3.4-5}$$

Substituting into the first equation in (3.4-1), and setting $(\partial u_1/\partial t) = 0$, we obtain (after replacing u_1 by u)

$$\Delta u + u[a - bu - ch(u)] = 0. \tag{3.4-6}$$

This equation will play an important role in the construction of upper and lower bounds for $u_1(x,t)$. Substituting the upper and lower bounds for u_1 into the function h in (3.4-5), we will eventually obtain respectively lower and upper bounds for $u_2(x,t)$. Theorem 3.4-1 gives lower bound for u_1 and upper bound for u_2. It looks difficult to apply at first sight. However, applying the same technique, together with the additional assumption $ab^{-1} > ef^{-1}$, one obtains very convenient results in Theorem 3.4-2. Theorem 3.4-2 essentially gives simple sufficient conditions on the initial and boundary conditions, so that $u_2(x,t)$ becomes arbitrarily small for large t in the interior, except for "boundary layer" adjustments near the boundary.

We consider system (3.4-1) for $(x,t) \in \Omega \times [0,\infty)$, with initial boundary conditions:

$$\begin{aligned} &u_i(x,0) = \phi_i(x) \geq 0, \quad x \in \bar{\Omega},\ i = 1,2, \\ &u_i(x,t) = \theta_i(x) \geq 0 \quad (x,t) \in \delta\Omega \times [0,\infty),\ i = 1,2. \end{aligned} \tag{3.4-7}$$

Here $\phi_i(x) = \theta_i(x)$ for $x \in \delta\Omega$ and ϕ_i, θ_i satisfy the compatibilty conditions of order 1 on $\delta\Omega$ at $t = 0$, $i = 1,2$. Further, we assume that $\phi_i \in H^{2+\alpha}(\bar{\Omega})$ ans $\theta_i \in H^{2+\alpha,1+\alpha/2}(\delta\bar{\Omega}_T)$, so that there exist unique solution $(u_1(x,t), u_2(x,t))$ with components in $H^{2+\alpha,1+\alpha/2}(\bar{\Omega}_T)$, each $T > 0$. (cf. section 1.3, Theorem 3.3-6 or [127].)

We now use the following reduced problem to construct lower bound for $u_1(x,t)$ and upper bound for $u_2(x,t)$:

$$\begin{aligned} &\Delta y + yg(x,y) = 0, \quad x \in \Omega, \\ &y = \theta_1(x) \qquad \text{for } x \in \delta\Omega. \end{aligned} \tag{3.4-8}$$

where $a\text{-}bu\text{-}ch(y) \geq g(x,y) \overset{(def)}{=} a\text{-}by\text{-}ch(y)\text{-}c(2\delta+I(x)+L(x))$, $\delta > 0$ is a small constant, and $I(x)$, $L(x)$ are nonnegative functions chosen to adapt respectively to $\phi_2(x)$ and $\theta_2(x)$. Note that the first equation in (3.4-8) is a slight modification of (3.4-6).

Theorem 3.4-1. *Let $\delta > 0$ be an arbitrary small number,* $I(x)$ *and* $L(x)$ *be respectively nonnegative and positive functions in* $H^{2+\alpha}(\bar{\Omega})$, *and* $y(x)$ *be a nonnegative solution of the boundary value problem* (3.4-8) *above. Suppose that the nonnegative function* $h(y(x))$ *has a "smooth truncation"* $M^y(x)$ *in the following sense:*

(i) $M^y(x) \in H^{2+\alpha}(\bar{\Omega})$,

(ii) $M^y(x) = h(y(x))$ *if* $h(y(x)) > \delta$ (i.e. *if* $y(x) < f^{-1}(e - g\delta)$), *and*

(iii) $0 \leq M^y(x) \leq \delta$ *if* $0 \leq h(y(x)) \leq \delta$ (i.e. *if* $y(x) \geq f^{-1}(e - g\delta)$).

Then the solution of the initial boundary value problem (3.4-1), (3.4-7) *will satisfy*

$$u_1(x,t) \geq y(x), \quad \text{and} \tag{3.4-9}$$

$$0 \leq u_2(x,t) \leq M^y(x) + \delta + I(x)e^{-rt} + L(x) \tag{3.4-10}$$

for $(x,t) \in \bar{\Omega}\times[0,\infty)$ *provided that* $\varepsilon > 0$ *is small enough, and*

$$\phi_1(x) \geq y(x) \quad \text{for } x \in \bar{\Omega} \tag{3.4-11}$$

$$0 \leq \phi_2(x) \leq M^y(x) + \delta + I(x) + L(x) \quad \text{for } x \in \bar{\Omega}, \text{ and} \tag{3.4-12a}$$

$$0 \leq \theta_2(x) \leq M^y(x) + \delta + L(x) \quad \text{for } x \in \delta\Omega. \tag{3.4-12b}$$

Here r _is_ _any_ _constant_ _with_ $0 < r < \delta g$.

Remark 3.4-1. Since $\theta_1(x) \geq 0$, the zero function is a lower solution for the scalar problem (3.4-8). An upper solution for (3.4-8) is a constant function with a sufficiently large positive constant. Consequently, there exists a nonnegative solution $y(x)$ to (3.4-8), as stated in Theorem 3.4-1 (cf. section 1.4 or 5.1). Moreover, the smoothness of the nonlinear expression in (3.4-8) implies that $y \in H^{2+\alpha}(\bar{\Omega})$.

Remark 3.4-2. When one restricts to the case $x \in \bar{\Omega} \subset R^1$, i.e., $n = 1$, one can readily prove that $M^y(x)$, satisfying conditions (i) to (iii) as stated in Theorem 3.4-1, does exist.

Remark 3.4-3. In order to apply Theorem 3.4-1 effectively, one may choose $L(x)$ to be arbitrarily small for x outside a small neighborhood of $\delta\Omega$, and to grow quickly to slightly larger than $\theta_2(x)-\delta$ at $x \in \delta\Omega$. ($L(x)$ therefore plays the role of a "boundary layer" correction for u_2). Then one may choose $I(x) \geq \phi_2(x)-\delta-L(x)$ for $x \in \bar{\Omega}$, so (3.4-12a, b) are always satisfied. Inequality (3.4-10) will then imply that $u_2(x,t)$ is nearly dominated by $M^y(x)$, or $h(y(x))$, for x outside a small neighborhood of $\delta\Omega$, when t is sufficiently large. One can therefore use the reduced problem (3.4-8) to approximate the asymptotic behavior of the full problem (3.4-1), (3.4-7), as $t \to +\infty$, provided ε is small enough and (3.4-11), (3.4-12a, b) are satisfied. Consider those x outside a neighborhood of $\delta\Omega$, so that $L(x)$ is defined arbitrarily small. If $y(x) \geq e/f$, then $h(y(x))$ and $M^y(x)$ will be small, and $u_2(x,t)$ will tend to small values as $t \to +\infty$ (by means of (3.4-10)). In other words, those will be locations where u_2 becomes extinct in the long run. Theorem 3.4-2 below describes a variant of this situation when $h(y(x)) \equiv 0$, which is a simple, but important, case.

Remark 3.4-4. Let $F(u_1,u_2) \equiv u_2(e - fu_1 - gu_2)$. We have $F(ef^{-1},0) = 0$

and $(\partial F/\partial u_2)(ef^{-1},0) = 0$. The relation $F(u_1,u_2) = 0$ defines u_2 as two smooth functions of u_1 (namely, $u_2 = 0$ and $u_2 = g^{-1}(e - fu_1)$. These two functions coalesce when $(u_1,u_2) = (ef^{-1},0)$. This is usually the difficult case when one studies the full problem (3.4-1), (3.4-7) by means of the reduced problem through setting $\varepsilon = 0$. Theorem 3.4-1 essentially treats this situation when $u_2 = h(u_1)$ switches from one smooth choice of the implicit function defined by $F(u_1,u_2) = 0$ to another.

<u>Proof of Theorem</u> 3.4-1. We will use Theorem 1.2-7 (and its following remark) by constructing appropriate lower and upper solutions v_i, w_i, $i = 1,2$. Let $v_1(x,t) = y(x)$, $w_2(x,t) = M^y(x) + \delta + I(x)e^{-rt} + L(x)$, for $(x,t) \in \bar{\Omega}\times[0,\infty)$. We have

$$\begin{aligned}
&\Delta v_1 + v_1[a - bv_1 - cw_2] - \frac{\partial v_1}{\partial t} \\
&= \Delta y + y[a - by - cM^y(x) - c\delta - cI(x)e^{-rt} - cL(x)] \\
&\geq \Delta y + y[a - by - c(h(y(x) + \delta) - c(\delta + I(x)e^{-rt} + L(x))] \\
&\geq \Delta y + y[a - by - ch(y) - c(2\delta + I(x) + L(x))] = 0
\end{aligned}$$

for $(x,t) \in \bar{\Omega}\times[0,\infty)$. On the other hand,

$$\begin{aligned}
&\varepsilon\Delta w_2 + w_2[e - fv_1 - gw_2] - \frac{\partial w_2}{\partial t} \\
&= \varepsilon\Delta w_2 + w_2[e - fy - gM^y(x) - g\delta - gI(x)e^{-rt} - gL(x)] + rI(x)e^{-rt} \\
&= \varepsilon\Delta w_2 + w_2[e - fy - gM^y(x)] - w_2[g\delta + gL(x)] + I(x)e^{-rt}(r - w_2g),
\end{aligned}$$

which is less than zero for ε sufficiently small (because $e - fy(x) - gM^y(x) = e - fy(x) - gh(y(x)) = 0$, if $y(x) < (e - g\delta)f^{-1}$, and $e - fy(x)$ $gM^y(x) \leq e - fy(x) \leq g\delta$ if $y(x) \geq (e - g\delta)f^{-1}$), for $(x,t) \in \bar{\Omega}\times[0,\infty)$.

We next let $v_2(x,t) \equiv 0$, $w_1(x,t) \equiv C$ where C is a large positive constant, $C > \max\{a/b, \max\{\phi_1(x): x \in \bar{\Omega}\}\}$. Clearly

$$\varepsilon\Delta v_2 + v_2[e - fw_1 - gv_2] - \frac{\partial v_2}{\partial t} \geq 0 ,$$

and

$$\Delta w_1 + w_1[a - bw_1 - cv_2] - \frac{\partial w_1}{\partial t} \leq 0$$

for $(x,t) \in \bar{\Omega}\times[0,\infty)$.

Finally, conditions (3.4-11) and (3.4-12a,b) imply that $v_i(x,0) \leq u_i(x,0) \leq w_i(x,0)$ for $x \in \Omega$, and $v_i(x,t) \leq u_i(x,t) \leq w_i(x,t)$ for $(x,t) \in \delta\Omega\times[0,\infty)$, $i = 1,2$. Theorem 1.2-7 therefore asserts that $y(x) \leq u_1(x,t) \leq C$ and $0 \leq u_2 \leq M^y(x) + \delta + I(x)e^{-rt} + L(x)$, for all $(x,t) \in \bar{\Omega}\times[0,\infty)$.

The following theorem is analogous to Theorem 3.4-1. We make the additional assumption $a/b > e/f$, and find conditions on the initial and boundary data which will imply extinction of u_2 for all x at the boundary, as $t \to +\infty$. One can compare the results of Theorem 3.4-2 with the case of the system of ordinary differential equations:

$$\frac{du_1}{dt} = u_1[a - bu_1 - cu_2] ,$$

$$\frac{du_2}{dt} = u_2[e - fu_1 - gu_2] .$$

If $a/b > e/f$, phase plane analysis easily shows that for $u_1(0) > 0$, $u_2(0) \geq 0$, $(u_1(0),u_2(0))$ close to $(a/b,0)$, one has $(u_1(t),u_2(t))$ $(a/b,0)$ as $t \to \infty$. To avoid excessive technicalities, we restrict to the case $x \in R^1$. In Theorem 3.4-2, we therefore assume $\Omega = (a_0,b_0)$, $a_0<b_0$, $\delta\Omega = \{a_0,b_0\}$.

<u>Theorem</u> 3.4-2. <u>Suppose that</u> $a/b > e/f$. <u>Let</u> R <u>be a number satisfying</u> $0 < R < a/b - e/f$. <u>Assume that the initial conditions satisfy</u>

(3.4-13) $\qquad \phi_1(x) \geq (a/b) - R \qquad$ <u>for all</u> $x \in [a_0,b_0]$,

(3.4-14) $\qquad \phi_2(x) < c^{-1}bR \qquad$ <u>for all</u> $x \in [a_0,b_0]$.

<u>Let</u> $\bar{a}$, $\bar{b}$ <u>be arbitrary numbers satisfying</u> $a_0 < \bar{a} < \bar{b} < b_0$. <u>Then for any arbitrary small</u> $\sigma > 0$, <u>the solution</u> $(u_1(x,t), u_2(x,t))$ <u>of</u> (3.4-1), (3.4-7) <u>will satisfy</u>

(3.4-15) $$0 \le u_2(x,t) < \sigma$$

<u>for all</u> $(x,t) \in [\bar{a},\bar{b}]x[K,\infty)$ <u>for large enough</u> $K > 0$, <u>provided</u> $\varepsilon > 0$ <u>is sufficiently small.</u>

<u>Proof.</u> We first proceed to construct L(x), I(x) for $x \in [a_0,b_0]$ by a procedure similar to that described in Remark 3.4-3. Let $\delta > 0$ be such that $\delta < \min\{c^{-1}bR, \sigma/2\}$ and $\max\{\phi_2(x): x \in [a_0,b_0]\} + 4\delta < c^{-1}bR$. For $x = a_0$ or b_0, define L(x) and I(x) be arbitrary numbers satisfying

(3.4-16) $$\begin{gathered}\max\{\phi_2(a_0), \phi_2(b_0)\} + \delta < L(x) < c^{-1}bR - 3\delta, \\ 0 < I(x) < \delta,\end{gathered}$$

We therefore have

(3.4-17) $$\max\{\phi_2(a_0), \phi_2(b_0)\} + 2\delta < I(x) + L(x) + \delta < c^{-1}bR - \delta$$

for $x = a_0$ or b_0. We will now define I(x), L(x) as functions in $H^{2+\alpha}(\bar{\Omega})$ by the following procedures so that

(3.4-18) $$\phi_2(x) + \delta < I(x) + L(x) + \delta < c^{-1}bR$$

for all $x \in [a_0,b_0]$. Let $\bar{\sigma}$ be a number satisfying $0 < \bar{\sigma} < \min\{\sigma - 2\delta, \delta\}$. Define L(x) and I(x) in $[\bar{a},\bar{b}]$ as any functions in $H^{2+\alpha}(\bar{D})$, $\bar{D} = [\bar{a},\bar{b}]$ satisfying $0 < L(x) < \bar{\sigma}$, $\phi_2(x) + \delta < I(x) < c^{-1}bR - 2\delta - \bar{\sigma}$, for $x \in [\bar{a},\bar{b}]$. If we let $h(x) = I(x) + L(x)$ for $x \in \{a_0,b_0\} \cup [\bar{a},\bar{b}]$, we clearly have for such x, the inequalities

(3.4-19) $$\phi_2(x) + 2\delta < h(x) + \delta < c^{-1}bR - \delta.$$

Extend h(x) to be a function in $H^{2+\alpha}(\bar{\Omega})$, $\bar{\Omega} = [a_0, b_0]$, so that

(3.4-20) $$\phi_2(x) + \delta < h(x) + \delta < c^{-1}bR,$$

for all $x \in [a_0, b_0]$. We next extend the definition of L(x) to $(a_0, \bar{a}) \cup (\bar{b}, b_0)$ so that L(x) is in $H^{2+\alpha}(\bar{\Omega})$ and $0 < L(x) < h(x)$ on $(a_0, \bar{a}) \cup (\bar{b}, b_0)$. Finally, set $I(x) = h(x) - L(x)$ on $(a_0, \bar{a}) \cup (\bar{b}, b_0)$. We therefore have inequalities (3.4-18) valid for all $x \in [a_0, b_0]$.

As in the proof of Theorem 3.4-1, we now contruct appropriate lower and upper solutions v_i, w_i, $i = 1,2$ and apply Theorem 1.2-7. Define $v_1(x,t) = a/b - R$, $w_2(x,t) = \delta + I(x)e^{-nt} + L(x)$, where $0 < n < \delta g$, for all $(x,t) \in [a_0, b_0]x[0,\infty)$. We have

$$\begin{aligned}\Delta v_1 + v_1[a - bv_1 - cw_2] - \frac{\partial v_1}{\partial t} &= v_1[bR - c(\delta + I(x)e^{-nt} + L(x))]\\ &\geq v_1[bR - c(\delta + I(x) + L(x))]\\ &> v_1[bR - cc^{-1}bR] = 0,\end{aligned}$$

for $(x,t) \in [a_0, b_0]x[0,\infty)$, by (3.4-17). On the other hand,

$$\begin{aligned}&\varepsilon\Delta w_2 + w_2[e - fv_1 - gw_2] - \frac{\partial w_2}{\partial t}\\ &= \varepsilon\Delta w_2 + w_2[e - f(\tfrac{a}{b} - R) - g\delta - gI(x)e^{-nt} - gL(x)] + nI(x)e^{-nt}\\ &< \varepsilon\Delta w_2 + w_2[e - f(\tfrac{a}{b} - R)] + I(x)e^{-nt}[-w_2 g + n]\end{aligned}$$

which is less than zero provided ε is sufficiently small (because $e - f(a/b - R) < 0$, $-w_2 g + n < -\delta g + n < 0$), for all $(x,t) \in [a_0, b_0]x[0,\infty)$.

We next set $v_2(x,t) \equiv 0$, $w_1(x,t) \equiv C$, where C is a large positive constant, $C > \max\{a/b, \max\{\phi_1(x): x \in [a_0,b_0]\}\}$. Clearly,

$$\varepsilon\Delta v_2 + v_2[e - fw_1 - gv_2] - \frac{\partial v_2}{\partial t} \ge 0$$

and

$$\Delta w_1 + w_1[a - bw_1 - cv_2] - \frac{\partial w_1}{\partial t} \le 0$$

for $(x,t) \in [a_0,b_0]x[0,\infty)$. Condition (3.4-13) and the choice of C imply that

$$v_i(x,0) \le u_i(x,0) \le w_i(x,0), \quad x \in [a_0,b_0], \tag{3.4-21}$$

and

$$v_i(x,t) \le u_i(x,t) \le w_i(x,t), \quad (x,t) \in \{a_0,b_0\}x[0,\infty), \tag{3.4-22}$$

for $i = 1$. Inequality (3.4-20) implies that (3.4-21) is valid for $i = 2$. Inequality (3.4-16) implies that (3.4-22) is valid for $i = 2$. Consequently, by Theorem 1.2-7, we have $u_2(x,t) \le w_2(x,t) = \delta + I(x)e^{-nt} + L(x)$ for all $(x,t) \in [a_0,b_0]x[0,\infty)$. Since by construction $L(x) < \bar{\sigma}$ for $x \in [\bar{a},\bar{b}]$ and $2\delta + \bar{\sigma} < \sigma$, we have inequality (3.4-15) for those (x,t) as stated in the theorem.

As a final comment on the subject in this section, we note that if $a/b > e/f$, we can use the solution $z(x)$ of the reduced problem:

$$\text{(3.4-23)} \quad \begin{aligned} &\Delta z(x) + z(x)k(z(x)) = 0 \qquad \text{for } x \in (a_0,b_0) \\ &z(a_0) = \theta_1(a_0), \quad z(b_0) = \theta_1(b_0) \end{aligned}$$

to obtain an upper bound for $u_1(x,t)$ in the form

$$u_1(x,t) \le z(x) \quad \text{for all } (x,t) \in [a_0,b_0]x[0,\infty),$$

and an approximate $N^z(x)$ for $h(z(x))$ to obtain a lower bound for $u_2(x,t)$ in the form

$$u_2(x,t) \geq N^z(x) \text{ for all } (x,t) \in [a_0,b_0]\times[0,\infty).$$

Here $k(z)$ in (3.4-23) is appropriately chosen so that

$$a - bz - ch(z) \leq k(z) \leq a - bz - ch(z) + \lambda$$

where λ is small, and the initial and boundary conditions of $(u_1(x,t), u_2(x,t))$ are assumed to satisfy appropriate conditions. Combining with Theorems 3.4-1 and 3.4-2 we can therefore obtain two sided bounds for $u_i(x,t)$, $i = 1,2$. For more details, see [139]. Other cases can of course be treated analougously.

Notes

Reaction-diffusion equations and systems with nonlinear boundary conditions had been studied in physical and biological problems by Mann and Wolf [153], and Thames and Elster [216] respectively. Rigorous mathematical treatments using comparison and monotone methods for such applied problems with nonlinear boundary conditions were made by Aronson and Peletier [15], Turner and Ames [221], Aronson [14], and many others. The presentation in section 3.2 for such problem follows the results in Leung [138]. Many models for nonlinear density-dependent diffusions and spatially varying environments were proposed and studied by Mimura, Nishiura and Yamaguti [160], Okubo [173] and Levin [150]. The materials in section 3.3 are gathered from Leung [141]. Singular perturbations of boundary value problems for elliptic equations (corresponding to small diffusion constant for steady state equations) were investigated by Fife [73], Howes [110],[111], and Wasow [225]. The parabolic case for the nonsteady state system, considered in section 3.4, is adapted from Leung [139]. We consider two competing species equations with one having small diffusion rate. Similar problem was investigated by deMottoni. Schiaffino and Tesei [64].

CHAPTER IV

Multigroup Fission Reactor Systems, Strongly Order-Preserving Systems

4.1 Introduction

In sections 4.2 to 4.4, we will consider the application of reaction-diffusion systems to the study of neutron fission reactors. We also discuss some ecological mutualist species interactions whose equations sometimes has similar structure. For the fission reactor theory, we investigate multigroup neutron-flux equations describing fission, scattering and absorption for n energy groups. The reactor core is represented by a bounded domain Ω in R^d, $d \geq 2$. The functions $u_i(x)$ or $\tilde{u}_i(x,t)$, $i = 1, \ldots, n$, $x = (x_1, \ldots, x_d) \in \Omega$ are the neutron flux of the ith energy group (decreasing energy for increasing i). T(x) is the core temperature above average coolant temperature. We will consider the system of nonlinear temperature feedback multigroup elliptic diffusion equations:

$$(4.1\text{-}1)\qquad \begin{aligned} &\Delta u_i + \sum_{j=1}^{n} H_{ij}(x,T)u_j = 0 \quad , \quad i = 1, \ldots, n \\ &\Delta T - c(x)T + \sum_{j=1}^{n} G_j(x,T)u_j = 0 \qquad \text{in } \Omega. \end{aligned}$$

Here $\Delta = \sum_{i=1}^{d} \partial^2/\partial x_i^2$; $c(x) > 0$ in $\bar{\Omega}$ (Ω closure) represents the cooling function. The functions determining interaction rates, $H_{ij}(x,T)$ and $G_j(x,T)$, are assumed to be functions of space and termperature. In nuclear engineering terminologies, H_{ij} describe fission, removal, group-transfer and absorption "cross sections", taking into account of the parameters of diffusion, neutron and energy release. For example:

$$H_{11}(x,T) \equiv \sigma_1^{-1}\,[\nu_1\Sigma_{f_1} - \Sigma_R]$$

(4.1-2) $$H_{1j}(x,T) \equiv \sigma_1^{-1}\,\nu_j\,\Sigma_{f_j}\quad,\qquad j = 2,\ \ldots,\ n$$

$$H_{ij}(x,T) \equiv \sigma_i^{-1}\,\Sigma_{S_{ji}}\qquad \text{for } i \neq j,\ i = 2,\ \ldots,\ n$$

$$H_{ii}(x,T) \equiv \sigma_i^{-1}\,\Sigma_{a_i}\qquad \text{for } i = 2,\ \ldots,\ n.$$

The symbols on the right of (4.1-2) are in conventional notation of nuclear engineering. The parameter σ_i is the diffusion coefficient of group i; Σ_{f_j} is the fission "macroscopic cross section" in group j (i.e., Σ_{f_j} is the probability per unit path length travelled that a neutron in energy group j will undergo fission); ν_j is the average number of neutron in group 1 released during fission induced by a neutron in group j. Note that in (4.1-2), we assume that only neutrons in group 1 are released during fission. Σ_R is the removal "macroscopic cross section" characterizing the probability that a neutron will be removed from group 1 (i.e. Σ_R is the probability per unit length travelled that a neutron in the 1st group will undergo a collision causing its own removal through absorption or slowing down to the other groups). $\Sigma_{S_{ji}}$ is the macroscopic group-transfer cross section (probability of collision causing transfer from group j to group i); Σ_{a_i} is the absorption macroscopic cross section for group i. Other hypotheses concerning the interactions between the various energy groups, will of course give rise to different formulas for H_{ij} in (4.1-2). However, there are certain basic properties which we will list later as conditions (C1) and (C2). The functions $G_j(x,T)$ can be assumed to be of the form:

(4.1-3) $$G_j(x,T) \equiv \theta_j\,\Sigma_{f_j}\quad,\qquad j = 1,\ \ldots,\ n$$

where θ_j is the effective energy released in each fission for group j. The detailed definitions can be found in [68, p. 288].

Various studies of models similar to (4.1-1) were made analytically and numerically, where the dependence on T are sometimes neglected, see e.g.

[118], [213], [51]. However, as temperature changes, materials in the core may contract, expand, or change phase, eventually causing a change in the macroscopic cross section. The advisability of a temperature-dependent nonlinear feedback model has been proposed and studied in [116], [21] and [177]. Here, we follow the development given in [143] and [144]. In practice, the multigroup equations are commonly applied in cases of four or more groups.

In this chapter, unless more specific conditions are imposed, the boundary $\delta\Omega$ of the bounded domain Ω is assumed to be C^2 smooth (i.e., can be locally represented as $x_i = \phi(x)$ for some i, ϕ with continuous second derivatives and independent of x_i). The functions H_{ij}, G_i, $i,j = 1, \ldots, n$ are continuous functions of $x \in \bar{\Omega}$, $T \geq 0$ and Lipschitz in T uniformly for $x \in \bar{\Omega}$; c(x) is continuous and positive in $\bar{\Omega}$. For convenience, let

$$\tilde{h}_{ij} = \inf\{H_{ij}(x,T) \mid x \in \bar{\Omega},\ T \geq 0\},$$
(4.1-4)
$$\bar{h}_{ij} = \sup\{H_{ij}(x,T) \mid x \in \bar{\Omega},\ T \geq 0\}$$

for $i,j = 1, \ldots, n$. Similarly, define $\tilde{g}_i$, $\bar{g}_i$ to be the corresponding inf. and sup. of G_i, $i = 1, \ldots, n$. We will always assume that

(C1) $$-\infty < \tilde{h}_{ij} \leq \bar{h}_{ij} < \infty,\ 0 \leq \tilde{g}_i \leq \bar{g}_i < \infty$$

for $i,j = 1, \ldots, n$. For $i \neq j$, H_{ij} describe group transfer and fissions of neutrons from other groups; while H_{ii} is affected by control rods and absorptions. Consequently, we always assume that

$$0 \leq \tilde{h}_{ij} \leq \bar{h}_{ij} < \infty \quad \text{each } i,j = 1, \ldots, n, \text{ with } i \neq j$$
(C2)
$$-\infty < \tilde{h}_{ii} \leq \bar{h}_{ii} < \infty,\ i = 1, \ldots, n.$$

Referring to (4.1-2), we see that conditions (C1) and (C2) are very reasonable and general assumptions for the reactor model. In some cases, we further assume that an energy group i always receives transfer of neutrons from some group j, $j < i$ (cf., condition (II) in Theorem 4.2-1'). In another case, a related "irreducible" condition is assumed (cf. condition (II^*) in Theorem 4.2-3). For later conveniences, we define:

(4.1-5) $\tilde{H}$ and $\bar{H}$ to be nxn square matrices whose $(i,j)^{th}$ entries are, respectively, $\tilde{h}_{ij}$ and $\bar{h}_{ij}$ for $1 \le i \le j \le n$.

Let $\lambda_1 > 0$ denote the first eigenvalue of the eigenvalue problem: $\Delta w + \lambda w = 0$ in Ω, $w = 0$ on $\delta\Omega$, where $w = \omega(x)$ is the corresponding normalized eigenfunction with $\max\{\omega(x) | x \in \bar{\Omega}\} = 1$. For positive integers r, $C^r(\Omega)$ and $C^r(\bar{\Omega})$ denote r-times continuously differentiable functions in Ω and $\bar{\Omega}$, respectively.

Finally, in section 4.6, we introduce some recent theories and developments in 'strongly order-preserving' systems. Abstract theorems as well as applications to systems of parabolic equations are described. Solutions of such systems are likely to tend to steady states. They are closely related to the systems considered sections 4.2 to 4.4, and are also applicable to competing and mutualistic species reaction-diffusion.

4.2 Blow-up and Decay Criteria for Temperature-Dependent Systems

In this section, we consider the system (4.1-1) and also its time-dependent analog when the right of (4.1-1) is replaced by $\text{col}(\frac{\partial u_1}{\partial t}, \ldots, \frac{\partial u_n}{\partial t}, \frac{\partial T}{\partial t})$. Hypotheses (C1) and (C2) will be assumed in the entire section. We will always impose nonnegative or zero Dirichlet boundary conditions on $\delta\Omega$. Theorem 4.2-1 gives some very readily applicable criteria when certain component of a nonnegative steady-state must be identically zero. A "sweeping" argument is used in the proof (cf. Theorem 1.4-3).

Theorem 4.2-1. *Suppose that the* $n\times n$ *matrix* $\tilde{H}-\lambda_1 I$ *has a square* $m\times m$ *submatrix* $A = (a_{ij})$, *(formed by deleting the* k_1^{th}, ..., k_{n-m}^{th} *rows and columns of* $\tilde{H}-\lambda_1 I$, $1 \le k_1 < k_2 \dots < k_{n-m} \le n$, $1 \le m \le n$) *with the properties that:*

(I) $A\vec{c} > 0$ *for some positive* m *vector* $\vec{c} > 0$; *and*

(II) *For each* $i = 2, \dots, m$ *at least one of* $a_{i1}, a_{i2}, \dots, a_{ii-1}$ *is positive.*

Then equation (4.1-1) *has no solution* $(\hat{u}_1(x), \dots, \hat{u}_n(x), \hat{T}(x))$ *with all the following three properties:*

(i) *each component is in* $C^2(\Omega) \cap C^1(\bar{\Omega})$,

(ii) $\hat{u}_i(x) \ge 0, i = 1, \dots, n$, $\hat{T}(x) \ge 0$ in $\bar{\Omega}$,

(iii) $\hat{u}_{s_1}(x) \not\equiv 0$ in $\bar{\Omega}$, *where* s_1 *is the first positive integer not included in* $k_1, \dots, k_{n-m}$.

(Roughly speaking, any solution of (4.1-1) satisfying (i) and (ii), must have its s_1^{th} component identically zero. When $m = n$, no k_i row or column is deleted, and $\tilde{H} - \lambda_1 I \equiv A$. A vector $\vec{v} > 0$ means that each component of $\vec{v}$ is positive).

Proof. Let $1 \le s_1 < s_2, \dots, < s_m \le n$ be integers not in $\{k_1, \dots, k_{n-m}\}$, and thus $a_{ij} = \tilde{h}_{s_i s_j} - \lambda_1 \delta_{ij}$, $1 \le i,j \le m$, where δ_{ij} is the Kronecker delta. Assume that $(\hat{u}_1(x), \dots, \hat{u}_n(x), \hat{T}(x))$ exists as described with properties (i), (ii) and (iii). We will construct a family of lower bounds for the functions $\hat{u}_{s_i}(x)$, $i = 1, \dots, m$, parametrized by $\delta > 0$. As $\delta \to \infty$, the lower bound will tend to ∞. For each $\delta > 0$, define $u^{\delta}_{s_i}(x) = c_i \delta \omega(x)$ for $x \in \bar{\Omega}$, $i = 1, \dots, m$ where col. $(c_1, \dots, c_n) = \vec{c}$ is the vector described in (I) above; define $u^{\delta}_{k_i}(x) \equiv 0$, $i = 1, \dots, n-m$ and $T^{\delta}(x) \equiv 0$. For each $i = 1, \dots, m$, when $u_j(x) \ge u^{\delta}_j(x)$ for all $j \ne s_i$, and $T(x) \ge T^{\delta}(x)$, we have

$$\Delta u^{\delta}_{s_i}(x) + H_{s_is_i}(x,T(x))\, u^{\delta}_{s_i}(x) + \sum_{j=1,\, j\neq s_i}^{n} H_{s_ij}(x,T(x))u_j(x)$$

$$\geq [-\lambda_1 + \tilde{h}_{s_is_i}]\, c_i\delta\omega(x) + \sum_{j=1,\, j\neq i}^{m} \tilde{H}_{s_is_j}\, u^{\delta}_{s_j} \tag{4.2-1}$$

$$= [-\lambda_1 + \tilde{h}_{s_is_i}]\, c_i\delta\omega + \sum_{j=1,\, j\neq i}^{m} a_{ij}\, c_j\, \delta\omega = (A\vec{c})_i\ \ \delta\omega(x) > 0$$

We now show that properties (i) to (iii) imply that $\hat{u}_{s_i}(x) > 0$ for $x \in \Omega$, $i = 1, \ldots, m$. Let $C > \max\{|\tilde{h}_{s_1s_1}|, |\bar{h}_{s_1s_1}|\}$, we have

$$\Delta\hat{u}_{s_1}(x) - C\hat{u}_{s_1}(x) = -[H_{s_1s_1}(x,\hat{T}(x)) + C]\hat{u}_{s_1}(x) - \sum_{j=1,\, j\neq s_1}^{n} H_{s_1j}(x,\hat{T}(x))\hat{u}_j(x) \leq 0$$

in Ω, $\hat{u}_{s_1} \geq 0$ in $\bar{\Omega}$. The maximum principle (Theorem 1.1-2) implies that $\hat{u}_{s_1}(x) > 0$ in Ω. Similarly, considering $\Delta\hat{u}_{s_i} - P\hat{u}_{s_i} \leq 0$ in Ω for large enough $P > 0$, we deduce from the maximum principle that $\hat{u}_{s_i} > 0$ in Ω or $\hat{u}_{s_i} \equiv 0$ in $\bar{\Omega}$, $i = 2, \ldots, m$. However, property (II) implies successively that the trivial function is not a solution of the s_i^{th} equation in $i = 2, \ldots, m$. Hence $\hat{u}_{s_i}(x) > 0$ for $x \in \Omega$. Moreover, the maximum principle at the boundary (Theorem 1.1-3) indicates that the outward normal derivatives $\partial\hat{u}_{s_i}/\partial\eta$ are negative at those boundary points where the corresponding function is 0.

From the above paragraph, we see that the set

$$J \equiv \{\tau > 0 \mid \hat{u}_{s_i}(x) > u^{\delta}_{s_i}(x),\ i = 1, \ldots, m \text{ for all } 0 \leq \delta < \tau,\ x \in \Omega\}$$

is nonempty. Suppose J has an upper bound, let its lub be $\bar{\delta}$. If there is a point at the boundary where $u^{\bar{\delta}}_{s_i} = \hat{u}_{s_i}$, some $i = 1, \ldots, m$, we deduce a contradiction to the definition of $\bar{\delta}$ by using the maximum principle at the boundary, with the inequality

$$\Delta(\hat{u}_{s_i} - u^{\bar{\delta}}_{s_i}) - P(\hat{u}_{s_i} - u^{\bar{\delta}}_{s_i})$$

$$(4.2\text{-}2) \quad = \{\Delta\hat{u}_{s_i} + \sum_{j=1}^{n} H_{s_i j}(x,\hat{T})\hat{u}_j\} - \{\Delta u^{\bar{\delta}}_{s_i} + H_{s_i s_i}(x,\hat{T})u^{\bar{\delta}}_{s_i} + \sum_{j=1, j\neq s_i}^{n} H_{s_i j}(x,\hat{T})\hat{u}_j\} - \{H_{s_i s_i}(x,\hat{T}) + P\}(\hat{u}_{s_i} - u^{\bar{\delta}}_{s_i}) \le 0$$

in Ω. (The last inequality is true for $P > \max\{|\tilde{h}_{s_i s_i}|, |\bar{h}_{s_i s_i}|\}$, and due to inequality (4.2-1), for $\hat{u}_j(x) \ge u^{\bar{\delta}}_j(x)$, $j \neq s_i$. $\hat{T} \ge T^{\bar{\delta}}$). Contradiction arises because (4.2-2) implies that $\partial\hat{u}_{s_i}/\partial\eta < \partial u^{\bar{\delta}}_{s_i}/\partial\eta$ at those points at the boundary where $\hat{u}_{s_i} = u^{\bar{\delta}}_{s_i}$, and thus $u^{\bar{\delta}+\varepsilon}_{s_i} < \hat{u}_{s_i}$ for all $x \in \Omega$, some small $\varepsilon > 0$. On the other hand, suppose that there is a point $\bar{x} \in \Omega$ where $u^{\bar{\delta}}_{s_i}(x) = \hat{u}_{s_i}(\bar{x})$, some $i = 1, \ldots, m$. Inequality (4.2-2) and the maximum principle imply that $u^{\bar{\delta}}_{s_i}(x) \equiv \hat{u}_{s_i}(x)$ in $\bar{\Omega}$. However, we consider in Ω:

$$(4.2\text{-}3) \quad \begin{aligned} 0 &= \Delta\hat{u}_{s_i} + H_{s_i s_i}(x,\hat{T})\hat{u}_{s_i} + \sum_{j=1, j\neq s_i}^{n} H_{s_i j}(x,\hat{T})\hat{u}_j \\ &= \Delta u^{\bar{\delta}}_{s_i} + H_{s_i s_i}(x,\hat{T})u^{\bar{\delta}}_{s_i} + \sum_{j=1, j\neq s_i}^{n} H_{s_i j}(x,\hat{T})\hat{u}_j > 0, \end{aligned}$$

which is a contradiction. The last inequality is true by letting $\delta = \bar{\delta}$ in (4.2-1).

The last paragraph shows that the set J is unbounded. However, as $\delta \to +\infty$, $u^{\delta}_{s_i}(x) \to +\infty$ for $x \in \Omega$, $i = 1, \ldots, m$. This proves the nonexistence of $(\hat{u}_1(x), \ldots, \hat{u}_n(x), \hat{T}(x))$.

Essentially Theorem 4.2-1 asserts that under assumptions (I), (II) and (i), (ii), the s_1^{th} component must be identically zero; otherwise it cannot be finite. The corresponding analog in the parabolic time-dependent case asserts further that all the $s_1^{th}, \ldots, s_m^{th}$ components must tend to $+\infty$ as $t \to +\infty$. This is the context of the following Theorem 4.2-2.

Remark. To avoid technical difficulties, we assume in Theorem 4.2-2 that $H_{ij}(x,T)$ and $G_j(x,T)$ are all extended to be defined for $T < 0$ in such a way that they are still Lipschitz in T uniformly for $x \in \bar{\Omega}$. We assume that (4.2-4) below has a solution with smoothness and initial-boundary conditions as described in Theorem 4.2-2. We will see that the solution will satisfy $\tilde{T}(x,t) \geq 0$ (cf., (4.2-10)), so that the extensions of H_{ij} and G_j are not really relevant for $T \leq 0$.

Theorem 4.2-2. Suppose that the nxn matrix $\tilde{H}-\lambda_1 I$ has a square mxm submatrix $A = (a_{ij})$, (formed as described in Theorem 4.2-1) with the property that:

(I) $$A\vec{c} > 0 \text{ for some positive m vector } \vec{c} > 0.$$

Let $(\tilde{u}_1(x,t), \ldots, \tilde{u}_n(x,t), \tilde{T}(x,t))$ be a solution of:

$$\frac{\partial \tilde{u}_i}{\partial t} = \Delta \tilde{u}_i + \sum_{j=1}^{n} H_{ij}(x,\tilde{T})\tilde{u}_j \quad , \ i = 1, \ldots, n$$

(4.2-4)

$$\frac{\partial \tilde{T}}{\partial t} = \Delta \tilde{T} - c(x)\tilde{T} + \sum_{j=1}^{n} G_j(x,\tilde{T})\tilde{u}_j$$

for $(x,t) \in \Omega\times(0,\infty)$, with each component function in $C^2(\Omega\times(0,\infty)) \cap C^1(\bar{\Omega}\times[0,\infty))$ and initial-boundary conditions satisfying:

$$\tilde{u}_i(x,0) > 0 \text{ for } x \in \Omega \text{ , and}$$

(4.2-5)

$$\frac{\partial \tilde{u}_i}{\partial \eta}(x,0) < 0 \text{ for } x \in \delta\Omega, \ i = s_1, \ldots, s_m$$

(Recall that $s_1, \ldots, s_m$ are those integers between 1 to n not included in

$\{k_1, \ldots, k_{n-m}\}$ described in Theorem 4.2-1), and

(4.2-6)
$$u_i(\tilde{x}, t) \geq 0, \ i = 1, \ldots, n, \ T(x, \tilde{t}) \geq 0 \text{ for } (x,t) \in (\Omega \times \{\bar{0}\}) \cup (\delta\Omega \times (0,\infty)).$$

Then $\tilde{u}_i(x,t) \to +\infty$ for all $x \in \Omega$, $i = s_1, \ldots, s_m$ as $t \to +\infty$. (More precisely, for such x and i, we have $\tilde{u}_i(x,t) \geq \varepsilon_1 \omega(x) e^{\varepsilon_2 t}$ for some positive constants ε_1, ε_2 and all $t \in [0,\infty)$).

Proof. For convenience, let $S = \{s_1, \ldots, s_m\}$. For $1 \leq i \leq n+1$, $i \notin S$, define $v_i \equiv 0$. In (I), denote $\vec{c} = \text{col.}\ (c_1, \ldots, c_m)$. Choose ε so that $0 < \varepsilon < \min\{(A\vec{c})_i \mid i = 1, \ldots, m\}$; and choose $\delta > 0$ so that $\delta c_i \omega(x) < \tilde{u}_i(x,0)$ for all $x \in \Omega$, $i \in S$. Define for $i \in S$: $v_i(x,t) = \delta c_i \omega(x) e^{\varepsilon t}$ for $(x,t) \in \bar{\Omega} \times [0,\infty)$.

Consider the set:

$$\mathcal{B} \equiv \{(x,t,z_1, \ldots, z_{n+1}) \mid (x,t) \in \Omega \times (0,\infty),\ z_i \geq v_i(x,t),\ i = 1, \ldots, n+1\}$$

Clearly, we have for each $i \in S$, $(x, t, z_1, \ldots, z_{n+1}) \in \mathcal{B}$:

$$\Delta v_i + H_{ii}(x, z_{n+1}) v_i + \sum_{j=1, j\neq i}^{n} H_{ij}(x, z_{n+1}) z_j - \frac{\partial v_i}{\partial t}$$

(4.2-7)
$$\geq [-\lambda_1 + \tilde{h}_{ii}]\, \delta c_i \omega(x) e^{\varepsilon t} + \sum_{j\in S, j\neq i} H_{ij}(x, z_{n+1}) \delta c_j \omega(x) e^{\varepsilon t} - \varepsilon \delta c_i \omega(x) e^{\varepsilon t}$$

$$\geq [-\lambda_1 + \tilde{h}_{ii} - \varepsilon] \delta c_i \omega e^{\varepsilon t} + \sum_{j\in S, j\neq i} \tilde{h}_{ij} \delta c_j \omega e^{\varepsilon t}$$

$$= \{(A\vec{c})_k - \varepsilon\} \delta \omega(x) e^{\varepsilon t} > 0$$

(here k is the integer where $s_k = i$). For $1 \leq i \leq n$, $i \notin S$ we have for $(x, t, z_1, \ldots, z_{n+1}) \in \mathcal{B}$:

(4.2-8)
$$\Delta v_i + H_{ii}(x, z_{n+1})v_i + \sum_{j=1, j\neq i}^{n} H_{ij}(x, z_{n+1})z_j - \frac{\partial v_i}{\partial t} \geq \sum_{j=1, j\neq i}^{n} \tilde{h}_{ij}z_j \geq 0$$

Finally, for $(x, t, z_1, \ldots, z_{n+1}) \in \mathcal{B}$, we have:

(4.2-9)
$$\Delta v_{n+1} - c(x)v_{n+1} + \sum_{j=1}^{n} G_j(x, v_{n+1})z_j - \frac{\partial v_{n+1}}{\partial t} \geq \sum_{j=1}^{n} \tilde{g}_j z_j \geq 0$$

Moreover, at $t = 0$ and for $x \in \delta\Omega$, v_i's satisfy

(4.2-10) $$v_i(x, t) \leq \tilde{u}_i(x, t), \; i = 1, \ldots, n, \; v_{n+1}(x, t) \leq \tilde{T}(x, t)$$

for $(x, t) \in (\bar{\Omega}\times\{0\}) \cup (\delta\Omega\times(0, \infty))$. From inequalities (4.2-7) to (4.2-10) we conclude from applying one side of Theorem 1.2-6 that (4.2-10) is true for $(x, t) \in \bar{\Omega}\times[0, \infty]$. Consequently, for $i = s_1, \ldots, s_m$, we have

$$\tilde{u}_i(x, t) \geq \delta c_i \omega(x)e^{\varepsilon t} \quad \text{for } (x, t) \in \bar{\Omega}\times[0, \infty).$$

<u>Remark</u>. In Theorem 4.2-2, suppose that (4.2-4) is modified by changing $\Delta\tilde{u}_i$ to $\sigma_i\Delta\tilde{u}_i$ for each $i = 1, \ldots, n$, with $\sigma_i > 0$, and $\Delta\tilde{T}$ to $\sigma\Delta\tilde{T}$, with $\sigma > 0$. Theorem 4.2-2 is true verbatim except with $\tilde{H} - \lambda_1 I$ changed to $\tilde{H} - \mathrm{diag}(\sigma_1\lambda_1, \sigma_2\lambda_1, \ldots, \sigma_n\lambda_1)$.

Theorem 4.2-3 below is a variant of Theorem 4.2-1 with a somewhat stronger hypothesis, under which the Perron-Frobenius theory on positive eigenvectors for nonnegative matrices can be applied. We first state some definitions and properties concerning square matrices. by a permutation of a square matrix M, we mean a permutation of the rows of M together with the same

permutation of the columns.

Definition. A square matrix M is called reducible if there is a permutation which transforms it into the form:

$$\begin{pmatrix} B & O \\ C & D \end{pmatrix}$$

where B and D are square matrices. Otherwise, A is call irreducible. A matrix is called non-negative (or positive) if all its elements are non-negative (or positive). The Perron-Frobenius Theorem says: An irreducible non-negative square matrix M always has a positive eigenvalue μ which is a simple root of the characteristic equation for M. The moduli of all the other eigenvalues of M cannot exceed μ. To the "maximal" eigenvalue μ there corresponds an eigenvector with all components positive (see e.g., [85] for reference).

Suppose that in Theorem 4.2-1, hypothesis (II) is modified to a more restrictive irreducible assumption, we can prove that more components of classical nonnegative solutions of (4.1-1) must be identically zero. Hypothesis (II) in Theorem 4.2-1 does not make any assumption on the entires above the diagonal of A, and one can thus readily find a reducible matrix A satisfying hypothesis (II). Theorem 4.2-3 below is therefore a somewhat more restrictive version of Theorem 4.2-1.

Theorem 4.2-3. Suppose that the nxn matrix $\tilde{H} - \lambda_1 I$ has a square submatrix $A = (a_{ij})$, (formed as described in Theorem 4.2-1) with the properties that:

(I) $A\vec{c} > 0$ for some positive m vector $\vec{c} > 0$; and

(II^*) A is irreducible.

Then, any solution $(\hat{u}_1(x), \ldots, \hat{u}_n(x), \hat{T}(x))$ of equation (4.1-1) with the properties that:

(i) each component is in $C^2(\Omega) \cap C^1(\bar{\Omega})$, and

(ii) $\hat{u}_i(x) \geq 0$, $i = 1, \ldots, n$, $\hat{T}(x) \geq 0$ in $\bar{\Omega}$, must satisfy:

$\hat{u}_{s_i}(x) \equiv 0$ in $\bar{\Omega}$, for $i = 1, \ldots, m$

(Recall that s_i's are those integers between 1 and n not in $\{k_1, \ldots, k_{n-m}\}$ described in Theorem 4.2-1).

Proof. Since the off diagonal entries of A are all nonnegative, property (II^*) implies by means of the Perron-Frobenius Theorem above that there is a positive m row vector $\vec{e}^t$ so that $\vec{e}^t A = r\vec{e}^t$ for a real number r. We have

$$r(\vec{e}^t \cdot \vec{c}) = (\vec{e}^t A) \cdot \vec{c} = \vec{e}^t (A\vec{c}) > 0$$

by Property (I). Consequently $r > 0$, because $\vec{e}^t$ and $\vec{c}$ are positive. Let $z(x) = \sum_{i=1}^{m} e_i \hat{u}_{s_i}(x)$, where $\vec{e}^t = (e_1, \ldots, e_m)$. We have for $x \in \Omega$:

$$\Delta z + \lambda_1 z = \sum_{i=1}^{m} e_i \sum_{j=1}^{n} - H_{s_i j}(x, \hat{T}(x))\hat{u}_j + \sum_{i=1}^{m} e_i \lambda_1 \hat{u}_{s_i}$$

$$\text{(4.2-11)} \qquad \leq - \sum_{i=1}^{m} e_i ([\tilde{H}-\lambda_1 I)\hat{u})s_i$$

$$\leq - \sum_{i=1}^{m} e_i (A\hat{\hat{u}})_i = -r\vec{e}^t \cdot \hat{\hat{u}} = -rz.$$

(here $\hat{u} \overset{(def)}{\equiv} \text{col}(\hat{u}_1(x), \ldots, \hat{u}_n(x))$, $\hat{\hat{u}} \overset{(def)}{\equiv} \text{col}(\hat{u}_{s_1}(x), \ldots, \hat{u}_{s_m}(x))$).
From (4.2-11) we have

$$0 \geq \int_\Omega \omega(x)\{\Delta+\lambda_1+r)z\, dx = \int_{\delta\Omega} (-z) \frac{\partial \omega}{\partial \eta} d\sigma + \int_\Omega z\Delta\omega dx + \int_\Omega (\lambda_1+r)\omega z dx$$

$$\geq \int_\Omega -\lambda_1 z\omega dx + \int_\Omega (\lambda_1+r)\omega z dx$$

$$= \int_\Omega rz\omega dx \geq 0$$

In order not to have a contradiction above, we must have $z(x) \equiv 0$ in $\bar{\Omega}$. Consequently, we have $\hat{u}_{s_i} \equiv 0$ in $\bar{\Omega}$ for $i = s_1, \ldots, s_m$.

The remaining part of this section discusses a condition when neutron density of each group will decay to zero for the time dependent parabolic model. This means that the only nonnegative steady state is the trivial one. Condition (4.2-12) in Theorem 4.2-4 is nearly the reverse of condition (I) in Theorem 4.2-1.

Theorem 4.2-4. *Suppose that the* nxn *matrix* $\bar{H} - \lambda_1 I$ *has the property that:*

(4.2-12) $\quad (\bar{H}-\lambda_1 I)\vec{y} < 0$ *for some positive* n *vector* $\vec{y} > 0$.

Then equation (4.1-1) *with boundary conditions*

$$u_i(x) = 0, \; i = 1, \ldots, n., \; T(x) = 0 \text{ for } x \in \delta\Omega$$

has the solution (0, ..., 0) *as the only solution with the properties that each component is in* $C^2(\Omega) \cap C^1(\bar{\Omega})$ *and nonnegative in* $\bar{\Omega}$. (Note that (4.2-12) implies that all the diagonal entries of $H - \lambda_1 I$ are negative, because all its off diagonal entries are nonnegative).

Proof: We will consider the related parabolic system (4.2-4) for $(x,t) \in \Omega \times (0,\infty)$, with boundary conditions

(4.2-13) $\quad \tilde{u}_i(x,t) = 0 \quad i = 1, \ldots, n, \quad \tilde{T}(x,t) = 0$

for $(x,t) \in \delta\Omega \times (0,\infty)$. Here $\tilde{u}_i$, $i = 1, \ldots, n$, $\tilde{T}$ are functions in $\bar{\Omega} \times [0,\infty)$. We will prove that all solutions of (4.2-4), (4.2-13) with components in

$C^2(\Omega \times (0,\infty)) \cap C^1(\bar{\Omega} \times [0,\infty))$ and initial conditions which are nonnegative for all $x \in \bar{\Omega}$, $t = 0$, will tend to zero as $t \to +\infty$. Consequently, the equilibrium solution as stated in the theorem can only be the trivial one.

Define $v_1 \equiv v_2 \equiv \ldots v_{n+1} \equiv 0$. Let $k > 0$ be a constant such that $kc_i\omega(x) \geq \tilde{u}_i(x,0)$, $i = 1, \ldots, n$ for each $x \in \bar{\Omega}$ (here c_i is the ith component of the vector $\vec{y}$ stated in the theorem). Let $d = \min\{c(x) | x \in \bar{\Omega}\}$ and σ a small enough constant with $0 < \sigma < d$ so that inequalities (4.2-12) is valid with $\bar{H} - \lambda_1 I$ replaced by $\bar{H} - (\lambda_1 - \sigma)I$ and $\vec{y}$ unchanged. Choose $c_{n+1} > 0$ so that $c_{n+1} > \max\{\max_{x\in\Omega} \tilde{T}(x,0), (d-\sigma)^{-1} \sum_{i=1}^{n} \bar{g}_i kc_i\}$. Finally, define $w_i = kc_i\omega(x)e^{-\sigma t}$, $i = 1, \ldots, n$ and $w_{n+1} = c_{n+1}e^{-\sigma t}$. Consider the set

$$J \equiv \{(x,t,z_1, \ldots, z_{n+1}) | (x,t) \in \Omega \times (0,\infty),\ v_i(x,t) \leq z_i \leq w_i(x,t), \text{ each } i = 1, \ldots, n+1\}.$$

Clearly, we have for each $i = 1, \ldots, n$

$$(4.2\text{-}14) \quad \Delta v_i + \sum_{\substack{j=1 \\ j\neq i}}^{n} H_{ij}(x,z_{n+1})z_j + H_{ii}(x,z_{n+1})v_i - \frac{\partial v_i}{\partial t} = \sum_{\substack{j=1 \\ j\neq i}}^{n} H_{ij}(x,z_{n+1})z_j \geq 0$$

$$(4.2\text{-}15) \quad \Delta v_{n+1} - c(x)v_{n+1} + \sum_{j=1}^{n} G_j(x,z_{n+1})z_j - \frac{\partial v_{n+1}}{\partial t} = \sum_{j=1}^{n} G_j(x,z_{n+1})z_j \geq 0$$

for all $(x,t,z_1, \ldots, z_{n+1}) \in J$. On the other hand, for all $(x,t,z_1, \ldots, z_{n+1}) \in J$:

$$(4.2\text{-}16) \quad \Delta w_i + \sum_{\substack{j=1 \\ j\neq i}}^{n} H_{ij}(x,z_{n+1})z_j + H_{ii}(x,z_{n+1})w_i - \frac{\partial w_i}{\partial t}$$

$$\leq k\omega(x)e^{-\sigma t} \{\sum_{\substack{j=1 \\ j\neq i}}^{n} c_j\bar{h}_{ij} + (-\lambda_1 + \bar{h}_{ii} + \sigma)c_i\} < 0, \quad i = 1, 2, \ldots, n.$$

$$\Delta w_{n+1}-c(x)w_{n+1} + \sum_{j=1}^{n} G_j(x,z_{n+1})z_j - \frac{\partial w_{n+1}}{\partial t}$$

(4.2-17)

$$\le e^{-\sigma t}\{(-d+\sigma)c_{n+1} + \sum_{j=1}^{n} \bar{g}_j k c_j \max_{x\in\bar{\Omega}} \omega(x)\} < 0,$$

because of the choice of c_j, $j = 1, \ldots, n+1$ and σ. Moreover, we have for $x \in \bar{\Omega}$:

$$v_i(x,0) \le \tilde{u}_i(x,0) \le w_i(x,0), \quad i = 1, \ldots, n, \text{ and}$$

(4.2-18)

$$v_{n+1}(x,0) \le \tilde{T}(x,0) \le w_{n+1}(x,0);$$

and for $(x,t) \in \delta\Omega \times [0,\infty)$,

$$v_i(x,t) \le \tilde{u}_i(x,t) \le w_i(x,t), \quad i = 1, \ldots, n, \text{ and}$$

(4.2-19)

$$v_{n+1}(x,t) \le \tilde{T}(x,t) \le w_{n+1}(x,t).$$

Therefore, if such a solution $(\tilde{u}_1(x,t), \ldots, \tilde{u}_n(x,t), \tilde{T}(x,t))$ exists in $\bar{\Omega} \times [0,\infty)$, it will satisfy (4.2-19) for all $(x,t) \in \bar{\Omega} \times [0,\infty)$, by inequalities (4.2-14) to (4.2-19) above. See e.g., Theorem 1.2-6.

Let $(\hat{u}_1(x), \ldots, \hat{u}_n(x), \hat{T}(x))$ be a solution of the boundary value problem described in the statement of the theorem. with properties as stated. It will be a solution of (4.2-4), (4.2-13) with the appropriate smoothness and nonnegative condition at $t = 0$. Letting

$$(\tilde{u}_1(x,t), \ldots, \tilde{u}_n(x,t), \tilde{T}(x,t)) = (\hat{u}_1(x), \ldots, \hat{u}_n(x), \hat{T}(x)) \quad \text{for } (x,t) \in \bar{\Omega} \times [0,\infty).$$

Inequality (4.2-19) for $(x,t) \in \bar{\Omega} \times [0,\infty)$ implies that

$$0 \le \hat{u}_i(x) \le kc_i\omega(x)e^{-\sigma t},\ i = 1, \ldots, n,\ 0 \le \hat{T}(x) \le c_{n+1}e^{-\sigma t}$$

for $(x,t) \in \bar{\Omega} \times [0,\infty)$. Consequently, $(\hat{u}_1, \ldots, \hat{u}_n, \hat{T}) \equiv (0, \ldots, 0)$.

We now observe a very direct consequence of Theorem 4.2-4.

Corollary 4.2-5. *Suppose that the* $n \times n$ *matrix* $\bar{H} - \lambda_1 I$ *has the properties that all its diagonal entries are negative, and it is diagonally dominant* (i.e., $|\bar{h}_{ii} - \lambda_1| > \sum_{\substack{j=1 \\ j \ne i}}^{n} \bar{h}_{ij}$, $i = 1, \ldots, n$). *Then the boundary value problem in Theorem* 4.2-4 *has the solution* $(0, \ldots, 0)$ *as the only solution with the properties that each component is in* $C^2(\Omega) \cap C^1(\bar{\Omega})$ *and non-negative in* $\bar{\Omega}$.

Proof. Choose $\vec{y} = \text{col.}(1, \ldots, 1)$ to satisfy hypothesis (4.2-12) and apply Theorem 4.2-4.

Remark. If $(\bar{H}-\lambda_1 I)\vec{y} = \gamma\vec{y}$ for some $\vec{y} > 0$, $\gamma < 0$, then we can clearly apply Theorem 4.2-4.

4.3 Prompt Feedback Fission Models and Mutualistic Species

Conditions for the existence of positive steady states had not been found for the general model (4.1-1) in the last section, under zero Dirichlet boundary condition. To make the analysis more tractable, we consider a slightly simplier model. We now assume that the reaction coefficients (i.e., cross sections) are functions of the neutron fluxes u_i directly. That is, the feedback is prompt, and does not have to be regulated through the change in T indirectly through the last equation in (4.1-1). More precisely, we have

$$\Delta u_i + \sum_{j=1}^{n} H_{ij}(u_1, \ldots, u_n)u_j = 0 \text{ in } \Omega,\ i = 1, \ldots, n, \tag{4.3-1}$$

(4.3-2) $$u_i(x) = 0,\ x \in \delta\Omega,\ i = 1, \ldots, n.$$

Define

$$h'_{ij} = \inf\{H_{ij}(u_1, \ldots, u_n) | u_k \geq 0,\ k = 1, \ldots, n\},$$

$$h''_{ij} = \sup\{H_{ij}(u_1, \ldots, u_n) | u_k \geq 0,\ k = 1, \ldots, n\},$$

$i, j = 1, \ldots, n$. The functions $H_{ij}(u_1, \ldots, u_n)u_j$ are assumed to belong to the class C^1 in the set $\{(u_1, \ldots, u_n) | u_k \geq 0,\ k = 1, \ldots, n\}$. We asume that $\delta\Omega$ belongs to $H^{2+\ell}$ (see Section 1.3 for details of symbols). The following conditions will be assumed:

(P1) $-\infty < h'_{ii} \leq h''_{ii} < \infty,\ i = 1, \ldots, n;$

$0 < h'_{1j},\ j = 2, \ldots, n;$

$0 \leq h'_{ij} \leq h''_{ij} < \infty,\ 1 \leq i, j \leq n,\ i \neq j;$

for each $i = 2, \ldots, n$ at least one of $h'_{i1}, h'_{i2}, \ldots, h'_{ii-1}$ is positive.

(P2) In the set $M \overset{(def)}{=} \{(u_1, \ldots, u_n) | u_k \geq 0,\ k = 1, \ldots, n\}$, $H_{11}(u_1, \ldots, u_n)$ is continuously differentiable with respect to $u_2, \ldots, u_n$; and $\left|\frac{\partial H_{11}}{\partial u_j}\right| < K$ for all $(u_1, \ldots, u_n) \in M$, $j = 2, \ldots, n$, where K is some positive constant.

(P3) There exist positive constants p and U^* such that $H_{11}(u_1, \ldots, u_n) \leq -p$ for all $(u_1, \ldots, u_n) \in M$ with $u_1 \geq U^*$.

Note: (P1) is analogous to (C1), (C2) in Section 4.2. However, $0 < h_{1j}$, $j = 2, \ldots, n$ is additional.

<u>Theorem</u> 4.3-1. <u>Let</u> $\hat{H}$ <u>be an</u> nxn <u>matrix</u>, $\hat{H} = (\hat{H}_{ij})$, $1 \le i, j \le n$, <u>where</u> $\hat{H}_{ij} \overset{(def)}{\equiv} h_{ij}$ <u>except</u> $(i,j) = (1,1)$, <u>and</u> $\hat{H}_{11} \overset{(def)}{\equiv} -p$. <u>Suppose that</u> (P1) <u>to</u> (P3) <u>are satisfied</u>. <u>Further, let</u>

$$H_{11}(0, \ldots, 0) > \lambda_1, \text{ and} \tag{4.3-3}$$

$$\hat{H}\vec{c} \le 0 \text{ for some positive n vector } \vec{c} > 0. \tag{4.3-4}$$

<u>Then the boundary value problem</u> (4.3-1) to (4.3-2) <u>has a solution</u> $(\hat{u}_1(x), \ldots, \hat{u}_n(x))$ <u>with components in</u> $H^{2+\ell}$ <u>and</u> $\hat{u}_i(x) > 0$ <u>in</u> Ω, $i = 1, \ldots, n$. ($\hat{H}\vec{c} \le 0$ means each of its component is ≤ 0).

<u>Proof</u>. We will construct upper and lower solutions for (4.3-1) to (4.3-2) and apply Theorem 1.4-1 to conclude the existence of a positive solution. By (4.3-3), there is a small $\alpha > 0$ so that $H_{11}(u, 0, \ldots, 0) > \lambda_1$ for $0 \le u < \alpha$. Choose $0 < \varepsilon < \min\{\alpha, K^{-1}h'_{12}, \ldots, K^{-1}h'_{1n}\}$. For each $i = 3, \ldots, n$, let $\hat{i}$ denote one of the j, $1 \le j \le i-1$ so that $h'_{i\hat{i}} > 0$ (such $h'_{i\hat{i}}$ exist by (P1)). Choose $0 < \delta_2 < h'_{21}\cdot\varepsilon\cdot[|h'_{22}-\lambda_1|]^{-1}$, $\delta_1 = \varepsilon$, and $0 < \delta_i < h'_{i\hat{i}}\,\delta_{\hat{i}}[|h'_{ii}-\lambda_1|]^{-1}$ for $i = 3, \ldots, n$. (For $i = 2, \ldots, n$, if $h'_{ii} - \lambda_1 = 0$, let $\delta_i > 0$ be arbitrary). Define lower solutions as

$$v_1(x) = \varepsilon\omega(x), \quad v_i(x) = \delta_i\omega(x), \quad i = 2, \ldots, n \tag{4.3-5}$$

for $x \in \Omega$. Define upper solutions as

$$w_i(x) \equiv c_i, \quad i = 1, \ldots, n, \quad x \in \bar{\Omega}, \tag{4.3-6}$$

where $\vec{c} = \text{col}(c_1, \ldots, c_n)$ as stated in (4.3-4). (Note that without loss of generality, we may assume $c_1 \ge U^*$, and $c_i > \delta_i$, $i = 2, \ldots, n$). We now check the appropriate inequalities for the v_i, w_i, $i = 1, \ldots, n$. We have

$$\Delta v_1 + H_{11}(v_1, u_2, \ldots, u_n)v_1 + \sum_{j=2}^{n} H_{1j}(v_1, u_2, \ldots, u_n)u_j$$

(4.3-7)

$$\geq \varepsilon\omega(x)[-\lambda_1+H_{11}(\varepsilon\omega(x), u_2, \ldots, u_n)] + \sum_{j=2}^{n} h'_{1j}u_j$$

for $u_j \geq 0$, $j = 2, \ldots, n$, $x \in \Omega$. However $[-\lambda_1 + H_{11}(\varepsilon\omega, 0, \ldots, 0)] > 0$ in Ω; and $F(s, x, u_2, \ldots, u_n) \overset{(def)}{\equiv} \varepsilon\omega(x)[-\lambda_1 + H_{11}(\varepsilon\omega(x), su_2, \ldots, su_n)] + \sum_{j=2}^{n} h'_{1j}su_j$ is an increasing function of $s \geq 0$, for fixed $u_j \geq 0$, $j = 2, \ldots, n$, each $x \in \Omega$ (by the choice of ε). Consequently, we have

(4.3-8)

$$\Delta v_1(x) + H_{11}(v_1(x), u_2, \ldots, u_n)v_1 + \sum_{j=2}^{n} H_{1j}(v_1(x), u_2, \ldots, u_n)u_j > 0$$

for all $u_j \geq 0$, $j = 2, \ldots, n$, $x \in \Omega$. For $i = 3, \ldots, n$, we have

$$\Delta v_i(x) + H_{ii}(u_1, \ldots, u_{i-1}, v_i, u_{i+1}, \ldots, u_n)v_i +$$

(4.3-9)

$$\sum_{\substack{j=1 \\ j\neq i}}^{n} H_{ij}(u_1, \ldots, v_i, \ldots, u_n)u_j$$

$$\geq \delta_i\omega(x)[-\lambda_1 + h'_{ii}] + h'_{i\hat{i}}u_{\hat{i}} \geq \omega(x)\{-\delta_i|h'_{ii}-\lambda_1| + h'_{i\hat{i}}\delta_{\hat{i}}\} \geq 0$$

for $v_j(x) \leq u_j \leq w_j(x)$, $j \neq i$, $x \in \Omega$ (by the choice of δ_i). For the case $i = 2$, all the inequalities in (4.3-9) is true, with $h'_{i\hat{i}}$, $u_{\hat{i}}$ and $\delta_{\hat{i}}$ replaced respectively by h'_{21}, u_1 and ε. For the upper solutions, we have

$$\Delta w_i(x) + H_{ii}(u_1, \ldots, u_{i-1}, w_i, u_{i+1}, \ldots, u_n)\, w_i(x) +$$

(4.3-10)

$$\sum_{\substack{j=1 \\ j\neq i}}^{n} H_{ij}(u_1, \ldots, w_i, \ldots, u_n)u_j$$

$$\leq \sum_{j=1}^{n} \hat{H}_{ij}c_j \leq 0$$

for each $i = 1, \ldots, n$, $v_j(x) \leq u_j \leq w_j(x)$, $j \neq i$, $x \in \Omega$ (by the choice of c_j, (4.3-4) and (4.3-6)). Theorem 1.4-1, (4.3-8) to (4.3-10) imply that there exists a solution $(\hat{u}_1(x), \ldots, \hat{u}_n(x))$ as described in the statement of theorem with $v_i(x) \leq \hat{u}_i(x) \leq w_i(x)$, $i = 1, \ldots, n$, $x \in \bar{\Omega}$. Consequently, $\hat{u}_i(x) > 0$ in Ω, $i = 1, \ldots, n$.

The prompt-feedback reactor problem has some similarity with various particular mutualism models in ecology, in the sense that the presence of each component contributes to faster growth rate of all other components (cf. (P1)). Although the equations are generally quite different, it is instructive to investigate the kind of results one might obtain for a two species mutualism model. One obtains the existence of positive steady states for the Volterra-Lotka diffusive system under very simple conditions. We consider the boundary value problem:

$$\text{(4.3-11)} \qquad \begin{aligned} \Delta u_1 + u_1(a - bu_1 + cu_2) &= 0, \\ \Delta u_1 + u_2(e + fu_1 - gu_2) &= 0, \qquad \text{in } \Omega; \\ u_1 = u_2 &= 0 \text{ on } \delta\Omega. \end{aligned}$$

Here a, b, c, e, f, g are nonnegative constants. The u and v are concentrations of two mutualistic (i.e., cooperating) species.

<u>Theorem</u> 4.3-3. <u>Suppose that</u>:

$$\text{(4.3-12)} \qquad a > \lambda_1 \text{ and } e > \lambda_1.$$

<u>The constants</u> b, g <u>are positive and</u> c, f <u>are nonnegative satisfying</u>

(4.3-13) $\qquad bg > cf.$

<u>Then the boundary value problem</u> (4.3-11) <u>has a solution</u> $(\bar{u}_1(x), \bar{u}_2(x))$ <u>with</u> $\bar{u}_i(x) > 0$, $i = 1, 2$ <u>for</u> $x \in \Omega$.

<u>Proof</u>. From (4.3-12), we can choose small positive constants ε_i, $i = 1,2$ so that:

$$a - \lambda_1 - b\varepsilon_1\omega(x) > 0 \text{ and } e - \lambda_1 - g\varepsilon_2\omega(x) > 0 \tag{4.3-14}$$

for $x \in \bar{\Omega}$. Condition (4.3-13) implies that we can choose arbitrarily large positive constants $M, N > \varepsilon_i\omega(x)$, $x \in \bar{\Omega}$, $i = 1,2$ so that:

$$a - bM + cN < 0 \quad \text{and} \quad e + fM - gN < 0. \tag{4.3-15}$$

Let $v_i = \varepsilon_i\omega(x)$, $i = 1,2$ and $w_1(x) \equiv M$, $w_2(x) \equiv N$. From (4.3-14), we have

$$\Delta v_1 + v_1[a-bv_1+cu_2] = \varepsilon_1\omega(x)[a-\lambda_1-b\varepsilon_1\omega(x)+cu_2] > 0,$$
$$\Delta v_2 + v_2[e+fu_1-gv_2] = \varepsilon_2\omega(x)[e-\lambda_1+fu_1-g\varepsilon_2\omega(x)] > 0$$

for all $x \in \Omega$, $v_i(x) \le w_i$, $i = 1,2$. From (4.3-15) we have

$$\Delta w_1 + w[a-bw_1+cu_2] = M[a-bM+cu_2] < 0,$$
$$\Delta w_2 + w_2[e+fu_1-gw_2] = N[e+fu_1-gN] < 0$$

for all $v_i(x) \le u_i \le w_i$. By Theorem 1.4-1, there exists a solution $(\bar{u}_1(x), \bar{u}_2(x))$ to the boundary value problem (4.3-11) with $v_i(x) \le \bar{u}_i(x) \le w_i(x)$, $i = 1,2$. Consequently, $\bar{u}_i(x) > 0$ in Ω.

Another interesting case of mutualistic loop is considered in [140]. We consider three species A, B and C (with corresponding concentrations u_1, u_2

and u_3) where A eats C, C eats B and B eats A. The Dirichlet problem is

$$\Delta u_i + u_i[a_i + \sum_{j=1}^{3} \lambda_{ij}u_j] = 0 \quad \text{in } \Omega,$$

(4.3-16)

$$u_i = g_i \quad \text{on } \partial\Omega,$$

$i = 1, 2, 3$, where a_i, λ_{ij}, $1 \le i, j \le 3$ are constants, with $\lambda_{ii} < 0$, $i = 1, 2, 3$. The relationship between the species described above gives

(4.3-17) $$\lambda_{12} \le 0, \ \lambda_{13} \ge 0; \ \lambda_{21} \ge 0, \ \lambda_{23} \le 0, \ \lambda_{31} \le 0, \ \lambda_{32} \ge 0.$$

We assume that for each pair (i,j), λ_{ij} and λ_{ji} cannot be both 0 so that there is indeed interaction between the corresponding pair of species. (Otherwise, the situation reduces to that of food chain or less than three species interaction.) For simplicity, we assume that the boundary functions g_i, $i = 1, 2, 3$ can be extended to $\hat{g}_i \in H^{2+\ell}(\bar{\Omega})$ and $g_i(x) \ge 0$, $\not\equiv 0$ for $x \in \delta\Omega$.

It is shown in [140] that for the problem above, if

(4.3-18) $$0 < \lambda_{32}\lambda_{21}\lambda_{13} < |\lambda_{11}\lambda_{22}\lambda_{33}| ,$$

then (4.3-16) has a solution $(\bar{u}_1(x), \bar{u}_2(x), \bar{u}_3(x))$ with $\bar{u}_i(x)$ in $H^{2+\ell}(\bar{\Omega})$, $i = 1, 2, 3$. More elaborate cases will be considered in the next chapter. However, they are not as similar to the reactor model in this chapter.

4.4 Down Scattering, Supercriticality and Directed Coupled Scattering

In this section we consider special cases in fission reactor theory. In the temperature dependent model (4.1-1) we always assume hypotheses (C1) and (C2), while in the prompt feedback case we always assume (P1) to (P3). When more specific assumptions are made concerning the nature of scattering and fission together with the sizes of their cross-sections, one can readily apply the theorems in the last two sections to predict reactor behavior. Hence one can deduce the proper needed adjustment to control the reactor.

We will always further assume "down scattering" in this section. That is we assume that group transfer only from higher to lower energy groups (i.e., from group j to group i, i > j). On the other hand, fissions from each group produces neutron only in the first group. More specifically:

(4.4-1) $H_{1j} \geq 0$ for $j = 2, \ldots, n$.

(4.4-2) For $i > 1$: $H_{ij} \geq 0$ if $j < i$, $H_{ii} \leq 0$ and $H_{ij} \equiv 0$ if $j > i$.

All formulas are true for $x \in \bar{\Omega}$, $T \geq 0$ in (4.4-1) and (4.4-2). In the first row, $i = 1$, only H_{11} is not assumed nonnegative because of the removal term $-\Sigma_R$ (cf. (4.1-2)) which can be adjusted by control rods.

In Theorem 4.4-1 we consider the simplest example when the fission cross section of the first group is large compared with the removal cross section. In order to have a finite steady state without blow-up, one must have $u_1 \equiv 0$. Then, the nature of scattering and fission in (4.4-2) will imply successively that $u_2, u_3, \ldots, u_n$ are all 0.

Theorem 4.4-1 (Supercriticality in down scattering). Assume (C1), (C2) and down scattering conditions (4.4-1), (4.4-2). Suppose further that

(4.4-3) $$\tilde{h}_{11} \equiv \inf\{H_{11}(x,s) \mid x \in \bar{\Omega}.\ s \geq 0\} > \lambda_1.$$

Then equation (4.1-1) with boundary conditions

$$u_i(x) = 0 \text{ on } \delta\Omega, \; i = 1, \ldots, n, \; T(x) = 0 \text{ on } \delta\Omega,$$

has the trivial solution as the only solution with each component nonnegative and in $C^2(\Omega) \cap C^1(\bar{\Omega})$.

Proof. Apply Theorem 4.2-1, choosing A as the 1x1 matrix $[\tilde{h}_{11}-\lambda_1]$. Condition (I) in Theorem 4.2-1 is satisfied by choosing c = [1], and condition (II) is true vacuously. Consequently, nonnegative solutions with smoothness described above must have $u_1(x) \equiv 0$ in $\bar{\Omega}$. Conditions (4.4-2) indicate that $H_{23}, \ldots, H_{2n}$ are $\equiv 0$, and we can thus apply maximum (Theorem 1.1-2) to conclude that $u_2 \equiv 0$ in $\bar{\Omega}$. Successively, the down scattering conditions and maximum principle will imply $u_2, u_3, \ldots, u_n$ are all $\equiv 0$ in $\bar{\Omega}$.

Remark: We call condition (4.4-3) supercriticality because it causes blow up unless everything are identically zero.

Another commonly used simplifying assumption in scattering behavior is that neutrons in a given energy group i only scatter into the next lower energy group i + 1, as a special case of down scattering. This is known as directly coupled scattering. More precisely we replace (4.4-2) by the following:

(4.4-4) For $i > 1$: $H_{i,i-1} \geq 0$, $H_{ii} \leq 0$, and
$H_{ij} \equiv 0$ if $j > i$ or $j < i-1$.

Even if the supercritical condition (4.4-3) are not satisfied, it is still possible for blowup situation to occur. Theorem 4.4-2 gives sufficient conditions that neutron formation in the first m groups are fast enough to blowup. Consequently, no finite nontrivial steady states can exist.

Theorem 4.4-2 (Directly coupled scattering). Assume (C1), (C2), (4.4-1),

(4.4-4), $\tilde{h}_{1k} > 0$, $\tilde{h}_{k,k-1} > 0$ $k = 2, \ldots, n$ _and_

(4.4-5) $$\tilde{h}_{11} \equiv \{\inf H_{11}(x,s) \mid x \in \bar{\Omega},\ s \geq 0\} < \lambda_1.$$

Suppose that there exists positive constants $\delta_2, \ldots, \delta_m$, $2 \leq m \leq n$, _with_ $\sum_{i=2}^{m} \delta_i = 1$, _so that_

(4.4-6) $$\tilde{h}_{21}\tilde{h}_{12} > (\lambda_1 - \tilde{h}_{11})(\lambda_1 - \tilde{h}_{22})\,\delta_2\,,$$
$$\delta_{k-1}\tilde{h}_{k,k-1}\tilde{h}_{1k} > (\lambda_1 - \tilde{h}_{kk})\,\tilde{h}_{1,k-1}\,\delta_k\quad,\ k = 3, \ldots, m.$$

Then equation (4.1-1) _with prescribed boundary conditions_

(4.4-7) $$u_i(x) = u_i^0(x) \geq 0,\ i = 1, \ldots, n,\ T(x) = T^0(x) \geq 0,\ x \in \delta\Omega,$$

has no solution $(\hat{u}_1(x), \ldots, \hat{u}_n(x), \hat{T}(x))$ _with_ $\hat{u}_1(x) \not\equiv 0$ _in_ $\bar{\Omega}$, _and properties_ (i) _and_ (ii) _as described in Theorem_ 4.2-1.

Remark. Conditions (4.4-6) successively describe the sizes of scattering into the kth group and fission caused by the kth group, $k = 2, \ldots, m$, which will lead to blowup.

Proof. Let $\tilde{\tilde{h}}_{11} < \tilde{h}_{11}$ so that the first inequality in (4.4-6) is still valid with $\tilde{h}_{11}$ replaced by $\tilde{\tilde{h}}_{11}$. Apply Theorem 4.2-1 with A as the mxm matrix on the upper left hand corner, and $\vec{c}$ = col. $(1, \delta_2\tilde{h}_{12}^{-1}(\lambda_1 - \tilde{\tilde{h}}_1), \delta_3\tilde{h}_{13}^{-1}(\lambda_1 - \tilde{\tilde{h}}_1), \ldots, \delta_m\tilde{h}_{1m}^{-1}(\lambda_1 - \tilde{\tilde{h}}_1), 0, \ldots, 0)$. Conditions in (4.4-6) imply that (I) in Theorem 4.2-1 is satisfied. Theorem 4.2-1 implies that $u_1(x) \equiv 0$ in $\bar{\Omega}$.

Remark. If the boundary condition (4.4-7) is replaced by $u_i(x) = 0$, $i = 1$,

..., n, $T(x) = 0$ on $\delta\Omega$, then we can apply the maximum principle successively to deduce that $u_i(x) \equiv 0$, $i = 1, \ldots, n$, $T(x) \equiv 0$ in $\bar{\Omega}$.

More elaborate conditions concerning blow up for the general down scattering case can be found in [45].

Finally, we describe the prompt feedback case with directly coupled scattering to find some cases when nontrivial positive steady-state do indeed exist. Consider the boundary value problem (4.3-1), (4.3-2) under hypotheses (P1) to (P3) and the directed coupled scattering conditions:

(4.4-8) $$H_{1j}(u_1, \ldots, u_n) > 0, \quad j = 2, \ldots, n;$$

(4.4-9) $$\text{for } i > 1: H_{i,i-1}(u_1, \ldots, u_n) > 0, \; H_{ii}(u_1, \ldots, u_n) < 0, \\ H_{ij} \equiv 0 \text{ if } j > i \text{ or } j < i-1.$$

The formulas above are assumed for $\{(u_1, \ldots, u_n) | u_k \geq 0, k = 1, \ldots, n\}$. Under the following conditions (4.4-11) which describe the sizes of scattering into the kth group and fission caused by the kth group, $k = 2, \ldots, m$, one will have the existence of nontrivial positive steady state. Note that the strong fission for the 1st group (4.4-10) is balanced by the limit on scattering and fission of the other groups (4.4-11).

<u>Theorem</u> 4.4-3 (<u>Directly coupled scattering</u>). <u>Assume</u> (P1) <u>to</u> (P3) <u>and</u> (4.4-8), (4.4-9). <u>Suppose that</u>

(4.4-10) $$H_{11}(0, \ldots, 0) > \lambda_1 ,$$

<u>and there exist positive constants</u> β_i, $i = 2, \ldots, n$ <u>with</u> $\sum_{i=2}^{n} \beta_i \leq 1$ <u>such that</u>

$$h''_{21}h''_{12} < \beta_2|h''_{22}|p ,$$

(4.4-11)

$$\beta_{k-1}h''_{k,k-1}h''_{1k} < |h''_{kk}|\ h''_{1,k-1}\ \beta_k \quad , \ k = 3, \ldots, n.$$

Then the boundary value problem (4.3-1), (4.3-2) has a solution $(\hat{u}_1(x), \ldots, \hat{u}_n(x))$ with components in $H^{2+\ell}(\bar{\Omega})$ and $\hat{u}_i(x) > 0$ in Ω, $i = 1, \ldots, n$. (Here, the number p in (4.4-11) is from hypothesis (P3)).

Proof. Let $\hat{H}$ be as defined in Theorem 4.3-1. Let $\vec{c}$ = col. $(1, \beta_2 p/h''_{12}, \beta_3 p/h''_{13}, \ldots, \beta_n p/h''_{1n})$. Condition (4.4-11) implies that (4.3-4) is satisfied. Apply Theorem 4.3-1.

Other conditions for the existence of nontrivial positive steady states for the more general down scattering case can be found in [45].

4.5 Transport Systems

In the study of reactor engineering, there are also substantial theories which use transport equations rather than diffusion equations. We present some results which are readily obtainable by using methods in this book. We consider the linear time-independent multigroup neutron transport system in an m dimensional domain Ω which is convex and bounded, $m \geq 1$. Let $u_i(x,v)$ represents the density of neutron in energy group i, at position $x \in \Omega$ moving with velocity $v \in D \subset R^m$; ∇ is the spatial gradient operator in x and $\hat{\eta}(x)$ is the outward unit normal at $x \in \delta\Omega$. The combination of transport, absorption, fission and source give rise to the following system:

(4.5-1) $$v\cdot\nabla u_i + \sigma_i(x,v)u_i = \sum_{j=1}^{N} \int_{\Omega} \sigma_{ij}(x,v,v')\ u_j(x,v')dv' + q(x,v), \quad x\in\Omega, \ v\in D,$$

(4-5-2) $u_i(x,v) = 0$ for $x \in \delta\Omega$, $v \cdot \hat{\eta}(x) < 0$ (i.e. for v incoming at the boundary).

Here, we have divided the neutron density into N energy groups. The function $\sigma_i(x,v)$ is the total cross-section for the ith group at point x with velocity v; $\sigma_{ij}(x,v,v')$ is the group transfer cross-section, which is the probability of scattering for a neutron from a group j with velocity v' to a group i with velocity v. In this section, we always assume that Ω is open bounded and convex, D is bounded, $\sigma_i(x,\sigma)$ is positive and piecewise continuous on $\bar{\Omega} \times \bar{D}$, $\sigma_{ij}(x,v,v')$ and $q_i(x,v)$ are nonnegative and piecewise continuous on $\bar{\Omega} \times \bar{D} \times \bar{D}$ and $\bar{\Omega} \times \bar{D}$ respectively. Here $q_i(x,v)$ is the interior source term; and at each $x \in \delta\Omega$, we assume that there is a outward unit normal $\hat{\eta}(x)$ with the boundary conditions (4.5-2) stating that no neutron is coming in through the boundary.

One of the important problems in neutron transport theory is to determine conditions on the size of the transport region and the physical parameters of the medium so that problem (4.5-1), (4.5-2) with a given source q_i does or does not have any nonnegative bounded solutions. In other words, the problem is to determine the critical size of the transport region in terms of the physical parameters of the transport medium. In this section, we will use the monotone methods to analyze this problem. More systematic study of this method for the reaction-diffusion elliptic systems will be presented in Chapter V.

We first transform (4.5-1) into a system of integral equations. For each $(x,v) \in \bar{\Omega} \times D$ with $v \neq 0$, we start from x and move in the direction of (-v) until we first meet the boundary $\delta\Omega$. This point on the boundary can be denoted as $x^* = x - sv$ where $s(x,v) \geq 0$. Thus $s(x,v) = |x-x^*|/|v|$ and $s|v|$ is the distance between x and x^*. Note that for any such (x,v), the point $x - tv \in \bar{\Omega}$ for all $t \in [0,s]$, and for $x \in \delta\Omega$, $s(x,v) = 0$. For any bounded integrable function h(x,v) defined on $\Omega \times D$, and positive piecewise continuous positive function $\sigma(x,v)$ on $\bar{\Omega} \times \bar{D}$, let w(x,v) be a solution of

(4.5-3) $$v\cdot\nabla w(x,v) + \sigma(x,v)\, w(x,v) = h(x,v), \text{ for } (x,v) \in \Omega \times D,$$

(4.5-4) $$w(x,v) = 0, \text{ for } x \in \delta\Omega,\ v\cdot\hat{\eta}(x) < 0 \text{ (i.e, v incoming).}$$

(Here, for $x \in \delta\Omega$ $v\cdot\hat{\eta}(x) < 0$, we assume that $w(\tilde{x},\tilde{v})$ tends to $w(x,v)$ as $(\tilde{x},\tilde{v}) \to (x,v)$ with $\tilde{x} \in \bar{\Omega}$). Using the variable $x' = x - sv$, and writing (4.5-3) in the form,

$$\frac{d}{ds} w(x'+sv,v) + \sigma(x'+sv,v)\, w(x'+sv,v) = h(x'+sv,v),$$

one readily obtains after multiplying by an exponential factor and integrating, that any solution of (4.5-3), (4.5-4) satisfies

(4.5-5) $$w(x,v) = \int_0^{s(x,v)} \exp\Big(-\int_0^{\tau} \sigma(x-\xi v,v)d\xi\Big)\, h(x-\tau v,v)d\tau$$

for $(x,v) \in \Omega \times D$. For $x \in \delta\Omega$, v incoming, we have $s(x,v) = 0$ in (4.5-5) so that (4.5-4) is satisfied. If $v = 0$, (4.5-3) means $\sigma(x,0)\, w(x,0) = h(x,0)$, and $s(x,v) = \infty$ in (4.5-5); and $w(x,0)$ still satisfies (4.5-5).

It is therefore natural to consider the following integral equation as a more general formulation for problem (4.5-1), (4.5-2):

(4.5-6) $$u_i(x,v) = \int_0^{s(x,v)} \exp\Big(-\int_0^{\tau} \sigma_i(x-\xi v,v)d\xi\Big)\Big[\sum_{j=1}^{N} F_{ij}(u_j)(x-\tau v,v) + q_i(x-\tau v,v)\Big]d\tau$$

for $i = 1, \ldots, N$, $(x,v) \varepsilon\ \bar{\Omega} \times D$. Here $F_{ij}(u_j)$ are defined as:

(4.5-7) $$F_{ij}(u_j)(x,v) = \int_{\Omega} \sigma_{ij}(x,v,v')\, u_j(x,v')dv'$$

for $1 \le i,\ j \le N$, $(x,v) \in \bar{\Omega} \times D$.

We will see that the integral equation (4.5-6) has a nonnegative solution if and only if there exists an upper solution. Such solution is found by means of constructing a monotone sequence iteratively, starting from the upper solution. For convenience, we define $u = (u_1, \dots, u_N)$ and write $u \le w$ if $u_i \le w_i$ for $i = 1, \dots, N$.

Definition. A function $w(x,v) = (w_1(x,v), \dots, w_N(x,v))$ with each component bounded, integrable and nonnegative in $\bar{\Omega} \times D$ is called an upper solution of (4.5-6) if it satisfies

$$w_i(x,v) \ge \int_0^{s(x,v)} \exp\left(-\int_0^{\tau} \sigma_i(x-\xi v, v)d\xi\right)\left[\sum_{j=1}^{N} F_{ij}(w_j)(x-\tau v, v) + q_i(x-\tau v, v)\right]d\tau \tag{4.5-8}$$

for $i = 1, \dots, N$, $(x,v) \in \bar{\Omega} \times D$. The function is called a lower solution of (4.5-6) if it satisfies (4.5-8) with $\ge$ replaced by $\le$.

Let $z(x,v) = (z_1(x,v), \dots, z_N(x,v))$ be any function with each component bounded and integrable in $\bar{\Omega} \times D$, define

$$(Tz)(x,v) = ((Tz)_1(x,v), \dots, (Tz)_N(x,v)), \text{ where} \tag{4.5-9}$$

$$(Tz)_i(x,v) = \int_0^{s(x,v)} \exp\left(-\int_0^{\tau} \sigma_i(x-\xi v, v)d\xi\right)\left[\sum_{j=1}^{i-1} F_{ij}(Tz_j)(x-\tau v, v) + \sum_{j=i}^{N} F_{ij}(z_j)(x-\tau v, v) + q_i(x-\tau v, v)\right]d\tau \tag{4.5-10}$$

for $(x,v) \in \bar{\Omega} \times D$, successively for $i = 1, \dots, N$. Suppose that $\tilde{u}(x,v)$ is an upper solution for (4.5-6), define $\bar{u}^{(0)}(x,v) = \tilde{u}(x,v)$ in $\bar{\Omega} \times D$ and recursively let

$$\bar{u}^{(k)}(x,v) = (T\bar{u}^{(k-1)})(x,v) \tag{4.5-11}$$

for $(x,v) \in \bar{\Omega} \times D$, $k = 1, 2, \ldots$. On the other hand, let $\underline{u}^{(0)} \equiv 0$ and recursively define

$$\underline{u}^{(k)}(x,v) = (T\underline{u}^{(k-1)})(x,v) \tag{4-5-12}$$

for $(x,v) \in \bar{\Omega} \times D$, $k = 1, 2, \ldots$. We will refer to $\{\bar{u}^{(k)}\}$ and $\{\underline{u}^{(k)}\}$ as maximal and minimal sequences respectively, and their ith components will be denoted by $\bar{u}_i^{(k)}$, $\underline{u}_i^{(k)}$ respectively, $i = 1, \ldots, N$. The two sequences have the following properties.

Lemma 4.5-1. *The maximal sequence* $\{\bar{u}^{(k)}\}$ *is monotone nonincreasing*, i.e.,

$$\bar{u}^{(k+1)}(x,v) \le \bar{u}^{(k)}(x,v), \text{ for all } (x,v) \in \bar{\Omega} \times D,\ k = 0, 1, \ldots . \tag{4.5-13}$$

The minimal sequence $\{\underline{u}^{(k)}\}$ *is monotone nondecreasing* i.e., (4.5-13) *is satisfied with* "≤" *replaced by* "≥", *and* $\bar{u}$ *replaced by* $\underline{u}$.

Proof. We will only show that $\{\bar{u}^{(k)}\}$ is monotone nonincreasing, the other case is exactly analogous. Since $\tilde{u}$ is an upper solution, from (4.5-8) to (4.5-11), we have

$$(\bar{u}_1^{(0)} - \bar{u}_1^{(1)})(x,v) = \tilde{u}_1(x,v) - \int_0^{s(x,v)} \exp\left(-\int_0^{\tau} \sigma_1(x-\xi v,v)d\xi\right)\left[\sum_{j=1}^{N} F_{1j}(\tilde{u}_j)(x-\tau v,v) + q_1(x-\tau v,v)\right]d\tau \ge 0 \text{ in } \bar{\Omega} \times D.$$

Assume by induction that

$$\bar{u}_i^{(1)} \le \bar{u}_i^{(0)}, \quad i = 1,\ldots, m-1, \text{ for } (x,v) \in \bar{\Omega} \times D. \tag{4.5-14}$$

We have for $(x,v) \in \bar{\Omega} \times D$:

$$(\bar{u}_m^{(0)} - \bar{u}_m^{(1)})(x,v) = \tilde{u}_m(x,v) - \int_0^{s(x,v)} \exp(-\int_0^\tau \sigma_m(x-\xi v,v)d\xi)[\sum_{j=1}^{m-1} F_{mj}(\bar{u}_j^{(1)})(x-\tau v,v)$$

$$+ \sum_{j=m}^{N} F_{mj}(\bar{u}_j^{(0)})(x-\tau v,v) + q_m(x-\tau v,v)]d\tau \tag{4.5-15}$$

$$\geq \tilde{u}_m(x,v) - \int_0^{s(x,v)} \exp(-\int_0^\tau \sigma_m(x-\xi v,v)d\xi)[\sum_{j=1}^{N} F_{mj}(\tilde{u}_j)(x-\tau v,v)$$

$$+ q_m(x-\tau v,v)]d\tau \geq 0$$

since $[F_{mj}(\tilde{u}_j) - F_{mj}(\bar{u}_j^{(1)})](x,v) = \int_\Omega \sigma_{mj}(x,v,v')(\tilde{u}_j-\bar{u}_j^{(1)})(x,v')dv' \geq 0$ for $j = 1, \ldots, m-1$, by (4.5-14). (Note that $(x - \tau v) \in \bar{\Omega}$ for each τ in (4.5-15)). Hence, we have

$$\bar{u}_i^{(1)} \leq \bar{u}_i^{(0)} \text{ for } i = 1, \ldots, N, \ (x,v) \in \bar{\Omega} \times D.$$

Next, we inductively assume that $\bar{u}^{(r)} \leq \bar{u}^{(r-1)}$ in $\bar{\Omega} \times D$. From (4.5-10) and (4.5-11) we obtain:

$$(\bar{u}_1^{(r)} - \bar{u}_1^{(r+1)})(x,v) = \int_0^{s(x,v)} \exp(-\int_0^\tau \sigma_1(x-\xi v,v)d\xi) \ [\sum_{j=1}^{N} \{F_{1j}(\bar{u}_j^{(r-1)}) \ (x-\tau v,v) - F_{1j}(\bar{u}_j^{r}) \ (x-\tau v,v)\}d\tau \geq 0.$$

Continue using (4.5-10) and (4.5-11), we deduce step by step that $(\bar{u}_i^{(r)} - \bar{u}_i^{(r+1)})(x,v) \geq 0$ for $i = 1, \ldots, N$, $(x,v) \in \bar{\Omega} \times D$. Consequently, by induction, we obtain

$$\bar{u}^{(k+1)} \leq \bar{u}^{(k)} \text{ for } k = 0, 1, 2, \ldots, \ (x,v) \in \bar{\Omega} \times D.$$

On the other hand, we have

$$\underline{u}_1^{(0)} \equiv 0 \le \int_0^{s(x,v)} \exp(-\int_0^{\tau} \sigma_1(x-\xi v,v)d\xi)q_1(x-\tau v,v)d\tau = \underline{u}_1^{(1)}(x,v) ,$$

and we can inductively obtain $\underline{u}^{(0)} \le \underline{u}^{(1)}$, and eventually $\underline{u}^{(k)} \le \underline{u}^{(k+1)}$ in $\bar{\Omega} \times D$ for k = 0, 1, 2, ..., as before. This proves the lemma.

Definition. A nonnegative solution u^* of (4.5-6) is said to be maximal (or minimal) with respect to the upper solution $\tilde{u}$ if $u \le u^* \le \tilde{u}$ (or $u^* \le u \le \tilde{u}$) for every solution u of (4.5-6) satisfying $0 \le u \le \tilde{u}$.

Theorem 4.5-1. Suppose that there exists a nonnegative upper solution $\tilde{u}$ of (4.5-6) in $\bar{\Omega} \times D$. Then the corresponding maximal sequence $\{\bar{u}^{(k)}\}$ defined recursively by (4.5-11) converges pointwise to a maximal solution $\bar{u}$ with respect to $\tilde{u}$. Moreover, the minimal sequence defined recursively by (4.5-12) converges pointwise to a minimal solution $\underline{u}$ with respect to $\tilde{u}$. Furthermore, we have the following inequalities in $\bar{\Omega} \times D$:

$$(4.5\text{-}16) \quad 0 \le \underline{u}^{(1)} \le \underline{u}^{(2)} \le \dots \le \underline{u} \le \bar{u} \le \dots \le \bar{u}^{(2)} \le \bar{u}^{(1)} \le \bar{u}^{(0)} \equiv \tilde{u}.$$

Proof. Since $\tilde{u} \ge 0$, and $\underline{u}^{(0)} \overset{(def)}{=} 0$, $\bar{u}^{(0)} \overset{(def)}{=} \tilde{u}$, we have $\underline{u}^{(0)} \le \bar{u}^{(0)}$. Suppose that $\underline{u}^{(k-1)} \le \bar{u}^{(k-1)}$, a similar argument as in (4.5-14) and (4.5-15) gives $\underline{u}^{(k)} \le \bar{u}^{(k)}$. Thus from Lemma 4.5-1, the sequences $\{\bar{u}^{(k)}\}$ and $\{\underline{u}^{(k)}\}$ are monotone and bounded, and the limits

$$\bar{u}_i(x,v) = \lim_{k\to\infty} \bar{u}_i^{(k)}(x,v), \quad \underline{u}_i(x,v) = \lim_{k\to\infty} \underline{u}_i^{(k)}(x,v)$$

exists for $(x,v) \in \bar{\Omega} \times D$, i = 1, ..., N, satisfying relation (4.5-16). Furthermore, we conclude from the dominated convergence theorem that $\underline{u} = (\underline{u}_1,$

..., $\underline{u}_N$) and $\bar{u} = (\bar{u}_1, \ldots, \bar{u}_N)$ are both solutions of (4.5-6) by taking limits in (4.5-12) and (4.5-11) respectively.

In order to show that $\underline{u}$ and $\bar{u}$ are respectively maximal and minimal with respect to $\tilde{u}$, we consider an arbitrary nonnegative solution $u = (u_1, \ldots, u_N)$ of (4.5-6) with $0 \le u \le \tilde{u}$. Assume that $u \le \bar{u}^{(k-1)}$ in $\bar{\Omega} \times D$; from (4.5-6) and (4.5-11), we deduce that $(\bar{u}_1^{(k)} - u_1)(x,v) \ge 0$ in $\bar{\Omega} \times D$. Suppose we have obtained $\bar{u}_j^{(k)} - u_j \ge 0$ for $j = 1, \ldots, m-1$, then by writing

$$(\bar{u}_m^{(k)} - u_m)(x,v) = \int_0^{s(x,v)} \exp(-\int_0^{\tau} \sigma_m(x-\xi v, v)d\xi)\{\sum_{j=1}^{m-1} [F_{mj}(\bar{u}_j^{(k)}) - F_{mj}(u_j)](x-\tau v, v) + \sum_{j=m}^{N} [F_{mj}(\bar{u}_j^{(k-1)}) - F_{mj}(u_j)[(x-\tau v. v)\}d\tau,$$

we can deduce inductively that $\bar{u}_j^{(k)} - u_j \ge 0$ for $j = 1, \ldots, N$. By induction again, we conclude that $u \le \bar{u}^{(k)}$ in $\bar{\Omega} \times D$, for $k = 0, 1, 2, \ldots$. Similarly, we show inductively that $\underline{u}^{(k)} \le u$ for $k = 0, 1, 2, \ldots$. This completes the proof.

<u>Remark</u>. Since every nonnegative bounded solution of (4.5-6) is also an upper solution, Theorem 4.5-1 implies that equation (4.5-6) has a nonnegative bounded solution if and only if there exists an upper solution.

From Theorem 4.5-1, we see that in order to obtain a nonnegative solution, it suffices to find sufficient conditions so that an explicit upper solution can be constructed. This leads to an explicit relationship between the size of the transport region and the cross sections σ_i and σ_{ij}. For convenience, we define the following notations for $i,j = 1, \ldots, N$:

$$\underline{\sigma}_i = \inf\{\sigma_i(x,v): (x,v) \in \bar{\Omega} \times D\} \quad , \underline{\sigma}^t = \min\{\underline{\sigma}_i: i = 1, \ldots, N\},$$
$$\hat{\sigma}_{ij}(v') = \sup\{\sigma_{ij}(x,v,v'): (x,v) \in \bar{\Omega} \times D\},$$
$$\bar{\sigma}_{ij} = \sup\{ \int_D \sigma_{ij}(x,v,v')dv': (x,v) \in \bar{\Omega} \times D\},$$

(4.5-17)
$$\bar{\sigma}_i = \sup\{\sum_{j=1}^{N} \int_D \sigma_{ij}(x,v,v')dv' : (x,v) \in \bar{\Omega} \times D\},$$
$$\bar{\sigma}^s = \max\{\bar{\sigma}_{ij} : 1 \le i,\ j \le N\},$$
$$\bar{q}_i = \sup\{q_i(x,v) : (x,v) \in \bar{\Omega} \times D\},\ \bar{q} = \max\{\bar{q}_i : i = 1, \ldots, N\}.$$

Theorem 4.5-2. Suppose that

(4.5-18)
$$\hat{\rho}_i \overset{(def)}{=} \sup_{x\in\bar{\Omega}} \{\sum_{j=1}^{N} [\int_D \hat{\sigma}_{ij}(v')(1-e^{-\underline{\sigma}_j s(x,v')})dv'/\underline{\sigma}_j]\} < 1$$

for each $i = 1, \ldots, N$. Then problem (4.5-6) has at least one nonnegative bounded solution in $\bar{\Omega} \times D$.

Proof. By Theorem 4.5-1, it is sufficient to find an upper solution. By hypotheses, (4.5-18), there are positive constants $\delta_i < 1 - \hat{\rho}_i$, $i = 1, \ldots, N$. Define $\delta = \min\{\delta_i : 1 \le i \le N\}$. For $(x,v) \in \bar{\Omega} \times D$, let

(4.5-19)
$$\tilde{u}_i(x,v) = \frac{q}{\delta} \int_0^{s(x,v)} \exp(-\int_0^{\tau} \sigma_i(x-\xi v, v)d\xi)d\tau, \text{ for } i = 1, \ldots, N.$$

Note that σ_i are piecewise continuous and positive in $\bar{\Omega} \times D$, thus $\tilde{u}_i$ are bounded. From (4.5-17), we have

(4.5-20)
$$\sum_{j=1}^{N} F_{ij}(\tilde{u}_j)(x,v) + q_i(x,v) =$$
$$\sum_{j=1}^{N} \frac{\bar{q}}{\delta} \int_{\Omega} \sigma_{ij}(x,v,v')\{\int_0^{s(x,v')} \exp(-\int_0^{\tau} \sigma_j(x-\xi v', v')d\xi)d\tau\}dv' + q_i(x,v)$$
$$\le \frac{\bar{q}}{\delta} \sum_{j=1}^{N} \frac{1}{\underline{\sigma}_j} \int_{\Omega} \hat{\sigma}_{ij}(v')[1 - \exp(-\underline{\sigma}_j s(x,v'))] + \bar{q}$$
$$\le \frac{\bar{q}\hat{\rho}_i}{\delta} + \bar{q} = \bar{q}(\frac{\hat{\rho}_i}{\delta} + 1) \le \frac{\bar{q}}{\delta}.$$

The above inequalities implies that

$$\int_0^{s(x,v)} \exp\left(-\int_0^{\tau} \sigma_i(x-\xi v, v)d\xi\right)\left[\sum_{j=1}^{N} F_{ij}(\tilde{u}_j)(x-\tau v, v) + q_i(x-\tau v, v)\right]d\tau \le \tilde{u}_i(x,v)$$

in $\bar{\Omega} \times D$, for $i = 1, \ldots, N$. Therefore $\tilde{u} = (\tilde{u}_1, \ldots, \tilde{u}_N)$ is an upper solution of (4.5-6). By Theorem 4.5-1, there exists a (nonnegative) bounded solution of (4.5-6).

The following corollary is a more readily applicable variant of Theorem 4.5-2.

Corollary 4.5-1. *Let ℓ be the largest distance between any two points in $\bar{\Omega}$. If*

$$\bar{\rho}_i \overset{(def)}{=} \bar{\sigma}_i[1-\exp(-\underline{\sigma}^t \ell)]/\underline{\sigma}^t < 1 \tag{4.5-21}$$

for $i = 1, \ldots, N$. Then problem (4.5-6) has at least one nonnegative bounded solution.

Proof. There are positive constants $\delta_i < 1 - \bar{\rho}_i$, $i = 1, \ldots, N$. Let $\delta = \min\{\delta_i: 1 \le i \le N\}$; and define $\tilde{u}_i(x,v)$ by exactly the same formula as in (4.5-19) (with the definition of δ_i and δ changed here). From (4.5-17), we have

$$\begin{aligned}\sum_{j=1}^{N} F_{ij}(\tilde{u}_j)(x,v) + q_i(x,v) &\le \frac{\bar{q}}{\delta\underline{\sigma}^t}[1 - \exp(-\underline{\sigma}^t\ell)]\sum_{j=1}^{N}\int_\Omega \sigma_{ij}(x,v,v')dv' + \bar{q} \\ &\le \frac{\bar{q}}{\delta\underline{\sigma}^t}[1 - \exp(-\underline{\sigma}^t\ell)]\bar{\sigma}_i + \bar{q} \\ &= \frac{\bar{q}}{\delta}\bar{\rho}_i + \bar{q} \le \frac{\bar{q}}{\delta}\end{aligned}$$

since $\ell \ge s(x,v)$. The rest of the arguments are the same as those in Theorem 4.5-2.

With even stronger assumptions on σ_i and σ_{ij}, we have a unique solution as follows.

Theorem 4.5-3 (Subcritical). Let ℓ be the largest distance between any two points in $\bar{\Omega}$. Suppose that

$$\{N\bar{\sigma}^s \, [1 - \exp(-\underline{\sigma}^t \ell)]/\underline{\sigma}^t\} < 1. \tag{4.5-22}$$

Then problem (4.5-6) has a unique nonnegative bounded solution. (Note that $\bar{\sigma}^s$ involves only scattering cross-sections σ_{ij}, and $\underline{\sigma}^t$ involves only the total cross-sections σ_i).

The proof is similar to above, and will be omitted. For details, see [46].

4.6 Strongly Order-Preserving Dynamical Systems, Connecting Orbit and Stability

Fission reactor equations and mutualistic species interactions are typical of a large class of reaction-diffusion equations. Their solutions are likely to tend to equilibria rather than oscillatory. An important characteristic for such systems is that they are "strongly order-preserving". This characteristic also prevails in population biology when there are two competing subcommunities of species, while each community consists of species which are mutualistic among themselves (cf. [217]). Other examples include repressible cyclic gene systems, models for infecteous diseases etc. (cf. [210] for survey involving ordinary differtial equations).

For simplicity, we only consider the Neumann problem:

$$\frac{\partial u_i}{\partial t} = \sigma_i \Delta u_i + f_i(u_1, \ldots, u_n) \qquad \text{for } x \in \Omega, \ t>0,$$

$$(4.6\text{-}1)\qquad u_i(x,0) = \varphi_i^0(x) \qquad \text{for } x \in \Omega,$$

$$\frac{\partial u_i}{\partial \eta} = 0 \qquad \text{for } x \in \delta\Omega,$$

$i = 1,..,n$. Here, we assume that Ω is a bounded domain in R^m, $m>1$, and $\delta\Omega \in H^{2+\alpha}$. The functions f_i are assumed to have continuous first partial derivatives in R^m, and σ_i are positive constants. With further assumptions on f_i and φ_i^0, the theories in Chapter 1 and 2 lead to the natural mapping:

$$(4.6\text{-}2)\qquad \Phi(t)\colon \varphi^0 \to u(.,t;\varphi^0),$$

where $u(.,t;\varphi^0) = (u_1,....,u_n)$ is the solution of (4.6-1) with initial data $u_i(x,0) = \varphi_i^0$ where $\varphi^0 = (\varphi_1^0,...,\varphi_n^0)$. Extending the definition of the mapping $\Phi(t)$ by continuity, Mora [163] shows that the mapping can be defined appropriately for all φ^0 in $C(\bar{\Omega})x...xC(\bar{\Omega})$ and the mapping $t \to \Phi(t)\varphi^0$ is continuous from $[0,s(\varphi^0))$ into $C(\bar{\Omega})x...xC(\bar{\Omega})$ where

$$s(\varphi^0) = \sup.\{t:\varphi^0 \in \mathrm{Dom}(\Phi(t))\}.$$

Here, Dom represents domain. To be specific, we define a local semiflow Φ on a Banach space X as follows. $\Phi = \Phi(t,\varphi)$ is a continuous mapping from an open subset $\mathrm{Dom}(\Phi) \subset [0,\infty)xX$ into X with the properties:

(i) $\mathrm{Dom}(\Phi(t)) \overset{(def)}{\equiv} \{\varphi \in X\colon (t,\varphi) \in \mathrm{Dom}(\Phi)\}$ satisfies $\mathrm{Dom}(\Phi(0)) = X$ and $\mathrm{Dom}(\Phi(t_2)) \subseteq \mathrm{Dom}(\Phi(t_1))$ if $0 \le t_1 < t_2$.

(ii) For any $t,\tau \ge 0$, we have $\Phi(t,\varphi) \in \mathrm{Dom}(\Phi(\tau)) \Longleftrightarrow \varphi \in \mathrm{Dom}(\Phi(t+\tau))$.

(iii) $\Phi(0,\varphi) = \varphi$ for all $\varphi \in X$.

(iv) $\Phi(t,\Phi(\tau,\varphi)) = \Phi(t+\tau,\varphi)$ for any $t,\tau \ge 0$, $\varphi \in \mathrm{Dom}(\Phi(t+\tau))$.

Comparing with symbols before, we are identifying $\Phi(t,\varphi)$ with $\Phi(t)\varphi$. From the fact that $\mathrm{Dom}(\Phi)$ is open, we deduce that for any $t>0$, the set

$\{\varphi \in X: \ t < s(\varphi)\}$ is open in X, i.e. $s(\varphi)$ is lower semicontinuous in X.

Our theories in this section follows essentially the development in Matano [155]. Analogous results can also be found in Hirsch [105]. If one considers the problem (4.6-1) under the additional assumption

$$(4.6\text{-}3) \qquad \frac{\partial f_i}{\partial u_j} > 0 \quad \text{for all } (u_1,\dots,u_n),\ i \neq j,\ i,j = 1,\dots,n,$$

one can deduce from Theorem 1.2-5 and Remark 1.2-3 that if $\psi_i^0(x) \geq \phi_i^0(x)$ for $i = 1,\dots,n$, $x \in \bar{\Omega}$, then $u_i(x,t;\psi^0) \geq u_i(x,t;\phi^0)$ for $i = 1,\dots,n$, $x \in \bar{\Omega}$. On the other hand, if one considers problem (4.6-1) with n = 2 and

$$(4.6\text{-}4) \qquad \frac{\partial f_1}{\partial u_2} < 0 \quad \text{and} \quad \frac{\partial f_2}{\partial u_1} < 0 \qquad \text{for all } u_1, u_2 ,$$

one can deduce from Theorem 1.2-7 that if $\psi_1^0(x) \geq \phi_1^0(x)$ and $\psi_2^0(x) \leq \phi_2^0(x)$ in $\bar{\Omega}$, then we have $u_1(x,t;\psi^0) \geq u_1(x,t;\phi^0)$ and $u_2(x,t;\psi^0) \leq u_2(x,t;\phi^0)$ for $x \in \bar{\Omega}$ and $t > 0$ where $\Phi(t)\psi^0$ and $\Phi(t)\phi^0$ are both defined. From these two examples, we see that it is convenient to define a partial order $\leq$ in a Banach spacce X with respect to a closed convex cone $P \subset X$ as follows: we write

$$(4.6\text{-}5) \qquad u \leq v \iff v - u \in P$$

for $u, v \in X$. Moreover, we say

$$(4.6\text{-}6) \qquad u < v \quad \text{if } u \leq v \text{ and } u \neq v \text{ in } X.$$

(Recall that a subset $\hat{P}$ of a Banach space X is called a cone if it has all the properties: (i) if $a \in \hat{P}$, $b \in \hat{P}$ then $a + b \in \hat{P}$, (ii) if $\alpha > 0$, $a \in \hat{P}$ then $\alpha a \in \hat{P}$, (iii) if $a \in \hat{P}$, $-a \in \hat{P}$ then $a = 0$.) Note that if we let $\tilde{P} = \{(u_1, u_2) \in C(\bar{\Omega})xC(\bar{\Omega}) : u_1(x) \geq 0,\ u_2(x) \leq 0 \text{ in } \bar{\Omega}\}$, and let $\leq$ be the partial order with respect to $\tilde{P}$ in $C(\bar{\Omega})xC(\bar{\Omega})$, we can more readily rephrase the properties for $\Phi(t)$ in (4.6-2) under the conditions (4.6-4). That is, we can write: if $\phi^0 \leq \psi^0$, then we have $\Phi(t)\phi^0 \leq \Phi(t)\psi^0$. Moreover, using the results in [156], the maximum principle Theorem 1.1-6 and the continuity of Φ, we can further conclude that if ϕ^0, ψ^0 are in $C(\bar{\Omega})xC(\bar{\Omega})$ with $\phi^0 < \psi^0$ then for any $t > 0$ (for which both $\Phi(t)\phi^0$ and $\Phi(t)\psi^0$ are defined), there exist open sets V and W respectively containing ϕ^0 and ψ^0, so that

$$\Phi(t)\tilde{\phi} < \Phi(t)\tilde{\psi}$$

whenever $\tilde{\phi} \in V$ and $\tilde{\psi} \in W$. This motivates the following definition.

Definition 4.6-1 _Let X be a Banach space with norm_ $\|\ \|$, _and partial order_ $\leq$. _A local semiflow_ Φ _on X is said to be order-preserving if_ $\phi \leq \psi$ _implies_ $\Phi(t)\phi \leq \Phi(t)\psi$ _for_ $0 \leq t < \hat{s}$, _where_ $\hat{s} = \min.\{s(\phi),\ s(\psi)\}$. _It is called strongly order-preserving if for any_ $\phi < \psi$ _and any_ $t \in (0, \hat{s})$, _there exists a_ $\delta > 0$ _such that_

$$\Phi(t)\tilde{\phi} < \Phi(t)\tilde{\psi}$$

is satisfied, whenever $\|\tilde{\phi} - \phi\| < \delta$, $\|\tilde{\psi} - \psi\| < \delta$.

We will also need the following definitions.

Definition 4.6-2 _A set_ $B \subset X$ _is said to be order-bounded from below (or from above) if there exists a_ $w \in X$ _such that_

$$w \leq \psi \quad (\text{ or } \psi \leq w\ )$$

is true for all $\psi \in B$. B _is called order-bounded if it is both order-_

bounded from below and above. B *is norm-bounded if* $\sup_{\psi\in B} \|\psi\| < \infty$.

Definition 4.6-3 *A point* $z \in X$ *is an equilibrium point for a local semiflow* Φ *if*

$$\Phi(t)z = z$$

for all $t \in [0,\infty)$. *A point* $z \in X$ *is called a sub-equilibrium (or super-equilibrium) if*

$$\Phi(t)z \ge z \qquad (\text{ or } \Phi(t)z \le z\ )$$

is true for all $t \in [0,s(z))$. *A sub-equilibrium (or super-equilibrium) is a strict sub-equilibrium (or strict super-equilibrium) if it is not a equilibrium.*

Note that for problem (4.6-1) under conditions (4.6-3), if $\varphi_i^0(x) \in H^{2+\alpha}(\bar{\Omega})$, $0<\alpha<1$, $i = 1,\dots,n$, are "lower" solutions in the sense:

$$\begin{aligned} &\sigma_i\Delta\varphi_i^0 + f_i(\varphi_1^0,\dots,\varphi_n^0) \ge 0 && \text{in } \Omega, \\ & && i = 1,\dots,n, \\ &\frac{\partial\varphi_i^0}{\partial\eta} \le 0 && \text{on } \delta\Omega, \end{aligned}$$

then we can deduce from Theorem 1.2-5 and Remark 1.2-3 that $\Phi(t)\varphi^0 \ge \varphi^0$. Here $\Phi(t)$ is the local semiflow defined by (4.6-2) on $X = C(\bar{\Omega})x\dots xC(\bar{\Omega})$, $\varphi^0 = (\varphi_1^0,\dots,\varphi_n^0)$, and the partial order $\le$ on X is with respect to the cone $P = \{(u_1,\dots,u_n) \in C(\bar{\Omega})x\dots xC(\bar{\Omega}) : u_i(x) \ge 0 \text{ in } \bar{\Omega},\ i=1,\dots,n\}$. That is, the "lower" solution φ^0 is a sub-equilibrium in the sense of Definition 4.6-3.

For convenience, we now label a few hypotheses, which will be selectively or entirely assumed in the main Theorems in this section.

[H1] Any order-bounded set in X is normed-bounded.

[H2] Φ is strongly order-preserving.

[H3] For any norm-bounded set $B \subset X$, we have

$$s_B \overset{(def)}{\equiv} \inf_{\varphi \in B} s(\varphi) > 0,$$

and the set $\Phi(t)B \overset{(def)}{\equiv} \{\Phi(t)\varphi : \varphi \in B\}$ is relatively compact for each $t \in (0, s_B)$. (That is, any sequence of points in $\Phi(t)B$ has a convergent subsequence).

[H4] Any pair of points φ, ψ in X have a least upper bound, denoted by $\varphi \vee \psi$. Moreover we have the property

$$(\lim_{k\to\infty} \varphi_k) \vee (\lim_{k\to\infty} \psi_k) = \lim_{k\to\infty} (\varphi_k \vee \psi_k).$$

For discussion as to situations when [H3] is satisfied, the reader is referred to [155],[156]. A key lemma for the proof of Theorem 4.6-2, which is essential for proving the useful Theorems 4.6-3 and 4.6-4, is the following.

Lemma 4.6-1 *Let X be a Banach space with partial order $\leq$, and Φ be a local semiflow on X satisfying hypotheses [H2] and [H3]. Suppose that $z \in X$ satisfies*

$$(4.6\text{-}7) \qquad \Phi(t_0)z > z$$

for some $t_0 > 0$; and the positive half orbit for the point z, i.e.

$$(4.6\text{-}8) \qquad Orb^+(z) = \{\Phi(t)z : 0 \leq t < s(z)\},$$

is norm-bounded. Then

$$\lim_{t\to +\infty} \Phi(t)z = v,$$

<u>where</u> v <u>is</u> <u>an</u> <u>equilibrium</u> <u>point</u> <u>in</u> X <u>for</u> Φ. <u>Furthermore,</u> <u>we</u> <u>have</u> $\Phi(t)z < v$ <u>for</u> <u>all</u> $t\geq 0$.

<u>Proof</u>. For $k = 0,1,2,\ldots$, let

$$z_k = \Phi(kt_0)z.$$

By hypothesis [H2], the strongly order-preserving property, we have

(4.6-9) $$z = z_0 < z_1 < z_2 < \ldots..$$

Note that the first part of [H3] applied to the norm-bounded set $Orb^+(z)$ implies that $s(z) = \infty$. Further, the hypothesis [H3] also implies that the set

$$\{z_k\}_{k\geq 1} = \{\Phi(t_0)z_k\}_{k\geq 0}$$

is relatively compact. Hence, together with the monotone property (4.6-9), we conclude that $\lim_{k\to\infty} z_k = v$ for some $v \in X$. Applying hypothesis [H2] to the pair $z_0 < z_1$ and the fact that $\lim_{t\to 0^+}\Phi(t)z_0 = z_0$, $\lim_{t\to 0^+}\Phi(t)z_1 = z_1$, we deduce that

(4.6-10) $$\Phi(t_0+t)z_0 = \Phi(t_0)\Phi(t)z_0 < \Phi(t_0)z_1$$

(4.6-11) $$\Phi(t_0)z_0 < \Phi(t_0)\Phi(t)z_1 = \Phi(t_0+t)z_1$$

for any $t \in [0,t_1]$, if $t_1>0$ is sufficiently small. Applying the operations $\Phi((k-1)t_0)$ and $\Phi(kt_0)$ to both sides of the inequalities (4.6-10) and (4.6-11) respectively, we obtain from [H2] again that

(4.6-12) $\Phi(t)z_k \leq z_{k+1} \leq \Phi(t)z_{k+2}$

for all $t \in [0,t_1]$, $k = 2,3,\ldots$. Letting $k \to \infty$, we deduce

(4.6-13) $\Phi(t)v = v$

for $t \in [0,t_1]$. The property (iv) of the semiflow Φ thus implies that (4.6-13) is true for $t \geq 0$, i.e. v is an equilibrium point. Applying (4.6-12) successively and using the monotone convergence of z_k to v, we can deduce that

$$\lim_{t\to\infty} \Phi(t)z = v .$$

Similarly, we deduce that $\Phi(t)z < v$.

Before we state the main theorems in this section, we will need a few more terminologies.

<u>Definition</u> 4.6-4 (i) <u>An equilibrium point</u> $z \in X$ <u>for a local semiflow</u> Φ <u>is said to be stable if for any</u> $\varepsilon > 0$, <u>there exists a</u> $\delta > 0$ <u>such that for all</u> $t \geq 0$, $\phi \in X$, <u>we have</u>:

$$\|\Phi(t)\phi - z\| < \varepsilon, \text{ provided that } \|\phi - z\| < \delta.$$

(ii) <u>An equilibrium point is unstable if it is not stable.</u>

(iii) <u>An equilibrium point</u> $z \in X$ <u>is asymptotically stable, if it is stable, and in addition, there exists an open neighborhood</u> $W_0 \subset X$ <u>of</u> z <u>such that</u> $\lim_{t\to+\infty}\Phi(t)w = z$ <u>for all</u> $w \in W_0$.

<u>Definition</u> 4.6-5 (i) <u>A closed set</u> Y <u>in a Banach space</u> X <u>is called</u>

positively invariant for the local semiflow Φ on X if

$$\Phi(t)Y \subset Y \quad \text{for all } t \geq 0.$$

(ii) A positively invariant closed set Y is said to be stable as a set if for any open neighborhood V of Y there exists an open neighborhood W of Y such that

$$\Phi(t)W \subset V \quad \text{for all } t \geq 0.$$

(iii) A positively invariant closed set Y is said to be asymptotically stable as a set if it is stable; and in addition, there exists an open neighborhood W_0 of Y such that if V is any neighborhood of Y, and w is any point in W_0, then

$$\Phi(t)w \in V \quad \text{for all } t \geq T.$$

Here $T > 0$ depends on w and V.

Definition 4.6-6 (i) Let Φ be order-preserving. An equilibrium point z is said to be stable from above if it is stable in the set

$$X^+(z) \overset{(def)}{=} \{\phi \in X, \ \phi \geq z\}.$$

That is, in the definition 4.6-4 (i) above, X is replaced by $X^+(z)$. If z has the further property that there exists an open neighborhood $W_0 \subset X$ of z such that $\lim_{t\to\infty} \Phi(t)w = z$ for all $w \in W_0 \cap X^+(z)$, we say z is asymptotically stable from above.

(ii) Stable from below and asymptotically stable from below for an equilibrium point z in a order preserving Φ is defined as in part (i) *with* $X(z)^+$ *replaced by*

$$X^-(z) \overset{(def)}{=} \{\phi \in X:\ \phi \le z\}.$$

Definition 4.6-7 *Let* $Orb^+(z)$ *be defined for the point* $z \in X$ *as in* (4.6-8). *The* ω-limit *set of* z *is defined as*

$$\omega(z) = \bigcap_{t>0} \overline{Orb^+(\Phi(t)z)}.$$

Here, the overbar denotes closure.

The following theorem considers an equilibrium which is not stable from above. It will be used to prove the existence of connecting orbits between two equilibria under appropriate conditions in Theorem 4.6-3. In the remaining part of this section, we will always assume X to be a Banach space with partial order $\le$ and Φ is a local semiflow on X.

Theorem 4.6-1. *Let* X *and* Φ *satisfy hypotheses* [H2] *and* [H3]. *Assume that* v *is an equilibrium point which is not stable from above. Then there exists a function* $w : (-\infty, 0] \to X$ *with the following properties*:

(i) $w(t) > v$ *for* $t \in (-\infty, 0]$;

(ii) $\Phi(t)w(\tau) = w(t+\tau)$ *for any* $t \ge 0$, $\tau \le 0$ *with* $t+\tau \le 0$;

(iii) $\lim_{t\to-\infty} w(t) = v$.

If, in addition, [H4] *is satisfied, then there exists a strictly monotone increasing function* w(t) *with properties* (i) *to* (iii).

Proof. For the proof of the first part of the theorem, we need the following lemma, which is a ready consequence of the strongly order-preserving hypothesis [H2].

Lemma 4.6-2. *Let* X *and* Φ *satisfy hypotheses* [H2] *and* [H3], *and* B *and*

s_B be as described in [H3]. Then for any $t_0>0$, $t_1>0$, $\varepsilon>0$ with $t_0+t_1 < s_B$, there exists $\delta>0$ such that : $y, z \in \Phi(t_0)B$ with

$$y \le z, \quad \|y - z\| \ge \varepsilon,$$

will imply that

$$\Phi(t_1)\tilde{y} < \Phi(t_1)\tilde{z}$$

whenever $\tilde{y}, \tilde{z} \in X$ satisfy $\|\tilde{y} - y\| < \delta$, $\|\tilde{z} - z\| < \delta$.

By the instability from above property of v, there exists a $\varepsilon_0 > 0$ such that for any $\delta > 0$, there exist a point $\phi_\delta \in X$ and corresponding minimal $T_\delta > 0$ so that

$$\phi_\delta > v, \quad \|\phi_\delta - v\| < \delta,$$
$$\|\Phi(T_\delta)\phi_\delta - v\| \ge \varepsilon_0.$$

The minimal property of T_δ means that

$$\text{(4.6-14)} \quad \begin{aligned} &\|\Phi(t)\phi_\delta - v\| < \varepsilon_0 \quad \text{for all } t \in [0, T_\delta), \text{ and} \\ &\|\Phi(T_\delta)\phi_\delta - v\| = \varepsilon_0 \quad \text{for any } \delta \in (0, \varepsilon_0). \end{aligned}$$

For application of Lemma 4.6-2, we let

$$B \overset{(\text{def})}{=} \{\psi \in X : \psi \ge v, \ \|\psi - v\| \le \varepsilon_0\}.$$

(Hence, by [H3], $s_B = \inf_{\psi \in B} s(\psi) > 0$). We now proceed to construct the function w(t). Let t^* be any positive number with $t^* < s_B$. From (4.6-14) and Lemma 4.6-2, there exists a positive number $\delta^* \in (0, \varepsilon_0)$, such that we have

$$\Phi(t^*)\psi < \Phi(T_\delta + t^*)\phi_\delta \tag{4.6-15}$$

for any $\delta \in (0, \varepsilon_0)$ and any $\psi \in X$ satisfying $\|\psi - v\| \le \delta^*$. Let

$$t_\delta = \min \{t \in [0, s(\phi_\delta)) : \|\Phi(t)\phi_\delta - v\| \ge \delta^*\}.$$

One readily sees that $t_\delta \to \infty$ as $\delta \to 0^+$. Choose a sequence $\delta_1 > \delta_2 > \ldots \to 0$ with $t_{\delta_m} \ge m$, $m = 1, 2, \ldots$ and define

$$\phi_{mk} = \Phi(t_{\delta_m} - k)\phi_{\delta_m}$$

for $m = 1, 2, \ldots$ and $k = 0, 1, \ldots, m-1$. From [H3], we may assume without loss of generality that

$$\lim_{m\to\infty} \phi_{mk} \overset{(\text{def})}{=} w_{-k}$$

does exist for each $k = 0, 1, \ldots$. Thus we can define $w : (-\infty, 0] \to X$ by

$$w(-k+s) = \Phi(s)w_{-k} \qquad \text{for } s \in [0,1],\ k = 1, 2, \ldots. \tag{4.6-16}$$

Note that $\Phi(1)w_{-k} = w_{-k+1}$ for $k = 1, 2, \ldots$, and

$$w(t) > v, \quad \|w(t) - v\| \le \delta^*, \quad \|w(0) - v\| = \delta^* \tag{4.6-17}$$

for all $t \le 0$. Moreover, w satisfies property (ii) in the statement of the Theorem.

We next proceed to show w satisfies property (iii). First, from (4.6-15) and (4.6-17), we obtain

(4.6-18) $\quad w(T_\delta-\tau+t^*) < \Phi(T_\delta+t^*)\phi_\delta$

for any $\delta \in (0,\varepsilon_0)$, $\tau \geq T_\delta$. On the other hand, suppose that there exist $\varepsilon_1 > 0$ and a sequence

$$0< t_1< t_2 <\ldots \to +\infty \quad \text{such that} \quad \|w(-t_m) - v\| \geq \varepsilon_1, \quad m = 1,2,\ldots$$

(i.e. we assume that $w(t)$ does not tend to v as $t \to -\infty$). Then from Lemma 4.6-2, we have

(4.6-19) $\quad w(-t_m+t^*) > \Phi(t^*)\phi_\delta$

for $\delta > 0$ sufficiently small and all $m = 1.2,\ldots$. Let such a small $\delta > 0$ be chosen, then choose m sufficiently large so that $t_m \geq T_\delta + t^*$. The strongly order-preserving property of Φ and (4.6-19) gives

(4.6-20) $\quad w(T_\delta-t_m+t^*) > \Phi(T_\delta+t^*)\phi_\delta$.

Comparing (4.6-20) with (4.6-18), with τ chosen as $\tau = t_m$, we see that we have arrived at a contradiction. Consequently, we must have $w(t) \to v$ as $t \to -\infty$.

It remains to prove the final part of the theorem. For its proof, we need the following lemma.

<u>Lemma</u> 4.6-3. <u>Let</u> X <u>satisfies</u> <u>hypothesis</u> [H4]. Suppose that $K \subset X$ <u>is a compact set, then it has a least upper bound and a greastest lower bound. That is, there exists</u> y^-, $y^+ \in X$ <u>such that</u>

$$y^- \leq z \leq y^+ \qquad \text{for all } z \in K;$$

and moreover if $u \in X$ *with* $u \le z$ (*or* $z \le u$) *for all* $z \in K$, *then it satisfies* $u \le y^-$ (*or* $y^+ \le u$).

This Lemma can be proved by approximating K with a finite ε-net and using hypotheses [H4]. We will denote the least upper bound and greastest lower bound of K by l.u.b.(K) and g.l.b.(K) respectively. Let [H4] be satisfied as well as [H2] and [H3], w(t) be as defined above, and $\tilde{\phi}_m$ = l.u.b.$\{w(t): -\infty < t \le -m\}$. One can readily deduce that $\tilde{\phi}_m$ is a strict sub-equilibrium and $\Phi(t)\tilde{\phi}_m$ is strictly monotone increasing in t . One can follow the first part of the proof, with the role of the sequence $\{\phi_{\delta_m}\}$ replaced by $\{\tilde{\phi}_m\}$, to construct a new w(t). This new w(t) will satisfy properties (i) to (iii) and is strictly monotone increasing in t. This completes the proof of Theorem 4.6-1.

Theorem 4.6-1 considers the behavior near an equilibrium which is *unstable* from above. The next Theorem considers behavior near an equilibrium which is *stable* from above. Theorem 4.6-2 is an important theorem in this section, and is used in proving both Theorem 4.6-3 and Theorem 4.6-4..

Theorem 4.6-2. *Let* X *and* Φ *satisfy hypotheses* [H2] *and* [H3], *and* v *is an equilibrium point. Let*

$$S^+(v) \overset{(def)}{=} \{\phi \in S: \phi > v\},$$

where S *is the set of all equilibrium points of* Φ *in* X. *Assume that* v *is separated from* $S^+(v)$, i.e.

$$\inf_{\phi\in S^+(v)} \|\phi - v\| > 0 ,$$

if the equilibrium point v is stable from above, then it is asymptotically stable (That is, either v is asymptotically stable from above or it is not stable from above).

For the proof of Theorem 4.6-2, we need two additional lemmas.

Lemma 4.6-4. *Let X and Φ satisfy hypotheses [H2] and [H3], and $y < z$ in X with both $Orb^+(y)$ and $Orb^+(z)$ being norm-bounded. Suppose that the set $\omega(y)\cap\omega(z)$ is nonempty, then it only contains equilibrium point(s).*

Proof. By assumption, there are sequences $0< t_1< t_2 \to \infty$ and $0< \bar{t}_1< \bar{t}_2 \to \infty$ such that

$$\lim_{m\to\infty} \Phi(t_m)y = \lim_{m\to\infty} \Phi(\bar{t}_m)z = w \in \omega(y)\cap\omega(z) \tag{4.6-21}$$

We now show that we also have

$$\lim_{m\to\infty} \Phi(t_m)z = w. \tag{4.6-22}$$

Suppose (4.6-22) is not true, then there exists an $\varepsilon > 0$ and an increasing subsequence $\{t_{m_k}\}$ of $\{t_m\}$ so that

$$\|\Phi(t_{m_k})y - \Phi(t_{m_k})z\| \geq \varepsilon \tag{4.6-23}$$

for $k = 1,2,\ldots$. From (4.6.21), (4.6-23) and Lemma 4.6-2, we deduce that

$$\Phi(\bar{t}_{m_k} + t_0)z = \Phi(t_{m_k} + t_0)y < \Phi(t_{m_k} + t_0)z$$

for any $t_0 > 0$, and m_k large enough. Thus, we have

$$\Phi(\hat{t})z > \Phi(\tilde{t})z$$

for some $\hat{t} > \tilde{t}$. We can now use Lemma 4.6-1 to assert that $\Phi(t)z \to w$ as $t \to +\infty$, contradicting (4.6-23). Consequently (4.6-22) must be valid.

Finally, we apply hypothesis [H2] to the pair $y < z$, and use the continuity of Φ to obtain

$$\Phi(t_m+s)y < \Phi(t_m)z, \quad \Phi(t_m)y < \Phi(t_m+s)z \tag{4.6-24}$$

for all $s \in [0,s_0]$, some $s_0 > 0$. Letting $m \to \infty$ in (4.6-24), we find $\Phi(t)w = w$ for all $s \in [0,s_0]$. This shows that w is an equilibrium, and completes the proof of the Lemma.

Lemma 4.6-5. _Let_ X _and_ Φ _satisfy hypotheses_ [H2] _and_ [H3]. _Suppose that_ Y _is a positively invariant (i.e._ $\Phi(t)y \in Y$, _for all_ $t \geq 0$ _if_ $y \in Y$) _norm-bounded closed set in_ X. _Then, the set_

$$G \overset{(def)}{=} \bigcup_{y \in Y} \omega(y)$$

is a non-empty subset of Y. _Further, it has the following properties:_

(i) G _has a maximal and a minimal point;_

(ii) _if_ y _is maximal (or minimal) in_ G, _then each point of_ $\omega(y)$ _is maximal (or minimal respectively) in_ G.

Proof (outline) The norm-boundedness of Y and hypothesis [H3] lead to the assertion that G is nonempty. The existence of a maximal and minimal point is proved by using Zorn's Lemma. In order to apply the Zorn's Lemma, we must show that any totally ordered subset of G has an upper bound in G. This property of G is shown by using the definition of ω-limit set and the strongly order-preserving property [H2]. Part (ii) of this lemma is proved by using [H2] and Lemma 4.6-4. For details, the reader is referred to [155].

Proof of Theorem 4.6-2. By the separation assumption, there exists an $\varepsilon_0 > 0$ such that any $\bar{v} \in S^+(v)$ must satisfy $\|\bar{v} - v\| \geq \varepsilon_0$. From Lemma 4.6-2, we can find a positive $\varepsilon_1 < \varepsilon_0$ and $t_0 > 0$ such that any $\phi \in X$ with

$$\phi > v, \quad \|\phi - v\| \leq \varepsilon_1$$

will satisfy

$$\Phi(t_0) < \bar{v} \tag{4.6-25}$$

for any $\bar{v} \in S^+(v)$.

Since v is stable from above, there exists $\delta > 0$ so that any $\theta \in X$ with

$$\theta > v, \quad \|\theta - v\| < \delta \tag{4.6-26}$$

will satisfy

$$\|\Phi(t)\theta - v\| \leq \varepsilon_1 \tag{4.6-27}$$

for all $t \geq 0$. Now, choose a fixed particular θ satisfying (4.6-26). Inequality (4.6-27) implies that $\bar{v}$ is not in $\omega(\theta)$. On the other hand, hypothesis [H3] implies that $\omega(\theta)$ is nonempty. Suppose that $\omega(\theta)$ contains v, then Lemma 4.6-4 and the separation assumption imply that $\Phi(t)\theta \to v$ as $t \to +\infty$. This leads to the conclusion that v is asymptotically stable from above.

The last conclusion follows from the fact that : under hypotheses [H2] and [H3], an equilibrium point which is stable from above will be asymptotically stable from above if there exists a $\theta \in X$ with $\theta > v$ and $\Phi(t)\theta \to v$ as $t \to +\infty$.

It remains to consider the case when v is not contained in $\omega(\theta)$. In this case, we have $\phi > v$ for any $\phi \in \omega(\theta)$. Let

$$B = \{w \in X : w \geq v, \ \|\Phi(t)w - v\| \leq \varepsilon_1 \text{ for any } t \geq 0\}, \text{ and}$$
$$Y = \{w \in B : w \leq \Phi(s)\phi \text{ for some } \phi \in \omega(\theta) \text{ and any } s \in [0,1]\}.$$

One readily sees that Y is norm-bounded and closed. The fact that both B and $\omega(\theta)$ are positively invariant implies that Y is positively invariant. Moreover, for some $t_0 \geq 0$ we obtain from Lemma 4.6-2 that $\Phi(t_0)w \in Y$ for any $w \in B$ sufficiently close to v. Hence, the set $Y \backslash \{v\}$ is not empty. Define

$$G = \bigcup_{w \in Y} \omega(w).$$

The set G has a maximal element, say $\bar{z}$, by Lemma 4.6-5. Since v is not contained in $\omega(\theta)$, inequalities (4.6-25) and (4.6-27) implies that $\omega(\theta)$ contains no equilibrium point. Thus, the relation $\bar{z} \leq \Phi(s)\phi$ $(0 \leq s \leq 1)$ leads to the fact that $\bar{z} < \tilde{\phi}$ for some $\tilde{\phi} \in \omega(\theta)$. Let $\bar{w} \in Y$ be such that $\bar{z} \in \omega(\bar{w})$. Applying hypothesis [H2] to the pair $\tilde{z} < \tilde{\phi}$, we find that there exist positive numbers t_1, t_2 and $\tilde{\delta}$ such that

$$\Phi(t_1)\bar{w} < \Phi(t_2)\bar{\theta} \tag{4.6-28}$$

is valid for any $\bar{\theta} \in X$ satisfying

$$\|\bar{\theta} - \theta\| < \tilde{\delta}. \tag{4.6-29}$$

Letting $\bar{\theta} = v + \alpha(\theta - v)$ where $\alpha \in (0,1)$ is chosen sufficiently close to 1 so that (4.6-29) is satisfied, we obtain

$$\bar{\theta} \in B, \quad \bar{\theta} < \theta.$$

Furthermore, from (4.6-28), we obtain a $\bar{\phi} \in \omega(\bar{\theta})$ so that

$$\bar{z} \le \bar{\phi}. \tag{4.6-30}$$

On the other hand, since $\omega(\theta)$ does not contain any equilibrium, we deduce from Lemma 4.6-4 that $\omega(\bar{\theta}) \cap \omega(\theta)$ is an empty set. From this, and the relation $\bar{\theta} < \theta$, we use Lemma 4.6-2 (successively if necessary) and the continuity of $\Phi(t)\theta$ in t to obtain the validty of

$$\Phi(t)\bar{\theta} < \Phi(t+s)\theta$$

for all large $t \ge 0$ and any $s \in [0,1]$. Since for all large t, $\Phi(t)\theta$ is within a neighborhood of $\omega(\theta)$, we can apply Lemma 4.6-2 to see that

$$\Phi(t)\bar{\theta} < \Phi(s)\phi \qquad (0 \le s \le 1)$$

for some $\phi \in \omega(\theta)$, t sufficiently large. That is, $\Phi(t)\bar{\theta} \in Y$ for some $t \ge 0$ sufficiently large; and consequently we must have $\bar{\phi}$ belonging to the set G. However, the maximality of $\bar{z}$ and (4.6-30) imply that $\bar{z} = \bar{\phi}$. This observation, together with (4.6-28) and Lemma 4.6-4 lead to the conclusion that $\bar{\phi}$ is an equilibrium point. Consequently, we have $\bar{\phi} = v$. This implies that we have the property:

$$\Phi(t)\bar{\theta} \to v \quad \text{as} \quad t \to +\infty.$$

From this property, we conclude that v is asymptotically stable from above, as is indicated in the first case.

The following two Theorems 4.6-3 and 4.6-4 are useful important consequences of Theorems 4.6-1 and 4.6-2.

Theorem 4.6-3. *Let* X *and* Φ *satisfy hypotheses* [H1] *to* [H4]. *Suppose that there exist two equilibrium points* v_1, v_2 *with* $v_1 < v_2$ *and the property that: there is no equilibrium point* v *satisfying* $v_1 < v < v_2$. *Then there exists a strictly montone entire orbit connecting* v_1 *and* v_2. *That is there is a function* $w : (-\infty,\infty) \to X$ *such that:*

(i) $\Phi(t)w(\tau) = w(t+\tau)$ *for any real numbers* t,τ;

(ii) $\lim_{t\to-\infty} w(t) = v_1$, $\lim_{t\to+\infty} w(t) = v_2$ (*or* $\lim_{t\to-\infty} w(t) = v_2$, $\lim_{t\to+\infty} w(t) = v_1$);

(iii) $w(t_1) < w(t_2)$ *if* $t_1 < t_2$. (*or* $w(t_1) > w(t_2)$ *if* $t_1 < t_2$).

Theorem 4.6-4. *Let* X *and* Φ *satisfy hypotheses* [H2] *and* [H3] *and* Y *is a positively invariant closed subset of* X. *Suppose that* Y *is norm-bounded and asymptotically stable as a set. Then* Y *contains a stable equilibrium.*

In order to prove Theorem 4.6-3, we need three further lemmas.

Lemma 4.6-6. *Let* X *and* Φ *satisfy hypotheses* [H1] *to* [H4], *and* $v_1 < v_2$ *are equilibrium points. Suppose that* v_1 *is stable from above and* v_2 *is stable from below. Then there exists at least one equilibrium point* v *satisfying* $v_1 < v < v_2$.

Lemma 4.6-7. *Let* X *and* Φ *satisfy hypotheses* [H2] *to* [H4]. *Suppose that* v *is asymptotically stable from above, then there exists a strict super-equilibriun* w *such that*

$$\lim_{t\to+\infty} \Phi(t)w = v.$$

Lemma 4.6-8. *Let* Y *be a retract of a Banach space, and* $g : Y \to Y$ *is a continuous map with the closure of* g(Y) *compact. Suppose that* Y_1 *and* Y_2 *are disjoint retracts of* Y *with* $g(Y_k) \subset Y_k$, $k = 1,2$. *Furthermore, if there are* U_k, $k = 1,2$ *nonempty open subsets of* Y *such that* $U_k \subset Y_k$, $k = 1,2$

so that g *has no fixed points in* $Y_k \backslash U_k$, $k = 1,2$. *Then the map* g *has at least three distinct fixed points* z_1, z_2 *and* z, *with* $z_k \in Y_k$. $k = 1,2$ *and* $z \in Y \backslash (Y_1 \cup Y_2)$.

Note that a nonempty subset A of a Banach space E is called a retract of E if there exists a continuous map $r : E \to A$ such that the restriction of r to A is the identity map. Lemma 4.6-8 is proved by considering the fixed point indices. Its proof can be found in Lemma 14.1 of Amann [8].

Proof of Lemma 4.6-7. Let $\psi > v$ be such that

$$\lim_{t \to \infty} \Phi(t)\psi = v. \tag{4.6-31}$$

Applying hypothesis [H2] to the pair $\psi > v$, and using (4.6-31), one readily obtain numbers $t_1 > t_0$ such that $\Phi(t_0)\psi > \Phi(t_1)\psi$. Letting $w =$ g.l.b.$\{\Phi(t)\psi : t_0 \le t \le t_1\}$ (cf. Lemma 4.6-3), then we can deduce that w is a strict super-equilibrium such that $\lim_{t \to +\infty} \Phi(t)w = v$.

Proof of Lemma 4.6-6 (Outline) In view of Theorem 4.6-2, if there is no equilibrium v, $v_1 < v < v_2$, then we must have both v_1 asymptotically stable from above and v_2 asymptotically stable from below. Let

$$\hat{Y} = \{\psi \in X: v_1 \le \psi \le v_2\} ,$$

and Y be the closure of $\Phi(t_0)\hat{Y}$ where t_0 is some fixed positive number. The set Y is positively invariant and compact. Let $t^* > 0$ be arbitrary, define a map $g : Y \to Y$ by

$$g(\psi) = \Phi(t^*)\psi.$$

From Lemma 4.6-7, there exist strict sub-equilibrium and super-equilibrium $\tilde{w}$ and $\bar{w}$ respectively with $v_1 < \bar{w} < \tilde{w} < v_2$. Let

$$Y_1 = \{\phi \in Y:\ g(\phi) \le \bar{w}\ \},\quad Y_2 = \{\phi \in Y:\ g(\phi) \ge \tilde{w}\ \},\ \text{and}$$

$$U_k = \{\phi \in Y:\ \|\phi - v_k\| < \delta\ \},\ k = 1,2,$$

where $\delta > 0$ is sufficiently small so that $U_k \subset Y_k$. It is clear that $g(\bar{w}) < \bar{w}$, $g(\tilde{w}) > \tilde{w}$, $g(Y_k) \subset Y_k$, and v_k is the only fixed point of g in Y_k, $k = 1,2$. By applying Lemma 4.6-8, we deduce that there exists at least one $\phi^* \in Y\setminus(Y_1 \cup Y_2)$ satisfying

$$\Phi(t^*)\phi^* = \phi^*.$$

Replacing t^* by $t^*/2^m$, $m = 1,2\ldots$ etc,. and considering the derivative of $\Phi(t)v$ with respect to t, we eventually deduce through an integral representation (cf. [103]) that there exist a v, ${}_1v < v <_2 v$, so that

$$\Phi(t)v = v$$

for all t.

<u>Proof of Theorem</u> 4.6-3. Since we assume that there is no equilibrium v satisfying $v_1 < v < v_2$, Lemma 4.6-6 implies that either v_1 is not stable from above or v_2 is not stable from below. In the first case, Theorem 4.6-1 and Lemma 4.6-1 imply the existence of w(t) satisfying (i) and the first alternative in (ii) and (iii). Similarly, in the second case, there must exist w(t) which satisfies (i) and the second alternative in (ii) and (iii).

Theorem 4.6-4 is proved by using Theorem 4.6-3, Lemmas 4.6-4 and 4.6-5 etc. The proof can be found in Matano [155], and will be omitted here. A corollary of Theorem 4.6-4 is the following.

Theorem 4.6-5. Let X and Φ satisfy hypotheses [H1] to [H3]. Suppose that there exist $w_1 < w_2$ in X such that

$$\Phi(t_1)w_1 > w_1 \quad \text{and} \quad \Phi(t_2)w_2 < w_2$$

for some positive numbers t_1, t_2. Then there exists at least one stable equilibrium point v satisfying $w_1 < v < w_2$.

This Theorem is analogous to Theorem 1.4-2 in Chapter 1. More general version of the theorem in be found in [155].

Theorem 4.6-5 had been used in [156] to study pattern formation for competing species under zero Neumann boundary condition. The following example is given in [156].

$$\begin{aligned} \frac{\partial u_1}{\partial t} &= \sigma_1 \Delta u_1 + u_1(r_1 - a_1u_1 - b_1u_2), \\ \frac{\partial u_2}{\partial t} &= \sigma_2 \Delta u_2 + u_2(r_2 - a_2u_2 - b_2u_2), \end{aligned} \qquad x \in \Omega,\ t > 0, \tag{4.6-32}$$

where σ_i, r_i, a_i, b_i, $i = 1,2$ are positive constants satisfying

$$a_1/a_2 < r_1/r_2 < b_1/b_2 , \tag{4.6-33}$$

with boundary conditions

$$\frac{\partial u_1}{\partial \eta} = \frac{\partial u_2}{\partial \eta} = 0 \qquad x \in \delta\Omega,\ t > 0. \tag{4.6-34}$$

Note that (4.6-32) satisfies (4.6-4) for (u_1, u_2) in the first open quadrant. Problem (4.6-32) to (4.6-34) has a pair of equilibrium solutions

$$p_1 = (r_1/a_1, 0) \quad \text{and} \quad p_2 = (0, r_2/b_2).$$

If we use the following partial order in $C(\bar{\Omega})xC(\bar{\Omega})$:

$$(\bar{u}_1(x),\ \bar{u}_2(x)) \geq (\tilde{u}_1(x),\ \tilde{u}_2(x)) \quad \Longleftrightarrow$$

$$\bar{u}_1(x) \geq \tilde{u}_1(x) \ \text{ and } \ \bar{u}_2(x) \leq \tilde{u}_2(x) \ \text{ for all } x \in \bar{\Omega},$$

we can use Theorem 4.6-5 to investigate the semiflow generated by (4.6-32) to (4.6-34). In [156], it is shown that one can construct a nonconvex domain Ω, which consists of two disjoint parts Ω_1, Ω_2 with a narrow connection between them, so that problem (4.6-32) to (4.6-34) has a stable <u>spatially-nonconstant</u> equilibrium solution. Theorem 4.6-5 is used; and in Ω_1 the solution is close to $(r_1/a_1,0)$, while in Ω_2 it is close to $(0,r_2/b_2)$. Note also that we have the relation $p_1 \geq p_2$ among the solutions p_1, p_2.

In Chapter 7, we will study more problems of the form (4.6-1), with zero Neumann boundary conditions. In section 7.3, many conditions are given which imply that only <u>spatially-constant</u> equilibrium solution is possible. Consequently problem (4.6-32) to (4.6-34) here will not satisfy those conditions in section 7.3, which uses a completely different method of analysis.

Notes

Nonlinear diffusion theory for nuclear reactor analysis had been initially proposed by Kastenberg and Chambrè [118], Cohen [50], Stakgold and Payne [213], etc. Temperature feedback models was suggested by Kastenberg [116], Pao [177], Belleni-Morante [21]. Substantial theory for reactor analysis using transport equations as in Section 4.5 had been developed by e.g. Case and Zweifel [41], Beals [19], Pao [77], Nelson [166] and others. Most of the materials in Sections 4.2 to 4.3 are obtained from Leung and Chen [143],[144] and in Sections 4.4-4.5 from Chen and Leung [45], [46]. Materials related to mutualist species in Section 4.3 can be found in Leung [140] and Korman and Leung [120]. On the other hand, stressing the time dynamical aspects of mutualism (cooperating species), recent work are done by Goh [91], Travis and Post [217], Freedman [78], Hirsch [104] and Smith [209]. The concept of strongly order-preserving system was clarified by Hirsch [105] and Matano [155]. The application of strongly order-preserving property to systems of parabolic partial differential equations was due to Matano [155], as described in section 4.6. The example of pattern formation in (4.6-32) for competing species under zero Neumann boundary condition was found in Matano and Mimura [156].

CHAPTER V

Monotone Schemes for Elliptic Systems, Periodic Solutions

5.1 Introduction

In the previous chapters, the major method for proving the existence of steady state solutions for elliptic systems is Theorem 1.4-2 of the type of intermediate value theorem. It essentially uses maximum principle and the homotopic invariance of degree. Another important technique for analyzing solutions of elliptic systems is the method of monotone schemes. Besides existence, it can be adapted to study uniqueness and stability for corresponding parabolic systems. Moreover, an analogous theory can be developed for finite difference systems. The corresponding monotone schemes provide numerical method for studying elliptic systems. The finite difference theory will be described in Chapter 6.

Before we consider the schemes for systems, let us briefly recall the well-known technique for the scalar case. Consider the boundary value problem:

$$\Delta u + f(x,u) = 0 \text{ in } \Omega \ , \ u(x) = g(x) \text{ for } x \in \delta\Omega \tag{5.1-1}$$

where Ω is a bounded domain in R^n, $n \geq 2$, with boundary $\delta\Omega$ in $H^{2+\ell}$, $0 < \ell < 1$. We assume f is H^{ℓ} in $x = (x_1, \dots, x_n)$ and $H^{1+\ell}$ in u, $g(x)$ can be extended to be a function in $H^{2+\ell}(\overline{\Omega})$. A function ϕ in $C^2(\overline{\Omega})$ is called an upper solution for problem (5.1-1) if

$$\Delta\phi + f(x,\phi) \leq 0 \text{ in } \Omega, \ \phi(x) \geq g(x) \text{ for } x \in \delta\Omega. \tag{5.1-2}$$

If both inequalities are reversed in (5.1-2), we say ϕ is a lower solution. The classical theorem goes as follows: Let w and v be respectively upper and lower solutions of problem (5.1-1) with $v \leq w$ in $\overline{\Omega}$, then there exists a solution for

(5.1-1) in the class $H^{2+\ell}(\bar{\Omega})$.

Let $P > 0$ be large enough so that $f(x,u) + Pu$ is an increasing function of u for $\min v \le u \le \max w$, each $x \in \Omega$. Define a transformation T by $z = Ty$ if

$$(\Delta-k)z = -[f(x,y) + ky] \text{ in } \Omega, \; z = g \text{ on } \delta\Omega. \tag{5.1-3}$$

Setting $z_1 = Tw$, we successively define $z_{i+1} = Tz_i$, $i = 1, 2, \ldots$. From the increasing property of $f(x,u) + ku$, maximum principle and the differential inequalities for w and v, one can show inductively that

$$v \le \ldots \ldots \le z_3 \le z_2 \le z_1 \le w \quad \text{in } \bar{\Omega} . \tag{5.1-4}$$

By dominated convergence theorem, the monotone limit $z = \lim_{i\to\infty} z_i$ is a limit in L^p, i.e. $\lim_{i\to\infty}(\int_\Omega |z_i-z|^p \, dx)^{1/p} = 0$. Using L^p estimate for solutions in the Sobolev space $W^{2,p}$ and then embedding theorem of Sobolev spaces into $H^\ell(\bar{\Omega})$, with $p > n$ large enough (see section A3 in the Appendix), we deduce that z_i actually converge in the space $H^\ell(\bar{\Omega})$. Finally, using Schauder estimate, Theorem 1.3-3, we conclude that z_i converge in the space $H^{2+\ell}(\bar{\Omega})$. Consequently, the montone limit z of z_i is actually in $H^{2+\ell}(\bar{\Omega})$ and taking limit in the following equations

$$\Delta z_{i+1} + f(x,z_i) = 0 \quad \text{in } \Omega, \quad z_i = g \text{ on } \partial\Omega,$$

we obtain a solution $z \in H^{2+\ell}(\bar{\Omega})$ for problem (5.1-1).

The details of the arguments above can be found in [7], [201]. Although we will be constructing monotone sequences as in (5.1-3) (5.1-4), however we will not be using L^p estimates and Sobolev spaces $W^{2,p}$ in the theory for systems in this chapter. Hence, we are not explaining these L^p estimates here in detail.

We will be defining iterative procedures for a system of 2 to m equations: $\Delta u_i + f_i(x, u_1, \ldots, u_m) = 0$ in Ω, $u_i = g_i$ on $\delta\Omega$. The functions f_i depends monotonically on the variables u_j, $j \ne i$. As $\partial f_i/\partial u_j$ can be of different signs for various i,j, we will find that it is more convenient to construct sequences such that every component is alternating monotone, i.e., for each i one has

(5.1-5) $$u_i^2 \le u_i^4 \le u_i^6 \le \dots \dots \le u_i^5 \le u_i^3 \le u_i^1 .$$

Since there are various possibilities for the different signs of $\frac{\partial f_i}{\partial u_j}$, the iterative schemes break down into many cases.

In sections 5.2 to 5.4, we emphasize on f_i which describe interacting population behavior and thus $f_i(x, u_1, \dots, u_{i-1}, 0, u_{i+1}, \dots u_m) = 0$. Each u_i^k is defined as a solution of a <u>nonlinear</u> <u>scalar</u> equation, using u_j^r, $j \ne i$, $r \le k$. One might say that the "diagonal" components are used "implicitly". To facilitate the understanding of the procedures, we restrict to $m \le 3$ in these three sections, and discuss the application of the schemes to uniqueness, stability and nonhomogeneous boundary conditions.

As the schemes essentially depend on $\frac{\partial f_i}{\partial u_j}$, $j \ne i$, one readily adapt the procedures to the case when f_i do not satisfy $f_i(x, u_1, \dots u_{i-1}, 0, u_{i+1}, \dots u_m) = 0$. This general situation for all m is discussed in section 5.5. Moreover, since the dependence of f_i on u_i does not have the special form as in sections 5.2 to 5.4, we modify the schemes so that the use of diagonal components are delayed one or two steps. Consequently, each u_i^k is defined as a solution of a <u>linear</u> <u>scalar</u> equation.

In section 5.6, we study parabolic systems which are periodic in time. They can be viewed as types of intermediate-value theorem, extending the theories in section 1.4-2. One obtains existence of time periodic solutions rather than steady states. Although this topic does not really concern monotone scheme, it can be combined with a corresponding monotone scheme theory for periodic parabolic systems. This part of the book is a natural place to include this topic.

In this chapter Ω always denotes a bounded domin in R^n, with its boundary $\delta\Omega \in H^{2+\ell}$, $0 < \ell < 1$. The number $\lambda = \lambda_1 > 0$ denotes the first (principal) eigenvalue for the eigenvalue problem $\Delta u + \lambda u = 0$ in Ω, $u = 0$ on $\delta\Omega$, and $u = \omega(x) > 0$ denotes a corresponding principal eigenfunction in Ω, with $\frac{\partial\omega}{\partial\eta} < 0$ on $\delta\Omega$.

5.2 Monotone Scheme for Prey-Predator Elliptic Systems

We consider the particular prey-predator Volterra-Lotka type diffusive system:

(5.2-1)
$$\begin{aligned} &\Delta u + u(a - bu - cv) = 0 \\ &\Delta v + v(e + fu - gv) = 0 \end{aligned} \quad \text{in } \Omega$$
$$u = v = 0 \quad \text{on } \delta\Omega.$$

where a, b, c, e, f, g are positive constants, with

(5.2-2) $\quad a > \lambda_1$ and $e > \lambda_1$.

The system (5.2-1) is not quasimonotone because the coefficient of uv in the two equations have opposite signs. For quasimonotone cases, one can simply construct purely monotone sequences as in (5.1-4), (cf. [200] or section 4.5). For the nonquasimonotone case here, we will see that the sequence constructed for each component is alternating monotone as in (5.1-5).

To insure the existence of positive solutions, we will make the following assumptions in the rest of this section:

$$\text{(5.2-3)} \qquad \begin{aligned} &cf < gb \\ &a > \frac{gb}{gb-cf}\left(\lambda_1 + \frac{ce}{g}\right). \end{aligned}$$

Lemma 5.2-1. *The boundary value problem* (5.2-1), *under hypotheses* (5.2-2) *and* (5.2-3) *has a solution* $(\bar{u},\bar{v})$, *where* $\bar{u},\bar{v}$ *are in* $H^{2+\ell}(\bar{\Omega})$ *and satisfy* $\delta\omega(x) < \bar{u}(x) < \frac{a}{b}$, $\delta\omega(x) < \bar{v}(x) < \frac{1}{g}(e + f\frac{a}{b})$ *for* $x \in \bar{\Omega}$. *Here,* δ *is a sufficiently small positive constant; consequently* $\bar{u}(x)$, $\bar{v}(x)$ *are positive for* $x \in \Omega$.

Proof. Hypotheses (5.2-3) implies that $(1 - \frac{cf}{gb})\, a > \lambda_1 + \frac{ce}{g}$, and hence $a - \lambda_1 > \frac{c}{g}(e + \frac{fa}{b})$. For each fixed v, $0 < v < \frac{1}{g}(e + \frac{fa}{b})$, we have $a - \lambda_1 - cv > 0$; hence $u = \delta\omega(x)$ is a lower solution for the scalar problem: $\Delta u + u(a - bu - cv) = 0$ in Ω, $u = 0$ on $\delta\Omega$, for $\delta > 0$ sufficiently small. Clearly for such fixed v, $u = a/b$ is an upper solution of this same scalar problem. On the other hand, for each u, $0 < u < a/b$, the function $v(x) = \delta\omega(x)$ is a lower solution of the scalar problem: $\Delta v + v(e + fu - gv) = 0$ in Ω, $v = 0$ on $\delta\Omega$, for $\delta > 0$ sufficiently small; and $v = \frac{1}{g}(e + \frac{fa}{b})$ is an upper solution. By Theorem 1.4-2, the boundary value problem (5.2-1) has a solution $(\bar{u},\bar{v})$ with properties as stated.

We will construct sequences which approximates $\bar{u}$ and $\bar{v}$. Each iterate of these sequences is a solution of an appropriate scalar boundary value problem. We first consider some existence, uniqueness and comparison results concerning related scalar equations.

Lemma 5.2-2. *Suppose that* $u_i \in H^{2+\ell}(\bar{\Omega})$, $i = 1,2$ *are solutions of*

$$\text{(5.2-4)} \qquad \Delta u + u[\ell(x) - pu] = 0 \ \text{ in } \Omega, \ u = 0 \ \text{ on } \delta\Omega,$$

where $\ell(x)$ *is in* $H^{\ell}(\bar{\Omega})$, p *is a positive constant, and* $u_i(x) > 0$ *in* Ω. *Then* $u_1(x) \equiv u_2(x)$ *in* $\bar{\Omega}$.

Proof. Let C and k be positive numbers satisfying $C > \max\{p^{-1}\sup_{\Omega}\ell(x), \sup_{\overline{\Omega}} u_1(x), \sup_{\overline{\Omega}} u_2(x)\}$, and $k > 2pC - \ell(x)$ for all $x \in \overline{\Omega}$. Let $w_0(x) = C$, $x \in \overline{\Omega}$ and inductively define $w_j(x)$, $j = 1, 2, \dots$ as solutions of

(5.2-5) $$\Delta w_j - kw_j = -w_{j-1}[\ell(x) - pw_{j-1}] - kw_{j-1} \text{ in } \Omega, \; w_j = 0 \text{ on } \delta\Omega.$$

From the choice of C and k, we have the function $f(w) \overset{(def)}{=} -w[\ell(x)-pw] - kw$ decreasing in w for $w \in [0,C]$, $x \in \overline{\Omega}$; and $w_0(x)$ is an upper solution for (5.2-4). Thus $\Delta w_0 - kw_0 \leq f(w_0)$ in Ω, and $w_1 - w_0$ satisfies

$$\Delta(w_1-w_0) - k(w_1-w_0) \geq f(w_0) - f(w_0) = 0 \quad \text{in } \Omega,$$
$$w_1 - w_0 < 0 \text{ on } \delta\Omega.$$

By the maximum principle, Theorem 1.1-2, we have $w_1 < w_0$ in Ω. By means of (5.2-5), maximum principle and the decreasing property of $f(w)$ we prove inductively that $w_j \leq w_{j-1}$ in Ω, $j = 2, 3, \dots$, consequently $w_0 \geq w_1 \geq w_2 \geq \dots$ in $\overline{\Omega}$. On the other hand, for each $i = 1,2$, we have

$$\Delta u_i - ku_i = f(u_i) \quad \text{in } \Omega \quad , \quad u_i = 0 \text{ on } \delta\Omega.$$

For each $j \geq 1$,

$$\Delta(w_j-u_i) - k(w_j-u_i) = f(w_{j-1}) - f(u_i) \quad \text{in } \Omega.$$

Hence $\Delta(w_j-u_i) - k(w_j-u_i) \leq 0$ in Ω if $w_{j-1} \geq u_i$. We can prove inductively that $w_j \geq u_i$ for each $j = 1, 2, \dots$. Therefore, for each $i = 1,2$, we have

(5.2-6) $$u_i \leq \dots \leq w_2 \leq w_1 \leq w_0 \quad \text{for } x \in \overline{\Omega}.$$

The limit $\lim_{j\to\infty} w_j(x) \overset{(def)}{=} \hat{w}(x)$ will be in $H^{2+\ell}(\overline{\Omega})$ and $\hat{w}(x)$ satisfies problem (5.2-4) as explained in section 5.1. From (5.2-6), we have $\hat{w}(x) \geq u_i(x)$, $x \in \overline{\Omega}$ $i = 1,2$

Substituting $\hat{w}$ into (5.2-4), multiplying the equation by u_i, and then interchanging the role of $\hat{w}$, u_i, we obtain after substraction:

$$u_i\Delta\hat{w} - \hat{w}\Delta u_i = p\hat{w}u_i[\hat{w} - u_i] \quad \text{in } \Omega$$

By the Green's identity, we have for $i = 1,2$:

$$(5.2\text{-}7) \qquad \int_\Omega p\hat{w}u_i[\hat{w} - u_i]\,dx = \int_{\partial\Omega} u_i\frac{\partial\hat{w}}{\partial n} - \hat{w}\frac{\partial u_i}{\partial n}\,d\sigma = 0.$$

Since $\hat{w}(x) \geq u_i(x) > 0$ in Ω, (5.2-7) implies that $\hat{w}(x) \equiv u_i(x)$ in $\bar{\Omega}$, $i = 1,2$. This proves the lemma.

The following theorem is fundamental for the method in this section.

<u>Theorem</u> 5.2-1 (<u>Comparison</u>). <u>Let</u> $\ell_i(x)$, $i = 1,2$ <u>be functions in</u> $H^\ell(\bar{\Omega})$ <u>satisfying</u> $\ell_1(x) \geq \ell_2(x) > \lambda_1$ <u>for all</u> $x \in \bar{\Omega}$. <u>Suppose that</u> $u_i \in H^{2+\ell}(\bar{\Omega})$, $i = 1,2$ <u>are respectively solutions of</u>

$$\Delta u + u[\ell_i(x) - pu] = 0 \text{ in } \Omega, \qquad u = 0 \text{ on } \delta\Omega,$$

<u>with</u> $u_i > 0$ <u>in</u> Ω. <u>Then</u> $u_1(x) \geq u_2(x)$ <u>for all</u> $x \in \Omega$. (<u>Here</u> p <u>is a positive constant</u>).

<u>Proof</u>. The function $w = u_1$ is an upper solution for the problem $\Delta w + w[\ell_2(x) - pw] = 0$ in Ω, $u = 0$ on $\delta\Omega$, because $\Delta u_1 + u_1[\ell_2 - pu_1] = u_1[-\ell_1 + pu_1 + \ell_2 - pu_1] \leq 0$ in Ω. Furthermore, $w = \delta\omega$, is a lower solution for this same problem, for sufficiently small $\delta > 0$, because $\Delta(\delta\omega) + \delta\omega[\ell_2 - p\delta\omega] = \delta\omega[-\lambda_1 + \ell_2 - p\delta\omega] > 0$ in Ω for small $\delta > 0$. From section 5.1, we have a solution $w = \tilde{u}_2 \in H^{2+\ell}(\bar{\Omega})$ to this problem, with $\delta\omega \leq \tilde{u}_2 \leq u_1$ in $\bar{\Omega}$. Consequently, $\tilde{u}_2 > 0$ in Ω, and by lemma 5.2-2 $u_2 = \tilde{u}_2$ in $\bar{\Omega}$. This proves the Theorem.

We now construct monotone sequences closing in to u^* and v^*. First, let $u_1 \in H^{2+\ell}(\bar{\Omega})$ be the unique strictly positive function in Ω, satisfying

(5.2-8) $$\Delta u_1 + u_1(a - bu_1) = 0 \text{ in } \Omega, \quad u_1 = 0 \text{ on } \delta\Omega.$$

Note that such solution exists because $\delta\omega(x)$ is a lower solution for small enough positive δ, and large positive constant functions are upper solutions; it is unique by Lemma 5.2-2. Similarly, let $v_1 \in H^{2+\ell}(\bar{\Omega})$, $v_1 > 0$ in Ω be the unique solution of

(5.2-9) $$\Delta v_1 + v_1(e + fu_1 - gv_1) = 0 \quad \text{in } \Omega, \quad v_1 = 0 \text{ on } \delta\Omega.$$

<u>Lemma</u> 5.2-3. $v_1(x) < c^{-1}(a-\lambda_1)$ <u>for</u> <u>all</u> $x \in \bar{\Omega}$.

<u>Proof</u>. Since all constant functions larger than a/b are upper solutions of $\Delta u_1 + u_1(a - bu_1) = 0$ in Ω, $u_1 = 0$ on $\delta\Omega$, we have $u_1(x) < a/b$ for all $x \in \Omega$, and $e + fu_1(x) < e + \frac{fa}{b}$ for all $x \in \Omega$. Consequently, the constant function $w = \frac{1}{g}(e + \frac{fa}{b})$ is an upper solution for $\Delta w + w(e + fu_1 - gw) = 0$ in Ω, $w = 0$ on $\delta\Omega$, and $w = \delta\omega$ is a lower solution for $\delta > 0$ sufficiently small. By Lemma 5.2-2, we have $\delta\omega(x) < v_1(x) < \frac{1}{g}(e + \frac{fa}{b})$, $x \in \bar{\Omega}$. On the other hand hypotheses (5.2-3) implies that $(1 - \frac{cf}{gb})\, a > \lambda_1 + \frac{ce}{g}$, hence $a - \lambda_1 > \frac{c}{g}(e + \frac{fa}{b})$, which proves the lemma.

With the aid of the following lemma, we now define u_i, v_i to be strictly positive functions in Ω, $i = 2,3 \ldots$ inductively as follows:

(5.2-10) $$\begin{aligned} \Delta u_i + u_i(a - bu_i - cv_{i-1}) &= 0 \\ \Delta v_i + v_i(e + fu_i - gv_i) &= 0 \end{aligned} \quad \text{in } \Omega$$

$u_i = v_i = 0$ on $\delta\Omega$, $u_i, v_i \in H^{2+\ell}(\bar{\Omega})$.

<u>Lemma</u> 5-2.4. <u>For each</u> $i = 1, 2, \ldots$, $u_i > 0$, $v_i > 0$ <u>in</u> Ω, <u>and</u> $u_i < u_1$, $v_i < v_1$ <u>in</u> Ω. <u>They are uniquely defined in</u> $H^{2+\ell}(\bar{\Omega})$.

<u>Proof</u>. The lemma is clearly true for $i = 1$. Assume that it is true for $i = j-1$. We have $a - cv_{j-1} > a - cv_1$, which is $> \lambda_1$ for $x \in \bar{\Omega}$ by lemma 5.2-3.

Therefore, the functions $w = \delta\omega$ is a lower solution for the scalar problem: $\Delta w + w(a - bw - cv_{j-1}) = 0$ in Ω, $w = 0$ on $\delta\Omega$, for sufficiently small $\delta > 0$. On the other hand, $w = u_1(x)$ is an upper solution. From lemma 5.2-2, $u_j(x)$ is uniquely defined as a solution of this scalar problem and $0 < \delta\omega(x) < u_j(x) < u_1(x)$ in Ω. Next, consider the scalar problem: $\Delta z + z(e + fu_j - gz) = 0$ in Ω, $z = 0$ on $\delta\Omega$. The function $z = v_1$ is an upper solution, since $\Delta v_1 + v_1(e + fu_j - gv_1) = v_1 f(u_j - u_1) < 0$ in Ω; and $z = \delta\omega$ is a lower solution for the problem, for $\delta > 0$ small enough. Again from lemma 5.2-2, $v_j(x)$ is uniquely defined in $H^{2+\ell}(\bar{\Omega})$ as a solution of this problem, and $0 < \delta\omega(x) < v_j(x) < v_1(x)$ in Ω. The lemma follows from induction.

We next observe some order relations among the sequences u_i, v_i, $i = 1, 2, \ldots$

Theorem 5.2-2. For each nonnegative integer n, the following are true for all $x \in \bar{\Omega}$:

$$u_{2n+2} < u_{2n+4} < u_{2n+3} < u_{2n+1}, \quad v_{2n+2} < v_{2n+4} < v_{2n+3} < v_{2n+1}. \tag{5.2-11}$$

Proof. We first consider the case when $n = 0$. Using the equations which u_1, u_3 satisfy and utilizing the comparison theorem 5.2-1, we conclude $u_3 < u_1$ in $\bar{\Omega}$. This in turn, using the same method, implies that $v_3 < v_1$ in $\bar{\Omega}$. Again, this implies that $u_2 < u_4$ and then $v_2 < v_4$ in $\bar{\Omega}$. To establish comparison between u_3 and u_4 or v_3 and v_4, we keep applying the comparison theorem 5.2-1 to first assert $u_2 < u_1$, then $v_2 < v_1$, then $u_2 < u_3$, then $v_2 < v_3$, then $u_4 < u_3$ and finally $v_4 < v_3$ (all inequalities being true in $\bar{\Omega}$). Inequalities (5.2-11) are thus true for $n = 0$. Assume that (5.2-11) are true for $n = j-1$, using the above method and $v_{2j} < v_{2j+2}$, we deduce the following in order: $u_{2j+3} < u_{2j+1}$, $v_{2j+3} < v_{2j+1}$, $u_{2j+2} < u_{2j+4}$, $v_{2j+2} < v_{2j+4}$. Then using $u_{2j+2} < u_{2j+1}$ we deduce the following in order: $v_{2j+2} < v_{2j+1}$, $u_{2j+2} < u_{2j+3}$, $v_{2j+2} < v_{2j+3}$, $u_{2j+4} < u_{2j+3}$, $v_{2j+4} < v_{2j+3}$ (All inequalities are true in $\bar{\Omega}$.) This proves (5.2-11) for $n = j$.

Theorem 5.2-2 clearly implies that:

$$(5.2\text{-}12)\quad \begin{array}{l} 0 < u_2 < u_4 < u_6 \cdots < u_5 < u_3 < u_1 \text{ , and} \\ 0 < v_2 < v_4 < v_6 \cdots < v_5 < v_3 < v_1 \text{ , for all } x \in \bar{\Omega}. \end{array}$$

Define $u^* = \lim_{n\to\infty} u_{2n+1}$, $u_* = \lim_{n\to\infty} u_{2n}$, $v^* = \lim_{n\to\infty} v_{2n+1}$, and $v_* = \lim_{n\to\infty} v_{2n}$. The functions u_*, u^*, v_*, v^* are in $H^{2+\ell}(\bar{\Omega})$, and $(w_1, w_2, w_3, w_4) = (u_*, u^*, v_*, v^*)$ is a classical solution for the problem:

$$(5.2\text{-}13)\quad \begin{array}{l} \Delta w_1 + w_1(a - bw_1 - cw_4) = 0, \quad \Delta w_2 + w_2(a - bw_2 - cw_3) = 0 \\ \Delta w_3 + w_3(e + fw_1 - gw_3) = 0, \quad \Delta w_4 + w_4(e + fw_2 - gw_4) = 0 \end{array}$$

in Ω, $w_i = 0$ on $\delta\Omega$, $i = i, \ldots 4$. (The regularity of the limiting function is proved as in section 5.1). Note that $w_i(x) > 0$ for all $x \in \Omega$.

We next investigate some relationship between problem (5.2-1) and (5.2-13).

Theorem 5.2-3. (i) $u_* \equiv u^*$ in $\bar{\Omega}$ iff $v_* \equiv v^*$ in $\bar{\Omega}$. (ii) Suppose that $u_* \equiv u^*$ in $\bar{\Omega}$, then the boundary value problem (5.2-1) has a unique nontrivial nonnegative solution with each component in $H^{2+\ell}(\bar{\Omega})$. More precisely, any such solution (u,v) with the property that $u \geq 0$, $v \geq 0$, both $\not\equiv 0$ in Ω will satisfy $(u,v) = (u_*,v_*) = (u^*,v^*)$ in $\bar{\Omega}$. Furthermore, $u > 0$, $v > 0$ in Ω. (iii) Suppose that the boundary value problem (5.2-13) has at most one solution with the property that all its components are positive in Ω, then $u_* \equiv u^*$ and $v_* \equiv v^*$ in $\bar{\Omega}$.

Proof. (i) Suppose $u_* \equiv u^*$ then v_* and v^* will be solutions to the same boundary value problem in the class $H^{2+\ell}(\bar{\Omega})$. By lemma 5.2-2, we conclude that $v_* \equiv v$ in $\bar{\Omega}$. Similarly, the converse is true.

(ii) Note that the existence of solution (u,v) to problem (5.2-1) with components in $H^{2+\ell}(\bar{\Omega})$ and nonnegative in Ω had already been established in lemma 5.2-1, even without the assumption that $u_* = u^*$ in $\bar{\Omega}$. It remains to prove that under this assumption we must have $(u,v) = (u_*,v_*)$. Since $v \geq 0$ in Ω, using the first equation in (5.2-1) and large constants as upper solutions, we conclude by means of Theorem 1.4-3 that $u \leq a/b$ in $\bar{\Omega}$. Subsequently, using the second equation in (5.2-1) we can conclude that $v \leq \frac{1}{g}(e + \frac{fa}{b})$ in $\bar{\Omega}$. Hypotheses (5.2-3) therefore implies that $a - cv > \lambda_1$ in $\bar{\Omega}$. We next show that $u > 0$ in Ω. Suppose not, and $u(\hat{x}) = 0$ for $\hat{x} \in \Omega$. Then the first equation in (5.2-1) implies that $\Delta u \leq 0$ in a neighborhood of $\hat{x}$, and the maximum principle implies $u \equiv 0$ in this neighborhood. This leads to the conclusion that $u \equiv 0$ in $\bar{\Omega}$, contradicting our assumption. Therefore $u > 0$ in Ω. Now, we can use the comparison Theorem 5.2-1 to conclude that $u \leq u_1$ in $\bar{\Omega}$; similarly, $0 < v \leq v_1$ in Ω. Again, applying Theorem 5.2-1, we have $u_2 \leq u$ and $v_2 \leq v$ in $\bar{\Omega}$. Repeated applications of Theorem 5.2-1 will, by induction, show that $u_{2n+2} \leq u \leq u_{2n+1}$, $v_{2n+2} \leq v \leq v_{2n+1}$, $x \in \bar{\Omega}$, for each nonnegative integer n. Clearly, if $u_* \equiv u^*$ (or $v_* \equiv v^*$), then $(u,v) = (u_*,v^*)$.

(iii) We have seen that $(w_1, w_2, w_3, w_4) = (u_*, u^*, v_*, v^*)$ is a solution of (5.2-13) with the property stated here. Moreover $(w_1, w_2, w_3, w_4) = (u^*, u_*, v^*, v_*)$ is also such a solution. Hence, the uniqueness assumption here implies that $u_* \equiv u^*$, $v_* \equiv v^*$ in $\bar{\Omega}$.

The proof of part (ii) of the last theorem also contains the proof of the following theorem.

Theorem 5.2-4. *Any solution* (u,v) *of boundary value problem* (5.2-1) *with both* u *and* v in $H^{2+\ell}(\bar{\Omega})$, ≥ 0, $\not\equiv 0$ *in* $\bar{\Omega}$, *must satisfy*:

$$u_* \leq u \leq u^* , \quad v_* \leq v \leq v^*$$

for all $x \in \bar{\Omega}$. (*Hence* $u > 0$, $v > 0$ *in* Ω.)

5.3 Application to Uniqueness and Stability

We continue our discussion on problem (5.2-1) with hypotheses (5.2-2) and (5.2-3) in this section. Let C,F be positive constants so that $c \leq C$, $f \leq F$ and

$$\text{(5.3-1)} \qquad CF < gb \quad , \quad a > gb(gb - CF)^{-1}(\lambda_1 + \frac{Ce}{g}) \; .$$

(Observe that if a, b, e, g are fixed and hypotheses (5.2-3) are satisfied, then (5.2-3) remains satisfied if c,f are reduced). Let $\hat{U}, \tilde{U}, \hat{V}, \tilde{V} \in H^{2+\ell}(\overline{\Omega})$ be strictly positive functions in Ω, which are solutions to the following:

$$\text{(5.3-2)} \qquad \begin{aligned} &\Delta\hat{U} + \hat{U}(a - b\hat{U}) = 0 \quad \text{in } \Omega, \quad \hat{U} = 0 \quad \text{on } \delta\Omega \\ &\Delta\hat{V} + \hat{V}(e + \frac{Fa}{b} - g\hat{V}) = 0 \quad \text{in } \Omega, \quad \hat{V} = 0 \quad \text{on } \delta\Omega \\ &\Delta\tilde{U} + \tilde{U}(a - b\tilde{U} - C\hat{V}) = 0 \quad \text{in } \Omega, \quad \tilde{U} = 0 \quad \text{on } \delta\Omega \\ &\Delta\tilde{V} + \tilde{V}(e - g\tilde{V}) = 0 \quad \text{in } \Omega, \quad \tilde{V} = 0 \quad \text{on } \delta\Omega \end{aligned}$$

Note that such $\hat{U}, \hat{V}, \tilde{V}$ exist because a, e, $e + F\frac{a}{b}$ are $> \lambda_1$; and $\hat{U}, \hat{V}, \tilde{V}$ are $\geq \delta\omega > 0$ in Ω, for sufficiently small $\delta > 0$. We can also prove as in section 5.2 that $\hat{V}(x) \leq \frac{1}{g}(e + \frac{Fa}{b})$, hence $a - C\hat{V}(x) \geq a - \frac{C}{g}(e + \frac{Fa}{b}) > \lambda_1$ for all $x \in \overline{\Omega}$, by hypotheses (5.3-1). Consequently, $\tilde{U} \geq \delta\omega > 0$ in Ω for $\delta > 0$ sufficiently small. By comparison theorem 5.2-1, we have $0 < \delta\omega \leq \tilde{U} \leq \hat{U}$, $0 < \delta\omega \leq \tilde{V} \leq \hat{V}$ for $x \in \Omega$, $\delta > 0$ sufficiently small. Since the outward normal derivatives of ω are negative on the boundary, there must exist a constant $K > 0$ such that

$$\text{(5.3-3)} \qquad \hat{U} \leq K\tilde{U}, \quad \hat{V} \leq K\tilde{V}, \quad \hat{U} \leq K\tilde{V}, \quad \hat{V} \leq K\tilde{U}$$

for all $x \in \overline{\Omega}$.

Let u_i, v_i, $i = 1, 2, \ldots$ be as defined in section 5.2. We have $u_1 = \hat{U}$, $v_1 < \hat{V}$, $\tilde{U} < u_2$, $\tilde{V} < v_2$ for $x \in \bar{\Omega}$, by theorem 5.2-1. Moreover

$$\text{(5.3-4)} \qquad \begin{array}{l} \tilde{U} < u_2 < u_4 < u_6 < \ldots < u_5 < u_3 < u_1 < \hat{U} \\ \tilde{V} < v_2 < v_4 < v_6 < \ldots < v_5 < v_3 < v_1 < \hat{V} \end{array}$$

for all $x \in \bar{\Omega}$. (Note that if c,f are reduced, such inequalities will be unchanged, and $\tilde{U}$, $\hat{U}$, $\tilde{V}$, $\hat{V}$ are unchanged). For $i > 1$, we have:

$$0 = \int_{\delta\Omega} (u_{2i+2} \frac{\partial u_{2i+1}}{\partial n} - u_{2i+1} \frac{\partial u_{2i+2}}{\partial n} d\sigma = \int_{\Omega} (u_{2i+2}\Delta u_{2i+1} - u_{2i+1}\Delta u_{2i+2}) \, dx$$

$$= -\int_{\Omega} u_{2i+1}u_{2i+2} \, [b(u_{2i+2} - u_{2i+1}) + c(v_{2i+1} - v_{2i})] \, dx$$

(here $\partial/\partial n$ means derivative with respect to outward normal), implying that

$$\text{(5.3-5)} \quad b\int_{\Omega} (u_{2i+1} - u_{2i+2}) \, u_{2i+1}u_{2i+2} \, dx = c\int_{\Omega} (v_{2i+1} - v_{2i}) \, u_{2i+1}u_{2i+2} \, dx$$

for each integer $i > 1$. Also for $i > 1$, we have

$$0 = \int_{\Omega}(v_{2i+1}\Delta v_{2i} - v_{2i}\Delta v_{2i+1}) \, dx$$

$$= -\int_{\Omega} v_{2i}v_{2i+1}[f(u_{2i} - u_{2i+1}) + g(v_{2i+1} - v_{2i})] \, dx$$

which implies that

$$\text{(5.3-6)} \qquad g\int_{\Omega} (v_{2i+1} - v_{2i})v_{2i}v_{2i+1} \, dx = f\int_{\Omega} (u_{2i+1} - u_{2i}) \, v_{2i}v_{2i+1} \, dx.$$

Moreover, for $i > 1$,

$$0 = \int_{\Omega} (u_{2i}\Delta u_{2i+1} - u_{2i+1}\Delta u_{2i}) \, dx$$

$$= -\int_{\Omega} u_{2i}u_{2i+1}[b(u_{2i} - u_{2i+1}) + c(v_{2i-1} - v_{2i})] \, dx,$$

$$0 = \int_\Omega (v_{2i-1}\Delta v_{2i} - v_{2i}\Delta v_{2i-1})\, dx$$

$$= -\int_\Omega v_{2i-1}v_{2i}[f(u_{2i} - u_{2i-1}) + g(v_{2i-1} - v_{2i})]\, dx$$

respectively gives

$$\text{(5.3-7)} \quad b \int_\Omega (u_{2i}u_{2i+1}\,(u_{2i+1} - u_{2i})\, dx = c \int_\Omega u_{2i}u_{2i+1}\,(v_{2i-1} - v_{2i})\, dx,$$

$$\text{(5.3-8)} \quad g \int_\Omega v_{2i-1}v_{2i}\,(v_{2i-1} - v_{2i})\, dx = f \int_\Omega v_{2i-1}v_{2i}\,(u_{2i-1} - u_{2i})\, dx.$$

Using (5.3-5), (5.3-6) and (5.3-3), (5.3-4) we deduce that:

$$\text{(5.3-9)} \quad \int_\Omega (u_{2i+1} - u_{2i+2})\, u_{2i+1}u_{2i+2}\, dx = \frac{c}{b}\int_\Omega (v_{2i+1} - v_{2i})\, u_{2i+1}u_{2i+2}\, dx$$
$$< \frac{c}{b}\int_\Omega K^2(v_{2i+1} - v_{2i})\, v_{2i}v_{2i+1}\, dx = K^2\,(\tfrac{c}{b})(\tfrac{f}{g})\int_\Omega (u_{2i+1} - u_{2i})\, v_{2i}v_{2i+1}\, dx.$$

Then, we use (5.3-7), (5.3-8) and (5.3-3), (5.3-4) again to obtain:

$$\text{(5.3-10)} \quad \int_\Omega (u_{2i+1} - u_{2i})\, u_{2i}u_{2i+1}\, dx = \frac{c}{b}\int_\Omega (v_{2i-1} - v_{2i})\, u_{2i}u_{2i+1}\, dx$$
$$< \frac{c}{b}\int_\Omega K^2\,(v_{2i-1} - v_{2i})\, v_{2i-1}v_{2i}\, dx = K^2\,(\tfrac{c}{b})(\tfrac{f}{g})\int_\Omega (u_{2i-1} - u_{2i})\, v_{2i-1}v_{2i}\, dx.$$

Combining (5.3-9), (5.3-10) and using (5.3-3), (5.3-4) once more, we obtain:

$$\text{(5.3-11)} \quad \int_\Omega (u_{2i+1} - u_{2i+2})\, u_{2i+1}u_{2i+2}\, dx < K^8\,(\tfrac{c}{b})^2(\tfrac{f}{g})^2 \int_\Omega (u_{2i-1} - u_{2i})\, u_{2i-1}u_{2i}$$

for each integer $i > 1$.

By means of (5.3-11), we conclude that if c,f with $0 < c < C$, $0 < f < F$ are such that

$$\text{(5.3-12)} \quad c^2f^2 < \frac{1}{K^8}\, b^2g^2\ ,$$

then $\lim_{i\to\infty}\int_\Omega (u_{2i+1} - u_{2i+2})\, u_{2i+1}u_{2i+2}\, dx = 0$. By dominated convergence theorem, and $\lim_{i\to\infty} u_{2i+1} = u^* > 0$ in Ω, $\lim_{i\to\infty} u_{2i+2} = u_* > 0$ in Ω, $\lim_{i\to\infty} (u_{2i+1} - u_{2i+2}) = u^* - u_* \geq 0$ in Ω, we conclude that $u^* = u_*$ for all $x \in \Omega$. (Note that (5.3-3), (5.3-4) imply that K is unchanged by reducing c and f). Applying theorem 5.2-4, we therefore arrive at the following:

Theorem 5.3-1. *Let hypothesis (5.2-3) be satisfied for boundary value problem (5.2-1). The solution (u,v) of (5.2-1) with both u and v in $H^{2+\ell}(\bar{\Omega})$, ≥ 0, $\not\equiv 0$ in $\bar{\Omega}$ is unique, provided that (5.3-12) is satisfied.*

To apply the monotone scheme in section 5.2 to a stability problem, we consider $(u(x,t), v(x,t))$ satisfying:

$$\text{(5.3-13)}\qquad \begin{aligned} \frac{\partial u}{\partial t} &= \Delta u + u(a - bu - cv) \\ \frac{\partial v}{\partial t} &= \Delta v + v(e + fu - gv) \end{aligned}$$

for $(x,t) \in \Omega \times (0,\infty)$, where Ω and a, b, c, e, f, g satisfy all the conditions as in section 5.2. Let $u_i(x)$, $v_i(x)$, $i = 1, 2, \ldots$ be as defined in section 5.2. For $T > 0$, let $\Omega_T = \Omega \times (0,T)$. $H^{2+\ell,(2+\ell)/2}(\bar{\Omega}_T)$ denotes the Banach space of all real-valued functions w having all derivatives of the form $D^\alpha D_t^r w$ (α is a multi-index, $r \geq 0$ is an integer, $D_t = \partial/\partial t$) with $2r + |\alpha| \leq 2$ continuous on $\bar{\Omega}_T$ and having finite norm $|w|^{(2+\ell)}_{\Omega_T}$ (as described in Section 1.3).

Theorem 5.3-2. *Let i be an arbitrary positive integer. Suppose that $(u(x,t), v(x,t))$ is a solution of (5.3-13) in $H^{2+\ell,(2+\ell)/2}(\bar{\Omega}_T)$, for each $T > 0$, satisfying:*

$$\text{(5.3-14)}\qquad u_{2i}(x) \leq u(x,0) \leq u_{2i-1}(x),\quad v_{2i}(x) \leq v(x,0) \leq v_{2i-1}(x)$$

for $x \in \bar{\Omega}$,

$$\text{(5.3-15)}\qquad u_{2i}(x) = u(x,t) = u_{2i-1}(x) = 0,\ v_{2i}(x) = v(x,t) = v_{2i-1}(x) = 0$$

<u>for</u> $(x,t) \in \delta\Omega \times [0,\infty)$. <u>Then</u> $(u(x,t), v(x,t))$ <u>will satisfy</u>:

$$(5.3\text{-}16) \qquad \begin{aligned} u_{2i}(x) &\leq u(x,t) \leq u_{2i-1}(x) \\ v_{2i}(x) &\leq v(x,t) \leq v_{2i-1}(x) \end{aligned}$$

<u>for all</u> $(x,t) \in \bar{\Omega} \times [0,\infty)$.

<u>Proof</u>. Observe that u_{2i}, u_{2i-1}, v_{2i}, v_{2i-1} satisfy the following inequalities:

$$\Delta u_{2i} + u_{2i}(a - bu_{2i} - cv_{2i-1}) - \frac{\partial u_{2i}}{\partial t} = 0 \geq 0,$$

$$\Delta u_{2i-1} + u_{2i-1}(a - bu_{2i-1} - cv_{2i}) - \frac{\partial u_{2i-1}}{\partial t}$$

$$\leq \Delta u_{2i-1} + u_{2i-1}(a - bu_{2i-1} - cv_{2i-2}) \leq 0 ,$$

$$\Delta v_{2i} + v_{2i}(e + fu_{2i} - gv_{2i}) - \frac{\partial v_{2i}}{\partial t} = 0 \geq 0,$$

$$\Delta v_{2i-1} + v_{2i-1}(e + fu_{2i-1} - gv_{2i-1}) - \frac{\partial v_{2i-1}}{\partial t} = 0 \leq 0$$

for all $(x,t) \in \bar{\Omega} \times [0,\infty)$ (here denote $v_0 \equiv 0$). Using Theorem 1.2-6, we obtain (5.3-16).

Theorem 5.3-2 gives a family of spatially dependent invariant regions closing in to the set $\{(x,u,v) \mid x \in \bar{\Omega},\ u_*(x) \leq u \leq u^*(x),\ v_*(x) \leq v \leq v^*(x)\}$. In case (5.3-12) is satisfied, the set becomes $\{(x, u_*(x), v_*(x)) \mid x \in \bar{\Omega}\}$, since $u_* = u^*$, $v_* = v^*$. For each $x \in \Omega$, let $I_i^x = \{(x,u,v) \mid u_{2i}(x) \leq u \leq u_{2i-1}(x),$ $v_{2i}(x) \leq v \leq v_{2i-1}(x)\}$, for each positive integer i. Clearly, Theorem 5.2-2 implies that $I_{i+1}^x \subseteq I_i^x$ for each $x \in \Omega$. The following remark will show that the set inclusion is proper.

<u>Remark</u> 5.3-1. In Theorem 5.2-1, suppose that we have the additional assumption that $\ell_1(x) \not\equiv \ell_2(x)$, $x \in \Omega$. Then $u_1(x) > u_2(x)$ for all $x \in \Omega$. Consequently,

our construction of monotone scheme will imply that the following inequalities are strict:

$$0 < u_2(x) < u_4(x) < \ldots < u_5(x) < u_3(x) < u_1(x)$$
$$0 < v_2(x) < v_4(x) < \ldots < v_5(x) < v_3(x) < v_1(x) \tag{5.3-17}$$

for each $x \in \Omega$. (Observe here that u_1, u_2 in (5.3-17) play different roles as those in the beginning of the remark.)

To prove the assertion of the above remark, we note that $w = u_1$ is an upper solution for: $\Delta w + w[\ell_2(x) - pw] = 0$ in Ω, $w = 0$ on $\delta\Omega$. We proved in Theorem 5.2-1 that $\delta\omega \leq u_2 \leq u_1$ for sufficiently small $\delta > 0$. We conclude the existence of such u_2 by starting to iterate from the upper solution $w = u_1$ (a procedure described in section 5.1). The property that the upper solution u_1 satisfies the boundary condition exactly at $\delta\Omega$ will imply that the next iterate is strictly less than the upper solution u_1 inside Ω (because of maximum principle), unless $\Delta u_1 + u_1[\ell_2(x) - pu_1] = 0$ in Ω. However $\Delta u_1 + u_1[\ell_2(x) - pu_1]$ $= u_1[-\ell_1(x) + \ell_2(x)] \not\equiv 0$ in Ω. Consequently, if $\ell_1(x) \not\equiv \ell_2(x)$ in Ω, we must have $u_2(x) < u_1(x)$ in Ω, in Theorem 5.2-1.

In view of (5.3-17) we observe that for each $x \in \Omega$, $I^x_{i+1} \subset$ interior of I^x_i for each integer $i \geq 1$, and

$$\{(x,u,v) \mid u_*(x) \leq u \leq u^*(x),\ v_*(x) \leq v \leq v^*(x)\} \subset \ldots \subset I^x_j \subset \ldots I^x_2 \subset I^x_1.$$

In view of Theorems 5.3-2 and 5.3-1, one might say that the positive equilibrium solution of (5.3-13) with $u = v = 0$ on $\delta\Omega \times [0,\infty)$ is <u>stable</u>, when (5.3-12) is satisfied.

5.4 More General Systems with Nonnegative Boundary Conditions

In this section we adapt the methods in the last two sections to consider systems of three equations which are more general than Volterra-Lotka interaction type, and has nontrivial nonnegative boundary conditions.

We consider systems of the form:

$$\begin{aligned} &\Delta u + u[a + f_1(u,v,w)] = 0 \\ (5.4\text{-}1) \quad &\Delta v + v[b + f_2(u,v,w)] = 0 \qquad \text{in } \Omega \\ &\Delta w + w[c + f_3(u,v,w)] = 0 \\ &u = g_1,\ v = g_2,\ w = g_3 \quad \text{on } \delta\Omega \end{aligned}$$

where a, b, c are constants, $f_i(u,v,w)$ for $i = 1, 2, 3$ have uniformly Hölder continuous first partial derivatives in compact sets of $\overline{R}^+ \times \overline{R}^+ \times \overline{R}^+$ ($\overline{R}^+$ denotes $[0,\infty)$). The functions g_i, $i = 1, 2, 3$ defined on $\delta\Omega$ are assumed to have extensions $\hat{g}_i \in H^{2+\ell}(\overline{\Omega})$, and $g_i(x) \geq 0$, $\not\equiv 0$ for $x \in \delta\Omega$. We will assume conditions on the signs of the first partial derivatives of f_i so that (5.4-1) can be used to study steady states for three interacting ecological species (or chemical interactions). The following "self-crowding" effects are always assumed:

$$(5.4\text{-}2) \qquad \frac{\partial f_1}{\partial u} < 0,\ \frac{\partial f_2}{\partial v} < 0,\ \frac{\partial f_3}{\partial w} < 0 \text{ on } \overline{R}^+ \times \overline{R}^+ \times \overline{R}^+ \quad ; \text{ and}$$

$$(5.4\text{-}3) \qquad \lim_{u\to+\infty} f_1(u,v,w) = -\infty,\ \lim_{v\to+\infty} f_2(u,v,w) = -\infty,\ \lim_{w\to+\infty} f_3(u,v,w) = -\infty$$

where the limits are uniform when the independent variables not tending to $+\infty$ are to remain in compact sets.

We will consider various cases for three species corresponding to food-chain, prey-predator and mutualist loop. In each case, we devise a scheme for constructing alternating monotone sequences closing into the solution, as in section 5.2. The rules will lead us to the understanding of the more general system of m equations in the next section when a more general modified scheme is developed.

We first consider a few lemmas which are analogous to the uniqueness lemma 5.2-2 and comparison Theorem 5.2-1 in section 5.2.

Lemma 5.4-1 (Uniqueness). *Let $h(x,z)$ be defined for (x,z) in $\bar{\Omega} \times [0,\infty)$ with Hölder continuous first partial derivatives in compact sets of $\bar{\Omega} \times [0,\infty)$. Suppose that $\frac{\partial h}{\partial z} < 0$ at each point in $\Omega \times (0,\infty)$ and there exists a constant $C > 0$ such that $h(x,z) < 0$ for all $x \in \bar{\Omega}$, $z > C$. Let $z_i \in H^{2+\ell}(\bar{\Omega})$, $i = 1,2$ be solutions of*

$$\Delta z + zh(x,z) = 0 , \quad x \in \Omega, \tag{5.4-4}$$

with the property $z_i(x) > 0$ for each $x \in \Omega$, $i = 1,2$, and $z_1(x) = z_2(x)$ for $x \in \delta\Omega$. Then $z_1(x) \equiv z_2(x)$ for all $x \in \bar{\Omega}$.

The proof is exactly analogous to that of lemma 5.2-2. We construct a solution of (5.4-4), with the same boundary condition as z_i, through iteration from a large constant upper solution. The resulting solution $\hat{z}$ will be $\geq z_i$, $i = 1,2$. Then, by using the Green's identity, we show that $\hat{z} \equiv z_i$, $i = 1,2$. The details can be found in [140].

Lemma 5.4-2. *Let $p(x,z)$ be a function defined for (x,z) in $\bar{\Omega} \times [0,\infty)$ which is Hölder continuous in compact sets of $\bar{\Omega} \times [0,\infty)$. Let $\tilde{z} \in H^{2+\ell}(\bar{\Omega})$, $\tilde{z} \geq 0$ in $\bar{\Omega}$, be a solution of*

$$\Delta z + zp(x,z) = 0 \ \text{ in } \Omega, \ z = g \text{ on } \delta\Omega$$

where $g \geq 0$, $\not\equiv 0$ on $\delta\Omega$. Then $\tilde{z}(x) > 0$ for all $x \in \Omega$. (Here, we assume that g has an extension $\hat{g} \in H^{2+\ell}(\bar{\Omega})$.)

Proof. Let γ be a positive number satisfying $\gamma > \max\{|p(x,0)|: x \in \bar{\Omega}\}$. Choose k, $0 < k < 1$ such that $|p(x,z) - p(x,0)| < \gamma$ for all z satisfying $0 \leq z \leq k \max\{g(x): x \in \delta\Omega\}$ and $x \in \bar{\Omega}$. For such (x,z) we clearly have $|p(x,z)| < 2\gamma$. Let v be a solution of $\Delta v - 2\gamma v = 0$ in Ω, $v = kg$ on $\delta\Omega$. From the maximum principle, we find that $0 < v(x) \leq k \max\{g(x): x \in \delta\Omega\}$ for all $x \in \Omega$. For each constant θ, $0 \leq \theta \leq 1$, we have $\Delta(\theta v) + (\theta v)\, p(x,\theta v) = \theta v[2\gamma + p(x,\theta v)] > 0$ for all $x \in \Omega$, and $\theta v = \theta kg \leq g$ for $x \in \delta\Omega$. By Serrin's sweeping principle, (Theorem 1.4-3), we conclude that $\tilde{z} \geq 1v$ for $x \in \bar{\Omega}$. Since $v > 0$ in Ω, we have $\tilde{z}(x) > 0$ for $x \in \Omega$.

Lemma 5.4-3 (Comparison). Let $h_i(x,z)$, $i = 1,2$ be functions defined on $\bar{\Omega} \times [0,\infty)$ satisfying all the assumptions concerning $h(x,z)$ in lemma 5.4-1. Further, suppose $h_1(x,z) \geq h_2(x,z)$ for all $(x,z) \in \bar{\Omega} \times [0,\infty)$. For each i=1, 2, let $z_i \in H^{2+\ell}(\bar{\Omega})$, $z_i(x) > 0$ for each $x \in \Omega$, satisfies

$$\Delta z_i + z_i h_i(x,z_i) = 0 \ \text{in}\ \Omega, \quad z_i = g \ \text{on}\ \delta\Omega \tag{5.4.5}$$

where $g \geq 0$, $\not\equiv 0$ on $\delta\Omega$ and has extension $\hat{g} \in H^{2+\ell}(\bar{\Omega})$. Then $z_1(x) \geq z_2(x)$ for all $x \in \Omega$.

Proof. The functions z_1, 0 are respectively upper and lower solutions for the problem $\Delta z + zh_2(x,z) = 0$ in Ω, $z = g$ on $\delta\Omega$. Hence, the problem has a solution $z = \bar{z}$ with $0 \leq \bar{z} \leq z_1$. By lemma 5.4-2, we have $\bar{z}(x) > 0$ in Ω. Then apply lemma 5.4-1 to conclude that $z_2 = \bar{z}$, thus $z_2 \leq z_1$.

(I) Food Chain

A species X is said to eat species Y if: (i) Y enhances the growth of X, and X inhibits or has no effect on the growth of Y; or (ii) Y has no effect on the growth of X, and X inhibits the growth of Y. In this case, we consider three species A, B, C (with corresponding concentrations u, v, w), where (i) C eats B,

(ii) B eats A, and (iii) C eats A or has no direct relation with A. This food chain condition can be mathematically summarized as

(5.4-6) $$\frac{\partial f_1}{\partial v} < 0, \quad \frac{\partial f_1}{\partial w} < 0 \quad ,$$

(5.4-7) $$\frac{\partial f_2}{\partial u} > 0, \quad \frac{\partial f_2}{\partial w} < 0 \quad ,$$

(5.4-8) $$\frac{\partial f_3}{\partial u} > 0, \quad \frac{\partial f_3}{\partial v} > 0 \quad ,$$

(5.4-9) $$\frac{\partial f_1}{\partial v} \text{ and } \frac{\partial f_2}{\partial u} \text{ cannot be both identically zero ,}$$

(5.4-10) $$\frac{\partial f_2}{\partial w} \text{ and } \frac{\partial f_3}{\partial v} \text{ cannot be both identically zero .}$$

Relations (5.4-6) to (5.4-10) are all considered in the region $\overline{R}^+ \times \overline{R}^+ \times \overline{R}^+$. Note that we may have both $\frac{\partial f_1}{\partial w}$ and $\frac{\partial f_3}{\partial u}$ being identically 0, corresponding to the situation when there is no direct relation between C and A. Case (II) below includes the situation when (5.4-9) is violated; and case (III) includes the situation when (5.4-10) is violated. When both (5.4-9) and (5.4-10) are violated, it will be two species interaction.

We now construct monotone sequences of functions closing in to solution(s) of our nontrivial, non-negative Dirichlet boundary value problem (5.4-1). First let $u_1 \in H^{2+\ell}(\overline{\Omega})$ be the unique strictly positive function in Ω, satisfying

(5.4-11) $$\Delta u_1 + u_1 [a + f_1(u_1, 0, 0)] = 0 \quad \text{in } \Omega, \; u_1 = g_1 \quad \text{on } \delta\Omega \; .$$

(Note that 0 and a large positive constant are respectively lower and upper solutions for this problem. Hence from section 5.1 and lemma 5.4-2, we have the existence of a solution which is positive in Ω. Then lemma 5.4-1 implies that u_1 is uniquely defined.) Similarly, let v_1 and w_1 be functions in $H^{2+\ell}(\overline{\Omega})$, $v_1 > 0$, $w_1 > 0$ in Ω respectively satisfying

$$(5.4\text{-}12)\qquad \Delta v_1 + v_1[b + f_2(u_1, v_1, 0)] = 0 \text{ in } \Omega,\ v_1 = g_2 \text{ on } \delta\Omega;$$

$$(5.4\text{-}13)\qquad \Delta w_1 + w_1[c + f_3(u_1, v_1, w_1)] = 0 \text{ in } \Omega,\ w_1 = g_3 \text{ on } \delta\Omega.$$

The existence and uniqueness of such functions follow again from the method of upper and lower solutions and lemmas 5.4-2 and 5.4-1.

We now define u_i, v_i, w_i, $i = 2,3$ to be strictly positive functions in Ω, inductively as follows:

$$(5.4\text{-}14)\qquad \Delta u_i + u_i[a + f_1(u_i, v_{i-1}, w_{i-1})] = 0 \text{ in } \Omega,\ u_i = g_1 \text{ on } \delta\Omega;$$

$$(5.4\text{-}15)\qquad \Delta v_i + v_i[b + f_2(u_i, v_i, w_{i-1})] = 0 \text{ in } \Omega,\ v_i = g_2 \text{ on } \delta\Omega;$$

$$(5.4\text{-}16)\qquad \Delta w_i + w_i[c + f_3(u_i, v_i, w_i)] = 0 \text{ in } \Omega,\ w_i = g_3 \text{ on } \delta\Omega.$$

The existence and uniqueness of such function in $H^{2+\ell}(\bar{\Omega})$ follow from exactly the same reasons as that for u_1, v_1, w_1.

Lemma 5.4-4. *For each* $i = 2, 3, \ldots,$ *the following are true:*

$$u_i < u_1,\ v_i < v_1,\ w_i < w_1 \text{ for all } x \in \bar{\Omega}.$$

Proof. Consider equations (5.4-11) and (5.4-14). We note the inequality $f_1(z,0,0) > f_1(z, v_{i-1}(x), w_{i-1}(x))$ for $x \in \Omega$, $z \in [0,\infty)$, for $i > 2$, because of conditions (5.4-6) and the positivity of v_{i-1}, w_{i-1} in Ω. Applying lemma 5.4-3, we conclude that $u_1 > u_i$ in $\bar{\Omega}$, for $i > 2$. Next, consider equations (5.4-12) and (5.4-13). Note the inequality $f_2(u_1(x), z, 0) > f_2(u_i(x), z, w_{i-1}(x))$ for $x \in \bar{\Omega}$, $z \in [0,\infty)$, $i > 2$, because of conditions (5.4-7), $w_{i-1} > 0$ and $u_1 > u_i$ in $\bar{\Omega}$. Applying lemma 5.4-3, we conclude that $v_1 > v_i$ in $\bar{\Omega}$, for $i > 2$. Finally, consider equations (5.4-13) and (5.4-16). Note the inequality

$f_3(u_1(x), v_1(x), z) > f_3(u_i(x), v_i(x), z)$ for $x \in \bar{\Omega}$, $z \in [0,\infty)$, $i > 2$, because of conditions (5.4-8), $u_1 > u_i$, $v_1 > v_i$ in $\bar{\Omega}$. Applying lemma 5.4-3 again, we conclude $w_1 > w_i$ in $\bar{\Omega}$, for $i > 2$.

We next deduce some more refined order relationships among the sequences u_i, v_i and w_i, $i = 1, 2, \ldots$.

Theorem 5.4-1. For each nonnegative integer n, the following are true:

(5.4-17i) $$u_{2n+2} < u_{2n+4} < u_{2n+3} < u_{2n+1} \quad ,$$

(5.4-17ii) $$v_{2n+2} < v_{2n+4} < v_{2n+3} < v_{2n+1} \quad ,$$

(5.4-17iii) $$w_{2n+2} < w_{2n+4} < w_{2n+3} < w_{2n+1} \quad ,$$

for all $x \in \bar{\Omega}$.

Proof. We first consider the case when $n = 0$. Observe the inequality $f_1(z, v_1(x), w_1(x)) < (f_1(z, v_2(x), w_2(x))$ for $x \in \bar{\Omega}$, $z \in [0,\infty)$, because of conditions (5.4-6) and $v_1 > v_2$, $w_1 > w_2$ in $\bar{\Omega}$ as proved in lemma 5.4-4. Applying lemma 5.4-3 to equations (5.4-14) for $i = 2$ and 3, we conclude that $u_2 < u_3$ in $\bar{\Omega}$. Next, observe that $f_2(u_2(x), z, w_1(x)) < f_2(u_3(x), z, w_2(x))$ for $x \in \bar{\Omega}$, $z \in [0,\infty)$, because of conditions (5.4-7) and $u_2 < u_3$, $w_2 < w_1$ in $\bar{\Omega}$ as proved above. Applying lemma 5.4-3 to equations (5.4-15) for $i = 2$ and 3, we conclude that $v_2 < v_3$ in $\bar{\Omega}$. Then using the inequalities just proved and (5.4-8) we check similarly that $f_3(u_2(x), v_2(x), z) < f_3(u_3(x), v_3(x), z)$ for $x \in \bar{\Omega}$, $z \in [0,\infty)$; and applying lemma 5.4-3 we obtain $w_2 < w_3$ in $\bar{\Omega}$.

We will continue to use (5.4-6) - (5.4-8) and lemma 5.4-3 repeatedly as above. $v_3 < v_1$ and $w_3 < w_1$ in $\bar{\Omega}$ imply that $f_1(z, v_1(x), w_1(x)) < f_1(z, v_3(x), w_3(x))$ for $x \in \bar{\Omega}$, $z \in [0,\infty)$. This leads to $u_2 < u_4$ in $\bar{\Omega}$. $v_2 < v_3$ and $w_2 < w_3$ in $\bar{\Omega}$ (established in the above paragraph) imply that $f_1(z, v_3(x), w_3(x)) <$

$f_1(z, v_2(x), w_2(x))$ for $x \in \bar{\Omega}$, $z \in [0,\infty)$. This leads to $u_4 < u_3$ in $\bar{\Omega}$. We have now proved (5.4-17i) for $n = 0$. Using the same techniques, we prove the following in order: $v_2 < v_4$, $v_4 < v_3$, $w_2 < w_4$, $w_4 < w_3$. This establishes (5.4-17ii) and (5.4-17iii) for $n = 0$.

Assume that the lemma is true for $n = j$, we then keep applying lemma 5.4-3 to establish the following inequalities in order:

$$u_{2j+4} < u_{2j+5} < u_{2j+3},\ v_{2j+4} < v_{2j+5} < v_{2j+3},\ w_{2j+4} < w_{2j+5} < w_{2j+3},$$

$$u_{2j+4} < u_{2j+6} < u_{2j+5},\ v_{2j+4} < v_{2j+6} < v_{2j+5},\ w_{2j+4} < w_{2j+6} < w_{2j+5}.$$

This proves that (5.4-17 i-iii) are true for $n = j + 1$, and thus the lemma.

Theorem 5.4-1 clearly implies that $0 < u_2 < u_4 < u_6 \cdots < u_5 < u_3 < u_1$, $0 < v_2 < v_4 < v_6 \cdots < v_5 < v_3 < v_1$ and $0 < w_2 < w_4 < w_6 \cdots < w_5 < w_3 < w_1$ for all $x \in \bar{\Omega}$. Define $u^* = \lim_{n\to\infty} u_{2n+1}$, $u_* = \lim_{n\to\infty} u_{2n}$, $v^* = \lim_{n\to\infty} v_{2n+1}$, $v_* = \lim_{n\to\infty} v_{2n}$, $w^* = \lim_{n\to\infty} w_{2n+1}$, $w_* = \lim_{n\to\infty} w_{2n}$. We have the following comparison theorem.

<u>Theorem</u> 5.4-2. <u>Any solution</u> (u,v,w) <u>of the boundary value problem</u> (5.4-1) <u>with</u> u, v, w <u>in</u> $H^{2+\ell}(\bar{\Omega})$, > 0, $\not\equiv 0$ <u>in</u> $\bar{\Omega}$ <u>must satisfy</u>:

$$u_* < u < u^*,\ v_* < v < v^*,\ w_* < w < w^* \tag{5.4-18}$$

<u>for all</u> $x \in \bar{\Omega}$. (<u>Here, we assume</u> (5.4-6) <u>to</u> (5.4-10).)

The theorem is proved by using lemma 5.4-2 and inducting with repeated application of lemma 5.4-3 as before (cf. Theorem 5.2-4).

If $u_* = u^*$, $v_* = v^*$, and $w_* = w^*$, we clearly have uniqueness of solution with properties as described in Theorem 5.4-2.

(II) Two Predators with One Prey

A species X is said to compete with species Y if: (i) X inhibits the growth of Y, and Y inhibits or has no effect on the growth of X; or (ii) X has no effect on the growth of Y, and Y inhibits the growth of X. In this subsection, we consider three species A, B, C (with corresponding concentrations u, v, w) where: (i) B eats A and/or C eats A (with at least one relation true), and (ii) B competes or has no direct relation with C. (If in (ii), B has no direct relation with C, we assume that both relations in (i) hold, otherwise there are only two interacting species). If the competition relation between B and C is changed to that of prey-predator, then the situation becomes food-chain, as considered in case (I).

The situation can be mathematically summarized as:

(5.4-19) $$\frac{\partial f_1}{\partial v} < 0, \ \frac{\partial f_1}{\partial w} < 0;$$

(5.4-20) $$\frac{\partial f_2}{\partial u} > 0, \ \frac{\partial f_2}{\partial w} < 0;$$

(5.4-21) $$\frac{\partial f_3}{\partial u} > 0, \ \frac{\partial f_3}{\partial v} < 0;$$

(5.4-22) If the condition: (a) $\frac{\partial f_1}{\partial v}$ and $\frac{\partial f_2}{\partial u}$ are both identically zero, holds, then the condition: (b) both $\frac{\partial f_1}{\partial w}$ and $\frac{\partial f_3}{\partial u}$ are identically zero, cannot hold. Also, if (b) holds, then (a) cannot hold.

(5.4-23) If both $\frac{\partial f_2}{\partial w}$ and $\frac{\partial f_3}{\partial v}$ are identically zero, then both (a) and (b) cannot hold.

Relations (5-4.19) to (5.4-23) are all considered in the region $\bar{R}^+ \times \bar{R}^+ \times \bar{R}^+$. Let $u_1 \in H^{2+\ell}(\bar{\Omega})$ be the unique strictly positive function in Ω, satisfying

(5.4-24) $\Delta u_1 + u_1[a + f_1(u_1,0,0)] = 0$ in Ω, $u_1 = g_1$ on $\delta\Omega$.

Similarly, let v_1 and w_1 be functions in $H^{2+\ell}(\bar{\Omega})$, $v_1 > 0$, $w_1 > 0$ in Ω respectively satisfying:

(5.4-25) $\Delta v_1 + v_1[b + f_2(u_1,v_1,0)] = 0$ in Ω, $v_1 = g_2$ on $\delta\Omega$;

(5.4-26) $\Delta w_1 + w_1[c + f_3(u_1,0,w_1)] = 0$ in Ω, $w_1 = g_3$ on $\delta\Omega$.

Define u_i, v_i, w_i, $i = 2, 3, \ldots$ to be strictly positive functions in Ω, inductively as follows:

(5.4-27) $\Delta u_i + u_i[a + f_1(u_i,v_{i-1},w_{i-1})] = 0$ in Ω, $u_i = g_2$ on $\delta\Omega$;

(5.4-28) $\Delta v_i + v_i[b + f_2(u_i,v_i,w_{i-1})] = 0$ in Ω, $v_i = g_3$ on $\delta\Omega$;

(5.4-29) $\Delta w_i + w_i[c + f_3(u_i,v_{i-1},w_i)] = 0$ in Ω, $w_i = g_3$ on $\delta\Omega$.

As in case (I) we can show similarly that $0 < u_2 < u_4 < u_6 \ldots < u_5 < u_3 < u$ and similar inequalities among v_i and w_i, $i = 1, 2, \ldots$ A theorem analogous to that of Theorem 5.4-2 is also true in this case.

(III) <u>One Predator with Two Prey</u>

In this subsection, we consider three species A, B, C (with corresponding concentrations u, v, w) where:

(i) A competes or has no direct relation with B,

(ii) C eats A and/or C eats B (with at least one relation hold)

(If in (i) A has no direction relation with B, we assume that both relations in (ii) hold, otherwise there are only two interacting species). If the competition relation between A and B is changed to that of prey-predator, then the situation becomes food-chain, as considered in Case (I).

The situation can be mathematically summarized as:

(5.4-30) $\frac{\partial f_1}{\partial v} < 0, \frac{\partial f_1}{\partial w} < 0$;

(5.4-31) $\frac{\partial f_2}{\partial u} < 0, \frac{\partial f_2}{\partial w} < 0$;

(5.4-32) $\frac{\partial f_3}{\partial u} > 0, \frac{\partial f_3}{\partial v} > 0$;

(5.4-33) If the condition: (a) $\frac{\partial f_1}{\partial w}$ and $\frac{\partial f_3}{\partial u}$ are both identically zero, holds, then the condition: (b) $\frac{\partial f_2}{\partial w}$ and $\frac{\partial f_3}{\partial v}$ are both identically zero, cannot hold. Also, if (b) holds, then (a) cannot hold.

(5.4-34) If both $\frac{\partial f_1}{\partial v}$ and $\frac{\partial f_2}{\partial u}$ are identically zero, then both (a) and (b) cannot hold.

Relations (5.4-30) to (5.4-34) are all considered in the region $\bar{R}^+ \times \bar{R}^+ \times \bar{R}^+$.

We now construct our corresponding monotone sequences for this section. Let $u_1, v_1, \in H^{2+\ell}(\bar{\Omega})$ be the unique strictly positive functions in Ω, satisfying:

(5.4-35) $\Delta u_1 + u_1[a + f_1(u_1,0,0)] = 0$ in Ω, $u_1 = g_1$ on $\delta\Omega$;

(5.4-36) $\Delta v_1 + v_1[b + f_2(0,v_1,0)] = 0$ in Ω, $v_1 = g_2$ on $\delta\Omega$.

Let $w_1 \in H^{2+\ell}(\bar{\Omega})$, $w_1 > 0$ in Ω be the function satisfying:

(5.4-37) $\Delta w_1 + w_1[c + f_3(u_1,v_1,w_1)] = 0$ in Ω, $w_1 = g_3$ on $\delta\Omega$.

Define u_i, v_i, w_i, $i = 2, 3, \ldots$ to be strictly positive functions in Ω, inductively as follows:

(5.4-38) $\Delta u_i + u_i[a + f_1(u_i,v_{i-1},w_{i-1})] = 0$ in Ω, $u_i = g_1$ on $\delta\Omega$;

(5.4-39) $\Delta v_i + v_i[b + f_2(u_{i-1},v_i,w_{i-1})] = 0$ in Ω, $v_i = g_2$ on $\delta\Omega$;

(5.4-40) $$\Delta w_i + w_i[c + f_3(u_i, v_i, w_i)] = 0 \quad \text{in } \Omega, \; w_i = g_3 \text{ on } \delta\Omega.$$

As in the two cases above we can show that $0 < u_2 < u_4 < u_6 \cdots < u_5 < u_3$ and similar inequalities among v_i and w_i, $i = 1, 2, \ldots$. A theorem analogous to that of theorem 5.4-2 is also true

(IV) Mutualistic Loop

Here, we consider three species A, B, C (with corresponding concentrations u, v, w) where (i) A eats C, (ii) C eats B and (iii) B eats A. We will need more restrictive conditions to obtain monotonic sequences which converge as in the previous cases. In order to simplify these conditions, we will only consider Volterra-Lotka type of interactions. The Dirichlet problem we consider in this section is the following:

$$\Delta u + u[a + \lambda_{11}u + \lambda_{12}v + \lambda_{13}w] = 0 \text{ in } \Omega, \; u = g_1 \text{ on } \delta\Omega$$

(5.4-41) $$\Delta v + v[b + \lambda_{21}u + \lambda_{22}v + \lambda_{23}w] = 0 \text{ in } \Omega, \; v = g_2 \text{ on } \delta\Omega$$

$$\Delta w + w[c + \lambda_{31}u + \lambda_{32}v + \lambda_{33}w] = 0 \text{ in } \Omega, \; w = g_3 \text{ on } \delta\Omega$$

where a, b, c, λ_{ij}, $1 < i, j < 3$ are constants, with λ_{11}, λ_{22} and λ_{33} being negative, $\lambda_{12} < 0$, $\lambda_{13} > 0$, $\lambda_{21} > 0$, $\lambda_{23} < 0$, $\lambda_{31} < 0$, $\lambda_{32} > 0$. We assume that for each pair (i,j), λ_{ij} and λ_{ji} cannot be both zero, so that there is indeed interaction between the corresponding pair of species. (Otherwise, the situation reduces to that of food chain or less than three species interaction.) The functions g_1, g_2, g_3 satisfy conditions as described before. We assume the following conditions which will insure the sequences we construct will be monotonic:

(5.4-42) $$\frac{\lambda_{32}\lambda_{21}\lambda_{13}}{|\lambda_{11}\lambda_{22}\lambda_{33}|} < 1,\ \lambda_{13} > 0,\ \lambda_{21} > 0,\ \lambda_{32} > 0$$

We now construct our monotone sequences as in the earlier cases. First, choose P to be a sufficiently large positive constant so that:

(5.4-43) $$\alpha \overset{(def)}{\equiv} (a + \lambda_{13}P)|\lambda_{11}|^{-1} > \max_{x \in \delta\Omega} g_1(x),$$

(5.4-44) $$\beta \overset{(def)}{\equiv} (b + \lambda_{21}\alpha)|\lambda_{22}|^{-1} > \max_{x \in \delta\Omega} g_2(x),$$

(5.4-45) $$\gamma \overset{(def)}{\equiv} (c + \lambda_{32}\beta)|\lambda_{33}|^{-1} > \max_{x \in \delta\Omega} g_3(x),$$

(5.4-46) $$\left[1 - \frac{\lambda_{32}\lambda_{21}\lambda_{13}}{|\lambda_{11}\lambda_{22}\lambda_{33}|}\right]P > \frac{c}{|\lambda_{33}|} + \frac{\lambda_{32}b}{\lambda_{33}\lambda_{22}} + \frac{\lambda_{32}\lambda_{21}a}{|\lambda_{33}\lambda_{22}\lambda_{11}|},$$

are all satisfied. Let $u_1 \in H^{2+\ell}(\bar{\Omega})$ be the unique strictly positive function in Ω satisfying

(5.4-47) $$\Delta u_1 + u_1[a + \lambda_{11}u_1 + \lambda_{13}P] = 0 \text{ in } \Omega,\ u_1 = g_1 \text{ on } \delta\Omega.$$

Similarly, let v_1 and w_1 be functions in $H^{2+\ell}(\bar{\Omega})$, $v_1 > 0$, $w_1 > 0$ in Ω respectively satisfying:

(5.4-48) $$\Delta v_1 + v_1[b + \lambda_{21}u_1 + \lambda_{22}v_1] = 0 \text{ in } \Omega,\ v_1 = g_2 \text{ on } \delta\Omega$$

(5.4-49) $$\Delta w_1 + w_1[c + \lambda_{32}v_1 + \lambda_{33}w_1] = 0 \text{ in } \Omega,\ w_1 = g_3 \text{ on } \delta\Omega.$$

Note that $0 < u_1(x) < \alpha$ for all $x \in \bar{\Omega}$, because the constant functions 0, α are respectively lower and upper solutions for (5.4-47) and lemma 5.4-1 applies. Similarly $0 < v_1(x) < \beta$ and $0 < w_1(x) < \gamma$ for all $x \in \bar{\Omega}$. Note that the constant function γ is an upper solution for w_1 in (5.4-49), and using (5.4-45), (5.4-44), (5.4-43), we obtain:

$$w_1 < \frac{c}{|\lambda_{33}|} + \frac{\lambda_{32}b}{\lambda_{33}\lambda_{22}} + \frac{\lambda_{32}\lambda_{21}a}{|\lambda_{33}\lambda_{22}\lambda_{11}|} + \frac{\lambda_{32}\lambda_{21}\lambda_{13}}{|\lambda_{33}\lambda_{22}\lambda_{11}|} P .$$

Using (5.4-46), we conclude that $w_1(x) < P$ for all $x \in \bar{\Omega}$. For convenience, we denote $w_0(x) \equiv 0$ for $x \in \bar{\Omega}$.

Define u_i, v_i, w_i, $i = 2, 3, \ldots$ to be strictly positive functions in Ω, inductively as follows:

(5.4-50) $\quad \Delta u_i + u_i[a + \lambda_{11}u_i + \lambda_{12}v_{i-1} + \lambda_{13}w_{i-2}] = 0$ in Ω, $u_i = g_1$ on $\delta\Omega$;

(5.4-51) $\quad \Delta v_i + v_i[b + \lambda_{21}u_i + \lambda_{22}v_i + \lambda_{23}w_{i-1}] = 0$ in Ω, $v_i = g_2$ on $\delta\Omega$;

(5.4-52) $\quad \Delta w_i + w_i[c + \lambda_{31}u_{i-1} + \lambda_{32}v_i + \lambda_{33}w_i] = 0$ in Ω, $w_i = g_3$ on $\delta\Omega$.

The existence and uniqueness of such functions can be shown as before. We can also show that $0 < u_2 < u_4 < u_6 < \ldots < u_5 < u_3 < u_1$ and similar inequalities among v_i and w_i. To show the existence of a solution of (5.4-41) with non-negative components in $H^{2+\ell}(\bar{\Omega})$, we first note that $u = u_1$ is an upper solution of the first equation of (5.4-41) for each $0 < v < v_1$, $0 < w < w_1$; $v = v_1$ is an upper solution of the second equation for each $0 < u < u_1$, $0 < w < w_1$; and $w = w_1$ is an upper solution of the third equation for $0 < u < u_1$, $0 < v < v_1$. On the other hand, $u = 0$, $v = 0$, $w = 0$ are respectively lower solutions for the corresponding equations in (5.4-41). Consequently, by Theorem 1.4-2, there exists a solution (u,v,w) for (5.4-41) with u, v, w in $H^{2+\ell}(\bar{\Omega})$, $0 < u < u_1$, $0 < v < v_1$, $0 < w < w_1 < P$. With u_i, v_i, w_i, $i = 2, 3, \ldots$ as defined in (5.4-50) to (5.4-52), and u_*, u^*, v_*, v^*, w^*, w_* defined as appropriate limiting functions as in case (I), we can show that a theorem analogous to that of theorem 5.4-2 is true. Applications of these sequences to the study of uniqueness and stability problem as in section 5.3 can also be made (cf. [140]).

5.5 General Scheme for a System of m Equations

In this section, we generalize the schemes in the last few sections to study the Dirichlet boundary value problem for elliptic systems:

$$\Delta u_i + f_i(x, u_1, \ldots, u_m) = 0, \quad x = (x_1, \ldots, x_n) \in \Omega,$$

(5.5-1)

$$u_i = g_i \quad \text{on } \delta\Omega, \qquad i = 1, \ldots, m$$

We will assume that each function f_i depends monotonically on the variables $u = (u_1, \cdot, \ldots, u_m)$, i.e. $\frac{\partial f_i}{\partial u_j} > 0$ or $\frac{\partial f_i}{\partial u_j} < 0$ for u varying in some region. We will not always visualize the system in an ecological context, hence we do not assume that f_i satisfy $f_i(x, u_1, \ldots, u_{i-1}, 0, u_{i+1}, \ldots u_m) = 0$ (Note that this is satisfied in sections 5.2 to 5.4, if we adapt the symbols there to the present situation). If we review the schemes in the last section (e.g. (5.4-38) to (5.4-40)), we notice that each compoenent of the kth iterate is obtained by solving a nonlinear scalar equation. The off-diagonal components make use of certain values already defined, while the diagonal components are used "implicitly", leading to a nonlinear equation. In our present situation, the method of utilizing previous iterates for off-diagonal components is analogous, and agrees with those in the last section when m = 3. The diagonal components now use those iterates two steps earlier, and hence our scheme is "explicit", with a linear scalar problem for the computation of a new iterate.

We will assume that:

(5.5-2) $\quad f_i(x, u_1, \ldots, u_m)$, $i = 1, \ldots, m$ are $H^{\ell}(\bar{\Omega})$ $0 < \ell < 1$ in x, and $H^{1+\ell}$ in each u_j.

for u in every compact set; the boundary functions g_i can be extended to be a function $\tilde{g}_i \in H^{2+\ell}(\bar{\Omega})$, for each $i = 1, \ldots, m$. We will also assume the existence of upper and lower solutions $w = (w_1, \ldots, w_m)$ and $v = (v_1, \ldots, v_m)$ respectively in $C^2(\bar{\Omega}) x \ldots x C^2(\bar{\Omega})$ to problem (5.5-1) in the following sense: For each i,

$$(5.5\text{-}3) \qquad v_i(x) \le w_i(x) \quad \text{for } x \in \bar{\Omega},$$

$$(5.5\text{-}4) \qquad \Delta w_i(x) + f_i(x, z_1, \ldots, z_{i-1}, w_i(x), z_{i+1}, \ldots, z_m) \le 0 \ ,$$

$$(5.5\text{-}5) \qquad \Delta v_i(x) + f_i(x, z_1, \ldots, z_{i-1}, v_i(x), z_{i+1}, \ldots, z_m) \ge 0 \ ,$$

for all $x \in \Omega$, $v_j(x) \le z_j \le w_j(x)$, $j \ne i$, and

$$(5.5\text{-}6) \qquad v_i(x) \le g_i(x) \le w_i(x) \quad \text{for } x \in \delta\Omega.$$

Theorem 5.5-1. *Let Ω and $\delta\Omega$ be as described at the end of section 5.1; and f_i, g_i satisfy the smoothness assumptions (5.5-2). Assume that problem (5.5-1) has upper and lower solutions w and v as described above; and for all i,j with $i \ne j$:*

$$(5.5\text{-}7) \qquad \frac{\partial f_i}{\partial u_j} \ge 0 \text{ or } \frac{\partial f_i}{\partial u_j} \le 0 \quad \text{for} \quad (x, u_1, \ldots, u_m) \text{ in the set}$$

$S \overset{(def)}{=} \{(x, u_1, \ldots u_m) \mid x \in \Omega,\ v_r \le u_r \le w_r,\ r = 1, \ldots, m\}$. *Let $u^{-1} = w$, $u^0 = v$ and u^k, $k = 1, 2, \ldots$ be defined recursively as solutions of:*

$$(5.5\text{-}8) \qquad \begin{aligned} \Delta u_1^k - P_1 u_1^k &= -f_1(x, u_1^{k-2}, u_2^{k+\ell_1(2)}, \ldots, u_m^{k+\ell_1(m)}) - P_1 u_1^{k-2} \\ \Delta u_2^k - P_2 u_2^k &= -f_2(x, u_1^{k+\ell_2(1)}, u_2^{k-2}, \ldots, u_m^{k+\ell_2(m)}) - P_2 u_2^{k-2} \\ &\ \vdots \\ \Delta u_m^k - P_m u_m^k &= -f_m(x, u_1^{k+\ell_m(1)}, u_2^{k+\ell_m(2)}, \ldots, u_m^{k-2}) - P_m u_m^{k-2} \end{aligned} \quad \text{in}$$

$$u_i^k = g_i \text{ on } \delta\Omega,\ i = 1, \ldots m,$$

where $P_i = \sup\{|\frac{\partial f_i}{\partial u_i}|: x \in \Omega \quad v_r \le u_r \le w_r, r = 1, \ldots, m\}$ *and for* $i \neq j$

(5.5-9)

$$\ell_i(j) = \begin{cases} -2 & \text{if } \partial f_i/\partial u_j \ge 0 \text{ in } S \\ -1 & \text{if } \partial f_i/\partial u_j \le 0 \text{ in } S \end{cases} \qquad \text{for } i+1 \le j \le n$$

$$\ell_i(j) = \begin{cases} 0 & \text{if } \partial f_i/\partial u_j \ge 0 \text{ in } S \\ -1 & \text{if } \partial f_i/\partial u_j \le 0 \text{ in } S \end{cases} \qquad \text{for } i \le j \le i-1$$

Then the boundary value problem (5.5-1) *has a solution* $u = (u_1, \ldots, u_m)$ *with* $v_i \le u_i \le w_i$, $u_i \in H^{2+\ell}(\bar{\Omega})$. *Moreover, any such solution satisfies the following inequalities.*

(5.5-10) $$v_i = u_i^0 \le u_i^2 \ldots \le u_i^{2r} \le \ldots \le u_i \le \ldots \le u_i^{2r-1} \le \ldots \le u_i^1 \le u_i^{-1}$$

for each $i = 1, \ldots, m$.

Proof. The existence of solution u to the boundary value problem (5.5-1) with components in $H^{2+\ell}(\bar{\Omega})$ between w and v follows from Theorem 1.4-2, even without hypotheses (5.5-7), it remains to prove (5.5-10). First, we proceed to show that

(5.5-11) $$v \equiv u_i^0 \le u_i^2 \le u_i \le u_i^1 \le u_i^{-1} \equiv w \quad \text{in } \bar{\Omega}$$

for each $i = 1, \ldots, m$. Note that with the additional terms $- P_i u_i$ on the right of (5.5-8), the dependence of the expression on the right side of the ith equation on u_i is nonincreasing. We have

(5.5-12)
$$\Delta u_1^1 - P_1 u_1^1 = -f_1(x, u_1^{-1}, u_2^{1+\ell_1(2)}, \ldots, u_m^{1+\ell_1(m)}) - P_1 u_1^{-1} \quad \text{in } \Omega,\ u_1^1 = g_1 \text{ on } \delta\Omega;$$

(5.5-13)

$$\Delta u_1^{-1} - P_1 u_1^{-1} \le -f_1(x, u_1^{-1}, u_2^{1+\ell_1(2)}, \ldots, u_m^{1+\ell_1(m)}) - P_1 u_1^{-1} \text{in } \Omega,\ u_1^{-1} \ge g_1 \text{ on } \partial\Omega;$$

(5.5-14)

$$\Delta u_1^0 - P_1u_1^0 \geq -f_1(x,u_1^0,u_2^{1+\ell 1(2)}, \ldots, u_m^{1+\ell 1(m)}) - P_1u_1^0 \quad \text{in } \Omega,\ u^0 \leq g_1 \text{ on } \delta\Omega$$

The right side of (5.5-12) is the same as that of (5.5-13), and is less or equal to that (5.5-14). From the maximum principle applied to differences of pairs of (5.5-12) to (5.5-14) we conclude that $u_1^0 \leq u_1^1 \leq u_1^{-1}$ in $\bar{\Omega}$. Suppose $u_j^0 \leq u_j^1 \leq u_j^{-1}$ in $\bar{\Omega}$ is true for $j < i \leq n$, we now show that

(5.5-15) $$u_i^0 \leq u_i^1 \leq u_i^{-1} \quad \text{in } \bar{\Omega}.$$

We have

(5.5-16)

$$\Delta u_i^1 - P_iu_i^1 = -f_i(x,u_1^{1+\ell i(1)}, \ldots, u_{i-1}^{1+\ell i(i-1)}, u_i^{-1}, u_{i+1}^{1+\ell i(i+1)}, \ldots, u_m^{1+\ell i(m)}) - P_iu_i^{-1} \quad \text{in } \Omega,\ u_i^1 = g_i \text{ on } \delta\Omega;$$

(5.5-17)

$$\Delta u_i^{-1} - P_iu_i^{-1} \leq -f_i(x,u_1^{1+\ell i(1)}, \ldots, u_i^{-1}, \ldots, u_m^{1+\ell i(m)}) - P_iu_i^{-1} \quad \text{in } \Omega,\ u_i^{-1} \geq g_i \text{ on } \delta\Omega;$$

(5.5-18)

$$\Delta u_i^0 - P_iu_i^0 \geq -f_i(x,u_1^{1+\ell i(1)}, \ldots, u_i^0, \ldots, u_m^{1+\ell i(m)}) - P_iu_i^0 \quad \text{in } \Omega,\ u_i^0 \leq g_i \text{ on } \delta\Omega.$$

The terms $u_j^{1+\ell i(j)}$ in (5.5-16) to (5.5-18) are all substituted in the same way. For $j < i$, $u_j^{1+\ell 1(j)}$ can possibly be u_j^1; however (5.5-17) and (5.5-18) are still valid because $u_j^0 \leq u_j^1 \leq u_j^{-1}$ is assumed valid for $j < i$. Applying maximum principle to differences as before, we deduce that (5.5-15) holds. Next we show that

(5.5-19) $$u_i^0 \leq u_i^2 \leq u_i^1 \quad \text{in } \bar{\Omega}$$

for each $i = 1, \ldots, m$. We have

(5.5-20)

$$\Delta u_1^2 - P_1 u_1^2 = -f_1(x, u_1^0, u_2^{2+\ell 1(2)}, \ldots, u_m^{2+\ell 1(m)}) - P_1 u_1^0 \text{ in } \Omega,\ u_1^2 = g_1 \text{ on } \delta\Omega.$$

(5.5-21)

$$\Delta u_1^0 - P_1 u_1^0 > -f_1(x, u_1^0, u_2^{2+\ell 1(2)}, \ldots, u_m^{2+\ell 1(m)}) - P_1 u_1^0 \text{ in } \Omega,\ u_1^0 < g_1 \text{ on } \delta\Omega.$$

Note that (5.5-21) is valid because $u_j^{2+\ell 1(j)}$ is either u_j^0 or u_j^1 which by (5.5-15) is always between v_j and w_j. Compare (5.5-20) with (5.5-21) and (5.5-12). Also note that if $\partial f_1/\partial u_j > 0$, then $1 + \ell_1(j) = -1$ and $u_j^{2+\ell 1(j)} < u_j^{1+\ell 1(j)}$; if $\frac{\partial f_1}{\partial u_j} < 0$, then $1 + \ell_1(j) = 0$ and $u_j^{1+\ell 1(j)} < u_j^{2+\ell 1(j)}$, by (5.5-15). Applying maximum principle to the differences of these equations as before, we obtain $u_1^0 < u_1^2 < u_1^1$. Assume that $u_j^0 < u_j^2 < u_j^1$ is valid for $j < i$, we now show that it is true for $j = i$. We have

(5.5-22)

$$\Delta u_i^1 - P_i u_i^1 = -f_i(u_1^{1+\ell i(1)}, \ldots, u_{i-1}^{1+\ell i(i-1)}, u_i^{-1}, u_{i+1}^{1+\ell i(i+1)}, \ldots, u_m^{1+\ell i(m)}) - P_i u_i^{-1} \text{ in } \Omega,\ u_i^1 = g_i \text{ on } \delta\Omega;$$

(5.5-23)

$$\Delta u_i^2 - P_i u_i^2 = -f_i(u_1^{2+\ell i(1)}, \ldots, u_{i-1}^{2+\ell i(i-1)}, u_i^0, u_{i+1}^{2+\ell i(i+1)}, \ldots, u_m^{2+\ell i(m)}) - P_i u_i^0 \text{ in } \Omega,\ u_i^2 = g_i \text{ on } \delta\Omega;$$

(5.5-24)

$$\Delta u_i^0 - P_i u_i^0 > -f_i(u_1^{2+\ell i(1)}, \ldots, u_{i-1}^{2+\ell i(i-1)}, u_i^0, u_{i+1}^{2+\ell i(i+1)}, \ldots, u_m^{2+\ell i(m)}) - P_i u_i^0 \text{ in } \Omega,\ u_i^0 < g_i \text{ on } \delta\Omega.$$

In (5.5-24), the inequality is valid when we substitute $u_j^{2+\ell i(j)}$ in the same way as in (5.5-23), although $u_j^{2+\ell i(j)}$ can possibly be u_j^2 for $j < i$. This is permissible due to the induction hypothesis concerning u_j^2. When we take the difference of (5.5-22) and (5.5-23), we need to compare $u_j^{1+\ell_i(j)}$ and $u_j^{2+\ell_i(j)}$. When $j > i$, the situation is like $j > 1$, $i = 1$ considered above. When $j < i$: we have $2 + \ell_i(j) = 2$ if $\partial f_i/\partial u_j > 0$, and $u_j^{2+\ell_i(j)} \leq u_j^{1+\ell_i(j)}$ by induction; while $2 + \ell_i(j) = 1$ if $\partial f_i/\partial u_j \leq 0$, and $u_j^{1+\ell_i(j)} \leq u_j^{2+\ell_i(j)}$ by (5.5-15). Subtracting, we obtain $\Delta(u_i^1 - u_i^2) \leq 0$ in Ω, $(u_i^1 - u_i^2) = 0$ on $\delta\Omega$, and we have $u_i^2 \leq u_i^1$ in $\bar{\Omega}$. Similarly we obtain $u_i^0 \leq u_i^2$ by subtracting (5.5-23) and (5.5-24). This proves (5.5-19). Thus, we have now

$$u_i^0 \leq u_i^2 \leq u_i^1 \leq u_i^{-1} \quad \text{in } \bar{\Omega}, \quad \text{and}$$

(5.5-25)

$$u_i^0 \leq u_i \leq u_i^{-1} \qquad \text{in } \bar{\Omega}$$

for each $i = 1, \ldots, m$. Use (5.5-25) and the exact equations (5.5-12) and (5.5-20) for u_1^1 and u_1^2 respectively, together with the equation

$$\text{(5.5-26)} \quad \Delta u_1 - P_1 u_1 = -f_1(x, u_1, \ldots, u_m) - P_1 u_1 \text{ in } \Omega, \; u_1 = g_1 \text{ on } \delta\Omega,$$

in order to apply maximum principle to the differences as before. This will lead to first $u_1 \leq u_1^1$ in $\bar{\Omega}$, and then $u_1^2 \leq u_1$ in $\bar{\Omega}$, i.e.,

$$\text{(5.5-27)} \qquad u_1^2 \leq u_1 \leq u_1^1 \quad \text{in } \bar{\Omega}$$

Next, assume that

$$\text{(5.5-28)} \qquad u_j^2 \leq u_j \leq u_j^1 \quad \text{in } \bar{\Omega}$$

is valid for $j < i \leq n$. Use the exact equations (5.5-22) and (5.5-23) for u_i^1 and u_i^2 respectively, together with the equation

$$\text{(5.5-29)} \quad \Delta u_i - P_i u_i = -f_i(x, u_1, \ldots, u_m) - P_i u_i \text{ in } \Omega, \; u_i = g_i \text{ on } \delta\Omega$$

in order to apply maximum principle to the differences as before. This will lead to first $u_i < u_i^1$ in $\bar{\Omega}$, and then $u_i^2 < u_i$ in $\bar{\Omega}$, i.e. (5.5-28) is true for $j = i$. Thus, inductively, we obtain the validity of (5.5-11) for each $i = 1, \ldots, m$.

The other inequalities in (5.5-10) will be proved by induction.

(A) First, suppose k is odd and we assume that

$$[H^o k] \quad u_i^0 < u_i^2 \ldots < u_i^{k-1} < u_i^{k-2} < \ldots < u^1 < u^{-1} \quad \text{in } \bar{\Omega}$$

is true for $i = 1, \ldots, m$. (Here, the superscript o in H^o designates oddness, and note that (5.5-11) is the same as $[H^o 3]$.) We now show that

$$u_i^{k-1} < u_i^k < u_i^{k-2} \quad \text{in } \bar{\Omega} \tag{5.5-30}$$

for $i = 1, \ldots, m$. The proof of (5.5-30) is by induction on rows.

(i) First we show that $u_1^{k-1} < u_1^k < u_1^{k-2}$ in $\bar{\Omega}$. Indeed, we have

$$\begin{aligned} \Delta u_1^k - P_1 u_1^k &= -f_1(x, u_1^{k-2}, u_2^{k+\ell_1(2)}, \ldots, u_m^{k+\ell_1(m)}) - P_1 u_1^{k-2}, \\ \Delta u_1^{k-1} - P_1 u_1^{k-1} &= -f_1(x, u_1^{k-3}, u_2^{k-1+\ell_1(2)}, \ldots, u_m^{k-1+\ell_1(m)}) - P_1 u_1^{k-3}, \\ \Delta u_1^{k-2} - P_1 u_1^{k-2} &= -f_1(x, u_1^{k-4}, u_2^{k-2+\ell_1(2)}, \ldots, u_m^{k-2+\ell_1(m)}) - P_1 u_1^{k-4}, \end{aligned} \tag{5.5-31}$$

each with boundary data g_1. If $\frac{\partial f_1}{\partial u_j} > 0$, then $k + \ell_1(j) = k - 2$ and hence, by $[H^o k]$,

$$u_j^{k-1+\ell_1(j)} < u_j^{k+\ell_1(j)} < u_j^{k-2+\ell_1(j)}, \quad \text{each } j > 1 . \tag{5.5-32}$$

If $\partial f_1/\partial u_j < 0$, then $k + \ell_1(j) = k - 1$ and hence by $[H^o k]$,

$$u_j^{k-2+\ell_1(j)} < u_j^{k+\ell_1(j)} < u_j^{k-1+\ell_1(j)}, \quad \text{each } j > 1 . \tag{5.5-33}$$

Applying the maximum principle to the differences of the equations in (5.5-31) we obtain $u_1^{k-1} < u_1^k < u_1^{k-2}$ in $\bar{\Omega}$.

(ii) Assuming that $u_j^{k-1} < u_j^k < u_j^{k-2}$ in $\bar{\Omega}$ for all j up to i - 1, we now show that (5.5-30) holds at component i. We have

$$\Delta u_i^k - P_i u_i^k = -f_i(x, u_1^{k+\ell_i(1)}, \ldots, u_i^{k-2}, \ldots, u_m^{k+\ell_i(m)}) - P_i u_i^{k-2},$$

(5.5-34) $$\Delta u_i^{k-1} - P_i u_i^{k-1} = -f_i(x, u_1^{k-1+\ell_i(1)}, \ldots, u_i^{k-3}, \ldots, u_m^{k-1+\ell_i(m)}) - P_i u_i^{k-3}$$

$$\Delta u_i^{k-2} - P_i u_i^{k-2} = -f_i(x, u_1^{k-2+\ell_i(1)}, \ldots, u_i^{k-4}, \ldots, u_m^{k-2+\ell_i(m)}) - P_i u_i^{k-4}$$

each with boundary conditions g_i. It is sufficient to consider the case when $i > j$, since for $i < j$ (on and above the diagonal) considerations are the same as in case (i) above. If $\partial f_i/\partial u_j > 0$, then $k + \ell_i(j) = k$ and (5.5-32) holds with $\ell_1(j)$ replaced by $\ell_i(j)$, for each $j < i$. If $\partial f_i/\partial u_j < 0$, then $k + \ell_i(j) = k - 1$ and (5.5-33) holds with $\ell_1(j)$ replaced by $\ell_i(j)$, for each $j < i$. By the maximum principle again, (5.5-30) holds at component i, through using equations (5.5-34).

(B) Next, suppose k is even and we assume that

$$[H^e k] \qquad u^0 < u^2 < \ldots < u^{k-2} < u^{k-1} < \ldots < u^1 < u^{-1} \quad \text{in } \bar{\Omega}$$

(Note that $[H^e 4]$ follows from $[H^o 3]$ and (5.5-30) with k = 3; e designates evenness.) We now show that

(5.5-35) $$u_i^{k-2} < u_i^k < u_i^{k-1} \quad \text{in } \bar{\Omega}$$

for i = 1, ..., m. The proof of (5.5-35) is by induction on rows.

(i) First we show that $u_1^{k-2} < u_1^k < u_1^{k-1}$. These iterates are again determined by (5.5-31).

If $\partial f_1/\partial u_j > 0$, then $k + \ell_1(j) = k-2$, and hence, by [$H^e k$]

$$u_j^{k-2+\ell_1(j)} < u_j^{k+\ell_1(j)} < u_j^{k-1+\ell_1(j)}, \text{ each } j > 1. \tag{5.5-36}$$

If $\partial f_1/\partial u_j < 0$, then $k + \ell_1(j) = k-1$, and hence, by [$H^e k$],

$$u_j^{k-1+\ell_1(j)} < u_j^{k+\ell_1(j)} < u_j^{k-2+\ell_1(j)}, \text{ each } j > 1. \tag{5.5-37}$$

In both cases (5.5-35) for $i = 1$ follows from (5.5-31).

(ii) Assume that $u_j^{k-2} < u_j^k < u_j^{k-1}$ in $\bar{\Omega}$ holds for all components j up to $i - 1$. We now show that (5.5-35) holds at component i. These iterates are determined by (5.5-34) and again we may assume $i > j$. If $\partial f_i/\partial u_j > 0$, then $k + \ell_i(j) = k$, and hence (5.5-36) holds with $\ell_1(j)$ replaced by $\ell_i(j)$, for each $j < i$. If $\partial f_i/\partial u_j < 0$, then $k + \ell_i(j) = k - 1$ and hence (5.5-37) holds with $\ell_1(j)$ replaced by $\ell_i(j)$, for each $j < i$. In both cases (5.5-35) for the ith component follows from (5.5-34).

Successive applications of parts (A) and (B) lead to the validity of [$H^o k+2$] and [$H^e k+2$], and the following is valid

$$u_i^0 < u_i^2 \dots < u_1^{2r} < \dots < u_i^{2r-1} < \dots < u_i^1 < u_i^{-1} \text{ in } \bar{\Omega} \tag{5.5-38}$$

for each positive integer r, $i = 1, \dots, m$.

Utilizing (5.5-38) up to $r = k$, and assume that we have

$$u_i^{2k-2} < u_i < u_i^{2k-3} \text{ in } \bar{\Omega} \text{ for } i = 1, \dots m, \tag{5.5-39}$$

we can prove that

$$u_i^0 < \dots < u_i^{2k-2} < u_i^{2k} < u_i < u_i^{2k-1} < u_i^{2k-3} \dots < u_i^{-1} \text{ in } \bar{\Omega} \tag{5.5-40}$$

for $i = 1, \dots, m$. (Note that (5.5-11) implies the validity of (5.5-39) for $k= 2$). The proof here is exactly analogous to proving (5.5-11) starting from

(5.5-25). We simply use the appropriate equations for the corresponding functions, take difference, and apply maximum principle. Inducting on the row i and omitting details, we finally obtain (5.5-40). The proof of (5.5-10) is completed by induction.

5.6 Periodic Solutions for Nonlinear Parabolic Systems

In this section we study periodic solutions for parabolic systems of equations. We will not discuss monotone schemes related to these systems. However, if we adapt the techniques of the previous sections in this chapter to the results of this section, one should be able to obtain analogous results.

For this section, let Ω be a bounded domain in R^n, with boundary $\delta\Omega$ in $H^{2+\alpha}$, $0 < \alpha < 1$, as described in section 5.1. Let $\Pi = \Omega x\ (-\infty,\infty)$ and Γ be its lateral boundary. For a scalar function $u(x,t)$ defined in $\overline{\Pi} = \overline{\Omega}\ x\ (-\infty,\infty)$, we define

$$|u|_0 = \sup_{\overline{\Pi}} |u|,$$

$$|u|_\alpha = |u|_0 + \sup \left\{ \frac{|u(P_1)-u(P_2)|}{[d(P_1,P_2)]^\alpha} : P_1, P_2 \in \overline{\Pi},\ P_1 \neq P_2 \right\},$$

$$|u|_{1+\alpha} = |u|_\alpha + \sum_{i=1}^{n} \left|\frac{\partial u}{\partial x_i}\right|_\alpha ,$$

$$|u|_{2+\alpha} = |u|_{1+\alpha} + \sum_{i=1}^{n} \left|\frac{\partial u}{\partial x_i}\right|_{1+\alpha} + \left|\frac{\partial u}{\partial t}\right|_\alpha ,$$

where $P_1 = (x',t')$, $P_2 = (x",t")$ and $d(P_1,P_2) = [|x'-x"|^2 + |t'-t"|]^{1/2}$. We say $u \in C^q(\overline{\Pi})$ if $|u|_q < \infty$ ($q = 0, \alpha, 1+\alpha, 2+\alpha$). For a vector function $u(x,t) = (u_1(x,t), \ldots, u_m(x,t))$ defined in $\overline{\Pi}$, let

$$|u| = [\sum_{j=1}^{m} (u_j)^2]^{1/2} ,$$

$$|u|_q = \sum_{j=1}^{m} |u_j|_q , \quad q = 0, \alpha, 1 + \alpha, 2 + \alpha ,$$

and $u \in C^q(\overline{\Pi})$ if $|u|_q < \infty$.

We first describe some preliminary results concerning linear theory for scalar parabolic equations. We consider the periodic boundary value problem:

$$Lu(x,t) \overset{(def)}{\equiv} \sum_{i,j=1}^{n} a_{ij}(x,t)u_{x_i x_j} + \sum_{i=1}^{n} b_i(x,t)\, u_{x_i} + c(x,t)u - u_t$$

$$(5.6\text{-}1) \qquad = f(x,t) \qquad \text{for } (x,t) \in \Pi,$$

$$u(x,t) = \psi(x,t) \qquad \text{for } (x,t) \text{ on } \delta\Omega \times (-\infty,\infty),$$

where $a_{ij}(x,t)$, $b_i(x,t)$, $c(x,t)$, $f(x,t)$ and $\psi(x,t)$ are all T-periodic functions of t for $x \in \Omega$.

<u>Theorem</u> 5.6-1. <u>Assume that</u>:

(i) <u>The functions</u> a_{ij}, b_i <u>and</u> c <u>are in</u> $C^{\alpha}(\overline{\Pi})$ <u>with</u> $\sum_{i,j=1}^{n} |a_{ij}|_{\alpha} + \sum_{i=1}^{n} |b_i|_{\alpha} + |c|_{\alpha} < k_1 < \infty$ <u>in</u> Π;

(ii) $\sum_{i,j=1}^{n} a_{ij}(x,t)\, \xi_i \xi_j > \sigma \sum_{i=1}^{n} \xi_i^2$ <u>for some constant</u> $\sigma > 0$, <u>for any real vector</u> $\xi = (\xi_1, \ldots, \xi_n)$, <u>and all</u> $(x,t) \in \overline{\Pi}$:

(iii) $c(x,t) < 0$ <u>in</u> $\overline{\Pi}$

(iv) f <u>is uniformly Hölder continuous in</u> Π; <u>that is, for any closed subset</u> S <u>of</u> Π, $|f(x,t) - f(x',t')| < k_2[|x-x'|^2 + |t-t'|]^{\alpha/2}$ <u>for any</u> (x,t), $(x',t) \in S$, <u>where</u> k_2 <u>is independent of</u> S.

(v) $\psi(x,t)$ *can be extended to a function* $\hat{\psi}(x,t) \in C^{2+\alpha}(\overline{\Pi})$, *such that* $\hat{\psi} = \psi$ *on* $\partial\Omega x(-\infty,\infty)$ *and* $\hat{\psi}$ *is periodic in* t *with period* T.

Then there exists a unique T-*periodic solution* $u(x,t)$ *in* $C^{2+\alpha}(\overline{\Pi})$ *for the linear boundary value problem* (5.6-1).

The proof of the above Theorem can be found in [72]. Further, there is also an estimate of the $|u|_{1+\nu}$ norm of the solution u in terms of the bound $|f|_0$. This is stated carefully in the next theorem, and it will be used in the proof of compactness for an appropriate operator in the study of the nonlinear problem later.

Theorem 5.6-2. *Consider the boundary value problem* (5.6-1) *with* $a_{ij}(x,t)$, $b_i(x,t)$, $c(x,t)$ *and* $f(x,t)$ *all* T-periodic *in* t *for all* $x \in \overline{\Omega}$, *and* $\psi(x,t) \equiv 0$. *Assume that*:

(i) *The functions* b_i, c *and* f *are continuous in* $\overline{\Pi}$, *and* a_{ij} are *in* $C^{\alpha}(\overline{\Pi})$, *with* $\sum_{i,j=1}^{n} |a_{ij}|_{\alpha} + \sum_{i=1}^{n} |b_i|_0 + |c|_0 \leq m_1 < \infty$.

(ii) $\sum_{i,j=1}^{n} a_{ij}(x,t)\, \xi_i\xi_j \geq \sigma \sum_{i=1}^{n} \xi_i^2$ *for some constant* $\sigma > 0$, *for any real vector* $\xi = (\xi_1, \ldots, \xi_n)$ *and all* $(x,t) \in \overline{\Pi}$.

(iii) $\sum_{i,j=1}^{n} |a_{ij}|_1^{\Gamma} \leq m_2 < \infty$, *where*

$$|a_{ij}|_1^{\Gamma} = \sup_{P\in\Gamma} |a_{ij}(P)| + \sup \left\{\frac{|a_{ij}(x,t)-a_{ij}(x',t')|}{|x-x'|+|t-t'|} : (x,t),(x',t') \in \Gamma,\ (x,t) \neq (x',t')\right\}$$

(*That is,* a_{ij} *is Lipschitz continuous on* Γ, *which is the lateral boundary of* Π.) *Suppose that* $u \in C^{2+\alpha}(\overline{\Pi})$ *is a* T-*periodic solution in* t, *then for any* $0 < \nu < 1$, *there exists a constant* K *depending on* m_1, σ, m_2 *and* ν *such that* $|u|_{1+\nu} \leq K|f|_0$.

We now describe the assumptions and notations of the nonlinear systems which we will consider. With the operator L as described in (5.6-1), we consider

(5.6-2)
$$Lu_i = f_i(x,t,u_1, \ldots u_m) \quad \text{for } (x,t) \in \Pi$$
$$u_i(x,t) = \psi_i(x,t) \quad \text{for } (x,t) \in \Gamma = \delta\Omega x(-\infty,\infty)$$

for $i = 1, \ldots, m$. We assume that:

(I) The functions a_{ij}, b_i, and c in the operator L satisfy conditions (i), (ii), (iii) in Theorem 5.6-1 and condition (iii) in Theorem 5.6-2.

(II) Each boundary function ψ_i satisfies the condition for ψ in (v) of Theorem 5.6-1.

(III) Each f_i is continuous for $(x,t) \in \overline{\Pi}$, $-\infty < u_j < \infty$, $j = 1, \ldots, m$, and periodic in t with period T. For fixed $u = (u_1, \ldots, u_m)$, each $f_i(x,t,u)$ is uniformly Hölder continuous in Π as described in (iv) of Theorem 5.6-1; and f_i satisfies local Lipschitz condition in u_j, $j = 1, \ldots, m$, uniformly with respect to $(x,t) \in \Pi$.

Theorem 5.6-3. *Consider the boundary value problem (5.6-2) under assumption (I) to (III) above. Let $\mathfrak{D}$ be an open bounded convex subset in R^m containing 0 such that $\psi(x,t) = (\psi_1(x,t), \ldots, \psi_m(x,t)) \in \mathfrak{D}$ for all $(x,t) \in \Gamma$. Further, assume that for each $u = (u_1, \ldots, u_m) \in \delta\mathfrak{D}$, there exists an outward normal $\eta(u) = (\eta_1(u), \ldots, \eta_m(u))$ to $\mathfrak{D}$ such that*

(5.6-3)
$$\sum_{i=1}^{m} \eta_i(u) f_i(x,t,u) > 0, \text{ for all } (x,t) \in \Pi.$$

Then the problem (5.6-2) has a T-periodic solution $u(x,t) = (u_1(x,t), \ldots, u_m(x,t))$ such that $u(x,t) \in \mathfrak{D}$ for all $(x,t) \in \overline{\Pi}$, and $u_i \in C^{2+\alpha}(\overline{\Pi})$, $i = 1, \ldots, m$.

Proof. Let $0 < \alpha' < \alpha$, and $E = \{u = (u_1(x,t), \ldots, u_m(x,t)) \mid u \in C^{\alpha'}(\bar{\Pi})$, each u_i periodic in t with period $T\}$. Define

$$\mathcal{O} = \{u \in E \mid u(x,t) \in \mathfrak{D} \text{ for all } (x,t) \in \bar{\Pi}\ ,\ |u|_{\alpha'} < N\}$$

where N is to be determined later. Thus $\mathcal{O}$ is an open bounded set in the Banach space E. From Theorem 5.6-1, for any $u \in \mathcal{O}$, there exists a unique $w = (w_1, \ldots, w_m) \in C^{2+\alpha'}(\bar{\Pi})$ which satisfies

$$\text{(5.6-4)} \qquad \begin{aligned} Lw_i &= f_i(x,t,u_1(x,t), \ldots, u_m(x,t)) \quad \text{for } (x,t) \in \Pi\ , \\ w_i(x,t) &= \psi_i(x,t) \quad \text{for } (x,t) \in \Gamma, \end{aligned}$$

w_i periodic in t with period T, $i = 1, \ldots, m$. Letting $\hat{\psi}_i$ to be the periodic extension of ψ_i to a function in $C^{2+\alpha}(\bar{\Pi})$, we have for $i = 1, \ldots, m$:

$$\text{(5.6-5)} \qquad \begin{aligned} L(w_i-\hat{\psi}_i) &= f_i(x,\ t,\ u_1(x,t), \ldots, u_m(x,t)) - L\hat{\psi}_i \quad \text{in } \Pi \\ w_i - \hat{\psi}_i &= 0 \text{ on } \Gamma \end{aligned}$$

with $w_i - \hat{\psi}_i$ as a T-periodic solution in $C^{2+\alpha'}(\bar{\Pi})$. By Theorem 5.6-2, there exists a constand K such that

$$|w_i|_{1+\alpha} \leq K[|f_i(x,t,u_1(x,t), \ldots, u_m(x,t))|_0 + |L\hat{\psi}_i|_0] + |\hat{\psi}_i|_{1+\alpha}\ .$$

Denote $w = Su$, then the set $S\mathcal{O}$ is a bounded subset of $C^{1+\alpha}(\bar{\Pi})$. Since $\alpha' < \alpha$, the map S is a compact operator from $\mathcal{O}$ into E. Further, suppose $\lim_{k\to\infty}|\tilde{u}^k - \tilde{u}|_{\alpha'} =$ with $\tilde{u}^k, \tilde{u} \in \bar{\mathcal{O}}$ and $\tilde{w}^k = S\tilde{u}^k$, $\tilde{w} = S\tilde{u}$, $k = 1, 2, \ldots$. Then

$$\begin{aligned} L(\tilde{w}^k-\tilde{w})_i &= f_i(x,t,\tilde{u}_1^k(x,t), \ldots, \tilde{u}_m^k(x,t)) - f_i(x,t,\tilde{u}_1(x,t), \ldots, \tilde{u}_m(x,t)) \text{ in} \\ (\tilde{w}^k-\tilde{w})_i &= 0 \quad \text{on } \Gamma \end{aligned}$$

From Theorem 5.6-2, we obtain

$$|\tilde{w}_i^k - \tilde{w}_i|_{1+\alpha} \leq K \mid f_i(x, t, \tilde{u}_1^k, \dots, \tilde{u}_m^k) - f_i(x, t, \tilde{u}_1, \dots, \tilde{u}_m)|_0$$

Hence $|w_i^k - w_i|_{\alpha'} \to 0$ as $k \to \infty$, and $S: \bar{\mathcal{O}} \to E$ is continuous. Let $H(u,\lambda) = u - \lambda Su$ for $u \in \bar{\mathcal{O}}$. $\lambda \in [0,1]$, then one can apply homotopy invariance of Leray-Schauder degree to $H: \bar{\mathcal{O}} \times [0,1] \to E$ (see Theorem 1.3-7). We have $H(u,0) = u$, and $H(u,1) = u - Su$, S compact and continuous. We need to deduce $H(u,\lambda) \neq 0$ for $\lambda \in [0,1]$, all $u \in \delta\mathcal{O}$ in order to obtain invariance. Clearly $H(u,0) \neq 0$ for $u \in \delta\mathcal{O}$. Suppose that there exists $\tilde{u} \in \delta\mathcal{O}$, with $H(\tilde{u},\tilde{\lambda}) = 0$, $0 < \tilde{\lambda} < 1$. We have $S\tilde{u} = \tilde{u}/\tilde{\lambda}$, that is for $i = 1, \dots, m$:

$$L\tilde{u}_i = \tilde{\lambda} f_i(x, t, \tilde{u}_1(x,t), \dots, \tilde{u}_m(t)) \quad \text{for } (x,t) \in \Pi ,$$

$$\tilde{u}_i = \tilde{\lambda}\psi_i(x,t) \quad \text{on } \Gamma$$

with $\tilde{u}(x,t) \in \bar{\Omega}$, periodic in t with period T. By Theorem 5.6-2 we obtain $|\tilde{u}_i|_{1+\alpha'} \leq \tilde{K}[|\tilde{\lambda} f_i(x, t, \tilde{u}_1, \dots, \tilde{u}_m)|_0 + |L\hat{\psi}_i|_0 + |\tilde{\lambda}\hat{\psi}_i|_{1+\alpha'}]$. Consequently, it is possible to choose N sufficiently large in the definition of $\mathcal{O}$, so that $|\tilde{u}|_{\alpha'} < N$. Thus $\tilde{u} \in \delta\mathcal{O}$ implies that there exist some $(x_0,t_0) \in \bar{\Pi}$ such that $\tilde{u}(x_0,t_0) \overset{(def)}{=} \tilde{u}_0 \in \delta\mathcal{D}$. By hypothesis, there exists outward $n(\tilde{u}_0)$ to $\mathcal{D}$ such that $\bar{\mathcal{D}} \subseteq \{y \in R^m | (y-\tilde{u}_0) \cdot n(\tilde{u}_0) \leq 0\}$. Here, $\cdot$ is the usual Euclidean dot product. Let $R(x,t) = (\tilde{u}(x,t) - \tilde{u}_0) \cdot n(\tilde{u}_0)$ for $(x,t) \in \bar{\Pi}$; we have $P(x,t) \leq 0$ in $\bar{\Pi}$. Suppose $R(x,t)$ has the maximum value 0 at $(x_0,t_0) \in \Gamma$ on the boundary. We have $\tilde{u}(x_0,t_0) = \tilde{\lambda}\psi(x_0,t_0)$ which is inside $\mathcal{D}$ by hypothesis, giving a contradiction. Thus the point x_0 is inside Ω. At the point of maximum (x_0,t_0) we have $R(x_0,t_0) = 0$, $R_{x_i}(x_0,t_0) = 0$ and $LR(x_0,t_0) \leq 0$. On the other hand

$$LR(x_0,t_0) = L(\tilde{u}(x,t) - \tilde{u}_0) \cdot n(\tilde{u}_0)|_{(x_0,t_0)}$$

(5.6-6)

$$= \sum_{i=1}^{m} [\tilde{\lambda} f_i(x_0,t_0,\tilde{u}_0) - c(x_0,t_0)\,\tilde{u}_{0i}]\, n_i(\tilde{u}_0) > \sum_{i=1}^{m} \tilde{\lambda} f_i(x_0,t_0,\tilde{u}_0)\, n_i(\tilde{u}_0) >$$

The inequalities above follow from $\tilde{u}_0 \cdot n(\tilde{u}_0) > 0$, $c(x_0,t_0) < 0$ and hypothesis (5.6-3). Inequality (5.6-6) gives rise to a contradiction, and thus there cannot be any $\tilde{u} \in \delta\mathcal{O}$, $0 < \tilde{\lambda} < 1$ such that $H(\tilde{u},\tilde{\lambda}) = 0$. From the homotopy invariance theorem of Leray-Schauder, Theorem 1.3-7 we conclude that the equation $Su = u$ has a solution. That is, the problem (5.6-2) has a solution $u = (u_1, \ldots, u_m)$ in E. From the definition of S, we have $u \in C^{2+\alpha'}(\overline{\Pi})$. Consequently $f_i(x, t, u_1(x,t), \ldots, u_m(x,t))$ is in $C^{\alpha}(\overline{\Pi})$, and moreover from Theorem 5.6-1, we conclude that $u \in C^{2+\alpha}(\overline{\Pi})$. This completes the proof of Theorem 5.6-3.

Remark 5.6-1. In Theorem 5.6-3, suppose $\mathfrak{D}$ is changed to take the form of rectangular shape: $\mathfrak{D} = \{u \in R^m |\ a_i < u_i < b_i,\ i = 1, \ldots, m\}$ with $0 \in \mathfrak{D}$. On the part of $\delta\mathfrak{D}$ where $u_k = b_k$, n becomes $(0, \ldots, 0, 1, 0, \ldots, 0)$, with 1 on the kth component only and (5.6-3) takes the form $f_k(x,t,u) > 0$. If $u = (u_1, \ldots, u_m) \in$ with $u_{k_i} = a_{k_i}$ or b_{k_i} $i = 1, \ldots r$, $0 < r < m$, then we replace (5.6-3) with the r natural conditions:

$$f_{k_i}(x,t,u) > 0 \quad \text{if } u_{k_i} = b_{k_i},\ u \in \delta\mathfrak{D},\ (x,t) \in \Pi\ ;$$

(5.6-7)

$$f_{k_i}(x,t,u) < 0 \quad \text{if } u_{k_i} = a_{k_i},\ u \in \delta\mathfrak{D},\ (x,t) \in \Pi\ ,$$

$i = 1, \ldots, r$. For rectangular regions containing 0, theorem 5.6-3 remains valid if (5.6-3) is replaced by (5.6-7). The proof is exactly analogous.

The following theorem is a very practical variant of the last theorem.

Theorem 5.6-4. *Consider the boundary value problem (5.6-2) under assumptions (I) to (III) above. Let* $\alpha(x,t)$ *and* $\beta(x,t)$ *be functions in* $C^2(\bar{\Pi})$ *which are periodic in* t *with period* T *for each* $x \in \bar{\Omega}$, *and have the following properties:*

(a) $\alpha_i(x,t) \leq \beta_i(x,t)$ *in* $\bar{\Pi}$ *for each component* $i = 1, \ldots, m$;

(b) $L\alpha_i(x,t) - f_i(x, t, u_1, \ldots, u_{i-1}, \alpha_i(x,t), u_{i+1}, \ldots, u_m) \geq 0$

$L\beta_i(x,t) - f_i(x, t, u_1, \ldots, u_{i-1}, \beta_i(x,t), u_{i+1}, \ldots, u_m) \leq 0$

for each $i = 1, \ldots, m$, *all* $(x,t) \in \bar{\Pi}$, $\alpha_j(x,t) \leq u_j \leq \beta_j(x,t)$, $j \neq i$.

(iii) $\alpha_i(x,t) \leq \psi_i(x,t) \leq \beta_i(x,t)$ *on* Γ, *for each* $i = 1, \ldots, m$.

Then the problem (5.6-2) has a T-*periodic solution* $u(x,t) \in C^{2+\alpha}(\bar{\Pi})$ *with* $\alpha_i(x,t) \leq u_i(x,t) \leq \beta_i(x,t)$ *in* $\bar{\Pi}$, *each* $i = 1, \ldots, m$.

Proof. For $(x,t) \in \bar{\Pi}$, $u \in R^m$, define for $i = 1, \ldots, m$,

$$\bar{f}_i(x,t,u) = \begin{cases} f_i(x,t,\bar{u}) + [u_i - \beta_i(x,t)] & \text{if } u_i > \beta_i(x,t) \\ f_i(x,t,\bar{u}) & \text{if } \alpha_i(x,t) \leq u_i \leq \beta_i(x,t) \\ f_i(x,t,\bar{u}) + [u_i - \alpha_i(x,t)] & \text{if } u_i < \alpha_i(x,t) \end{cases}$$

where $\bar{u} = (\bar{u}_1, \ldots, \bar{u}_m)$ is defined to be dependent on (x,t) by

$$\bar{u}_j = \begin{cases} \beta_j(x,t) & \text{if } u_j > \beta_j(x,t) \\ u_j & \text{if } \alpha_j(x,t) \leq u_j \leq \beta_j(x,t) \\ \alpha_j(x,t) & \text{if } u_j < \alpha_j(x,t) \end{cases}$$

for each $j = 1, \ldots, m$. Consider the problem

$$(5.6\text{-}8) \qquad \begin{aligned} Lu_i &= \bar{f}_i(x,t,u) \quad \text{in } \Pi \\ u_i &= \psi_i(x,t) \quad \text{on } \Gamma \end{aligned}$$

for $i = 1, \ldots, m$.

There exist constants a_i, b_i, $i = 1, \ldots, m$ such that $a_i < 0 < b_i$ and

$$a_i < \alpha_i(x,t) \leq \beta_i(x,t) < b_i \quad \text{for all } (x,t) \in \overline{\Pi} ,$$

$$a_i < \psi_i(x,t) < b_i \quad \text{for all } (x,t) \in \Gamma ,$$

$$\bar{f}_i(x, t, u_1, \ldots, u_{i-1}, a_i, u_{i+1}, \ldots, u_m) < 0$$

$$< \bar{f}_i(x,t,u_1, \ldots, u_{i-1}, b_i, u_{i+1}, \ldots, u_m)$$

for all $(x,t) \in \overline{\Pi}$, $a_j \leq u_j \leq b_j$, each $j \neq i$, $i = 1, \ldots, m$. The functions $\bar{f}_i$ satisfy condition (III) for f_i, and from remark 5.6-1, theorem 5.6-3 is applicable to the problem (5.6-8) with rectangular $\mathfrak{D}$. Thus there exists a T-periodic solution $u(x,t)$ of (5.6-8) in $C^{2+\alpha}(\overline{\Pi})$, with $a_i < u_i(x,t) < b_i$ for $(x,t) \in \overline{\Pi}$, $i = 1, \ldots, m$.

It remains to show that $\alpha_i(x,t) \leq u_i(x,t) \leq \beta_i(x,t)$ for $(x,t) \in \overline{\Pi}$. Suppose that $\theta > 0$ is the positive maximum of $u_i(x,t) - \beta_i(x,t)$ for $(x,t) \in \overline{\Pi}$, attained at the point (x_0,t_0). Since on Γ, we have $u_i = \psi_i \leq \beta_i$, the point (x_0,t_0) where positive maximum is attained must be inside Π. It follows that

$$(5.6\text{-}9) \quad L(u_i-\beta_i)(x_0,t_0) = \bar{f}_i(x_0,t_0,u(x_0,t_0)) - L\beta_i(x_0,t_0)$$

$$\geq f_i(x_0,t_0,\bar{u}) + u_i(x_0,t_0) - \beta_i(x_0,t_0) - f_i(x_0, t_0, \bar{u}_1, \ldots, \bar{u}_{i-1}, \beta_i(x_0,t_0), \bar{u}_{i+1}, \ldots, \bar{u}_m)$$

$$= \theta > 0$$

(Recall that $\bar{u}$ has each component satisfying $\alpha_j(x_0,t_0) \leq \bar{u}_j \leq \beta_j(x_0,t_0)$, and $\bar{u}_i = \beta_i(x_0,t_0)$ in the present situation). However, since $(u_i-\beta_i)(x_0,t_0)$ has a local maximum, we have $(u_i-\beta_i)_t(x_0,t_0) = (u_i-\beta_i)_{x_j}(x_0,t_0) = 0$, and

$$L(u_i-\beta_i)(x_0,t_0) \leq 0 .$$

This contradicts (5.6-9), and consequently $u_i(x,t) \leq \beta_i(x,t)$ in $\bar{\Pi}$. Similarly, we obtain $\alpha_i(x,t) \leq u_i(x,t) \leq \beta_i(x,t)$ in $\bar{\Pi}$ for each $i = 1, \ldots, m$. Within such region, we have $\bar{f}_i(x,t,u(x,t)) = f_i(x,t,u(x,t))$ for each i; and $u(x,t)$ which satisfies (5.6-8) actually satisfies (5.6-2). This proves the theorem.

Notes

The method described in Section 5.1 leading to Section 5.2 was developed by Keller [119], Cohen [51], Amman [7], Sattinger [200], Amman and Crandall [11]. Materials in Section 5.2 to 5.4 are gathered from Leung [137], [140], constructing sequences which are oscillating for each component (i.e., with even and odd monotone subsequences). These lead to generalization in Theorem 5.5-1 from Korman and Leung [120]. Work similar to Theorems 5.5-1 is also found in Ladde, Lakshmikantham and Vatsala [124], [125]. Theorems 5.6-1 and 5.6-2 are respectively due to Fife [72] and Friedman [82]. Theorems 5.6-3 and 5.6-4 are adapted from Tsai [219]. Other related work concerning applications to chemical reactions and others can be found in e.g. Aris and Zygourakis [13], Cohen and Laetsch [52], Noyes and Jwo [172], and Pao [181]. The work of Schmitt [206] and Tsai [220], applicable to existence theory described in Section 1.4 is also related to the subject to this chapter. Some recent work concerning periodic parabolic problems related to Section 5.6 is done by Ortega [175] and Hess [102].

CHAPTER VI

Systems of Finite Difference Equations, Numerical Solutions

6.1 Monotone Scheme for Finite Difference Systems of Elliptic Equations

In this chapter we adapt the monotone schemes method to find approximate solutions for semilinear elliptic systems. We combine finite difference method with the monotone procedures developed in the last chapter. Accelerated version of the schemes is also considered in Section 6.3. We will consider up to two dimensional domain in Section 6.4, the method can naturally extend to higher dimensions. We will be only concerned with positive solutions to systems with Volterra-Lotka type ecological interactions. The method can however carry over to other interactions with similar monotone properties (c.f. Section 5.3). Further, the acceleration method can be applied to nonlinear interactions, with the appropriate convexity property (c.f. Equation (6.3-4)).

In the first two sections, we only restrict to one space variable. Rigorous foundations for the finite difference numerical calculations are given. It is shown that the finite difference solutions converge uniformly to the solutions of the continuous equations.

We consider the boundary value problem:

$$\text{(6.1-1)}\qquad \begin{aligned} &u''(x) + u(x)[a-bu(x)-cv(x)] = 0, \\ &v''(x) + v(x)[e+fu(x)-gv(x)] = 0; \\ &u(\alpha) = u(\beta) = v(\alpha) = v(\beta) = 0, \end{aligned} \qquad \text{for } \alpha < x < \beta$$

where a, b, c, e, f, g are positive parameters satisfying:

(6.1-2) $cf < gb,$

(6.1-3) $e > \lambda_1 \overset{(def)}{=} \frac{\pi^2}{(\beta-\alpha)^2},$

(6.1-4) $a > \frac{gb}{gb-cf} (\lambda_1+\frac{ce}{g}).$

Here $\lambda = \lambda_1$ is the first eigenvalue for the boundary value problem $w''(x) + \lambda w(x) = 0$, $\alpha < x < \beta$, $w(\alpha) = w(\beta) = 0$. It was shown in Chapter 5 that Problem (6.1-1) under hypotheses (6.1-2) to (6.1-4) has C^2 solution (u,v) with the property that $u(x) > 0$ and $v(x) > 0$ for $\alpha < x < \beta$. Moreover, if $cf << gb$ (cf. (5.3-12)), such solution is unique.

Let $h_N = N^{-1}(\beta-\alpha)$, $x_k = \alpha + N^{-1}k(\beta-\alpha)$, $k = 0, \ldots, N$. The values for $u(x_k)$, $v(x_k)$ will be respectively approximated by $u_{N,k}$, $V_{N,k}$, $k = 0, \ldots, N$ with $u_{N,0} = u_{N,N} = v_{N,0} = v_{N,N} = 0$. The second derivative will be replaced by the operator Δ_h where

$$\Delta_h u_{N,k} = h_N^{-2}[u_{N,k-1} - 2u_{N,k} + u_{N,k+1}]$$

for $k = 1, \ldots, N-1$. The hypotheses (6.1-3), (6.1-4) will be respectively replaced by

(6.1-5) $e > \frac{2(1-\cos\frac{\pi}{N})N^2}{(\beta-\alpha)^2} \overset{(def)}{=} \lambda(N),$

(6.1-6) $a > \frac{gb}{gb-cf} (\lambda(N) + \frac{ce}{g}).$

Note that $\lambda(N) \to \frac{\pi^2}{(\beta-\alpha)^2} = \lambda_1$ as $N \to \infty$ After some preliminary lemmas concerning scalar elliptic differential equations, we will construct numerically computable sequences $u_{N,k}^{(i)}$, $v_{N,k}^{(i)}$, $k = 0, 1, \ldots, N$, $i = 1, 2, \ldots.$ Their piecewise linear extensions:

$$u_N^{(i)}(x) = u_{N,k}^{(i)} + h_N^{-1}(x-x_{N,k})(u_{N,k+1}^{(i)} - u_{N,k}^{(i)}),$$
$$v_N^{(i)}(x) = v_{N,k}^{(i)} + h_N^{-1}(x-x_{N,k})(v_{N,k+1}^{(i)} - v_{N,k}^{(i)}),$$

$x_{N,k} \le x \le x_{N,k+1}$, $k = 0, \ldots, N-1$ will be shown in Section 6.2 to have the property that $\lim_{N\to\infty} u_N^{(i)}(x) = u^{(i)}(x)$, $\lim_{N\to\infty} v_N^{(i)}(x) = v^{(i)}(x)$ uniformly for $\alpha \le x \le \beta$. The functions $u^{(i)}(x)$ and $v^{(i)}(x)$ correspond respectively to u_i and v_i described in Section 5.1. Subsequences of $u^{(i)}(x)$, $v^{(i)}(x)$, with odd or even indices are found in Section 5.1 to converge to u_*, u^*, v_*, v^*, so that all positive solutions of (6.1-1) will satisfy $u_* \le u \le u^*$, $v_* \le v \le v^*$. Consequently, the sequences $u_{N,k}^{(i)}$, $v_{N,k}^{(i)}$ are useful for approximating positive solutions of (6.1-1). If calculations indicate that $u_* \approx u^*$ or $v_* \approx v^*$ one might conjecture the uniqueness of positive solution of the problem (Note that although if $cf << gb$ is sufficient for uniqueness, the general uniqueness question remained unanswered for our prey-predator case).

For $N \ge 2$, consider the discrete version for the problem $w''(x) + w[p(x) - qw] = 0$, $\alpha < x < \beta$, $w(\alpha) = w(\beta) = 0$:

$$\Delta_h w_{N,k} + w_{N,k}[p_k - qW_{N,k}] = 0, \quad k = 1, \ldots, N-1;$$

(6.1-7)

$$w_{N,0} = w_{N,N} = 0,$$

where $p_k = p(x_k)$, q is a positive constant. For the analysis of (6.1-7) up to Theorem 6.1-1, N will be fixed; hence we abbreviate $w_{N,k}$, h_N respectively as w_k, h. Grid functions $\{w_k\}_{k=0}^{N}$ will be denoted simply by w. As in the differential equation case, we define a grid function ϕ to be an upper solution for the problem (6.1-7) if it satisfies

$$\Delta_h \phi_k + \phi_k[p_k - q\phi_k] \le 0, \quad k=1, \ldots, N-1;$$

(6.1-8)

$$\phi_0, \ \phi_N \ge 0.$$

If all inequalities in (6.1-8) are reversed, we call the grid function a lower solution for (6.1-7).

Lemma 6.1-1. *Suppose that*

(6.1-9) $$q > 0 \ \text{and} \ p_k > 0 \ \text{for} \ k = 1, \ldots, N-1.$$

Then the constant grid function ϕ *with* $\phi_k = \max_{0 \le k \le N} \{p_k/q\}$ *is an upper solution for* (6.1-7). *Furthermore, if*

(6.1-10) $$r = \min_{1 \le k \le N-1} p_k - 2h^{-2}(1-\cos\tfrac{\pi}{N}) > 0,$$

then for any δ *such that* $0 < \delta \le r/q$, *the grid function* ψ *with* $\psi_k = \delta \sin \frac{(x_k-\alpha)\pi}{(\beta-\alpha)} = \delta \sin \frac{k\pi}{N}$, $k = 0, \ldots, N$ *is a lower solution for* (6.1-7). (*Note*: $2h^{-2}(1-\cos\frac{\pi}{N}) = \lambda(N)$).

Proof. $\Delta\phi_k + \phi_k[p_k - q\phi_k] = \phi_k[p_k - \max_{0 \le k \le N} \{p_k\}] \le 0$ for $k = 1, \ldots, N-1$, and $\phi_0 = \phi_N > 0$; hence ϕ is an upper solution. For ψ, we have $\psi_0 = \psi_N = 0$; and for $k = 1, \ldots, N-1$:

$$\Delta_h\psi_k + \psi_k[p_k - q\psi_k]$$

$$= \delta\{[\sin \frac{(k-1)\pi}{N} - 2\sin \frac{k\pi}{N} + \sin \frac{(k+1)\pi}{N}]\, h^{-2} + p_k \sin \frac{k\pi}{N} - q\delta \sin^2 \frac{k\pi}{N}\}$$

$$= \delta \sin \frac{k\pi}{N} \{\frac{2}{h^2}(\cos\frac{\pi}{N} - 1) + p_k - q\delta \sin \frac{k\pi}{N}\}.$$

The last expression is ≥ 0 if $0 < \delta \le r/q$.

Theorem 6.1-1. (*Existence for scalar equation*). *Assume all conditions concerning* q, p_k, *and* r *in the last lemma. Let* $\hat{\phi}$ *be an arbitrary upper solution for* (6.1-7) with $\hat{\phi}_k > 0$, $k = 1, \ldots, N-1$, *then there exists a grid function* w *which is a solution of* (6.1-7) *such that:*

$$\delta \sin\frac{k\pi}{N} \le w_k \le \hat{\phi}_k \quad , \ k = 0, \ \dots, \ N.$$

Here δ *is a sufficiently small positive constant* $\le r/q$.

Proof. Let Q be a nonnegative number satisfying $Q > 2q \max_{1\le k\le N-1} \hat{\phi}_k - \min_{1\le k\le N-1} p_k$. The functions $f_k(w) = w(p_k - qw) + Qw$ will then be increasing functions of w for $0 \le w \le \max_{1\le k\le N-1} \hat{\phi}_k$, each $k = 1, \dots, N-1$. Define the transformation T by: $z_k = (Ty)_k$, $k = 0, \dots, N$ if

$$(\Delta_h - Q)z_k = -y_k[p_k - qy_k + Q], \ k = 1, \ \dots, \ N-1;$$
$$z_0 = z_N = 0.$$

We observe that T is well defined since the matrix $h^2(\Delta_h - Q)$ is nonsingular with all its eigenvalues lying in the interval $(-4-2Qh^2, -2Qh^2)$. By using the property that f_k is increasing and the maximum principle for the operator $\Delta_h - Q$, we can show that T is monotone: if $y_k^{(1)} \ge y_k^{(2)}$, $k = 0, \dots, N$, then $(Ty^{(1)})_k \ge (Ty^{(2)})_k$, (cf. Section 5.1). Further, if we define for $k = 0, \dots, N$, $z_k^{(1)} = (T\hat{\phi})_k$ and $z_k^{(i+1)} = (Tz^{(i)})_k$, $i = 1, 2, \dots$ we can show by the usual manner that $\delta \sin\frac{k\pi}{N} \le \dots \le z_k^{(3)} \le z_k^{(2)} \le z_k^{(1)} \le \hat{\phi}_k$, for $k = 0, \dots, N$, and $\delta > 0$ sufficiently small as described. As $i \to \infty$, $z_k^{(i)} \to w_k$ as stated.

Theorem 6.1-2 (*Uniqueness for scalar equation*). *Let* $q > 0$. *Then there is at most one solution of* (6.1-7) *with the property that* $w_k > 0$ for $k = 1, \dots, N-1$.

Proof. Suppose that (6.1-7) has two different solutions u, v with the property stated. Up to a switch of the roles of u and v, there must exist integers I, J with $0 \le I-1 < I \le J < J+1 \le N$, such that

(6.1-11) $$v_{I-1} \le u_{I-1};\ v_k > u_k > 0,\ k = I, \ldots, J;\ v_{J+1} \le u_{J+1}.$$

From the identity $\sum_{k=I}^{J} u_k(v_{k+1}-2v_k+v_{k-1}) = u_{J+1}(v_{J+1}-v_J) - u_I(v_I-v_{I-1}) - \sum_{k=I}^{J} (u_{k+1}-u_k)(v_{k+1}-v_k)$, and using equation (6.1-7) for v, we have

$$h^{-2}[u_{J+1}(v_{J+1}-v_J) - u_I(v_I-v_{I-1}) - \sum_{k=I}^{J}(u_{k+1}-u_k)(v_{k+1}-v_k)]$$
$$= -\sum_{k=I}^{J} p_k u_k v_k + q\sum_{k=I}^{J} u_k v_k^2.$$

Interchange the roles of u and v in the last equation and subtract the two equations we deduce that

(6.1-12) $$v_J u_{J+1} + v_I u_{I-1} - v_{J+1}u_J - u_I v_{I-1} = h^2 q \sum_{k=I}^{J} v_k u_k(u_k-v_k) < 0.$$

The last inequality above is true by means of (6.1-11). On the other hand, (6.1-11) also implies that

$$v_J u_{J+1} + v_I u_{I-1} - v_{J+1}u_J - u_I v_{I-1}$$
$$\ge v_J v_{J+1} + v_I v_{I-1} - v_{J+1}u_J - u_I v_{I-1}$$
$$= v_{J+1}(v_J-u_J) + v_{I-1}(v_I-u_I) \ge 0$$

From (6.1-12) and the last inequality, we have a contradiction.

We next prove a comparison lemma, by means of which we will construct monotone sequences of grid functions $u_{N,k}^{(2n)}$, $u_{N,k}^{(2n+1)}$, $v_{N,k}^{(2n)}$, $v_{N,k}^{(2n+1)}$, n = 1, 2, ..., k=1, ..., N. These grid functions will converge to the solutions of (6.1-1) or their bounds, as $N \to \infty$ and $n \to \infty$.

<u>Lemma</u> 6.1-2 (<u>Comparison</u>). <u>Let</u> $\ell_k^{(1)} \ge \ell_k^{(2)} > \lambda(N)$ <u>for each</u> k = 1, ..., N-1.

For $i = 1, 2,$ *let* $w^{(i)}_{N,0} = w^{(i)}_{N,N} = 0$, $w^{(i)}_{N,k}$, $k = 1, \ldots, N-1$ *be positive numbers* satisfying

$$\Delta w^{(i)}_{N,k} + w^{(i)}_{N,k}\,[\ell^{(i)}_k - q w^{(i)}_{N,k}] = 0$$

for $k = 1, \ldots, N-1$ (*Here* q *is a positive constant*). *Then* $w^{(1)}_{N,k} \geq w^{(2)}_{N,k}$ *for each* $k = 0, \ldots, N$.

Proof. The function $z_{N,k} = w^{(1)}_{N,k}$, $k = 0, \ldots, N$ is an upper solution for the problem:

$$\text{(6.1-13)}\quad \begin{aligned} &\Delta_h\, z_{N,k} + z_{N,k}[\ell^{(2)}_k - q z_{N,k}] = 0, \quad k = 1, \ldots, N-1; \\ &z_{N,0} = z_{N,N} = 0. \end{aligned}$$

Furthermore, since $\ell^{(2)}_k > \lambda(N)$, Theorem 6.1-1 implies the existence of a solution $z_{N,k} = \tilde{w}_{N,k}$ to the problem (6.1-13) with $\delta \sin \frac{k\pi}{N} \leq \tilde{w}^{(1)}_{N,k} \leq w^{(1)}_{N,k}$, $k = 0, \ldots, N$ (δ some positive constant). Consequently $\tilde{w}_{N,k} > 0$, for $k = 1, \ldots, N-1$, and by Theorem 6.1-2 we have $\tilde{w}_{N,k} = w^{(2)}_{N,k}$, $k = 0, \ldots, N$. Hence $w^{(1)}_{N,k} \geq w^{(2)}_{N,k}$.

The rest of this section will require the hypotheses (6.1-2), (6.1-5) and (6.1-6) concerning e, a and N etc. We assume these hypotheses and proceed to construct $u^{(i)}_{N,k}$, $v^{(i)}_{N,k}$. $i = 1, 2, \ldots$.

First, let $u^{(1)}_{N,k} > 0$, $k = 1, \ldots, N-1$ satisfy

$$\text{(6.1-14)}\quad \begin{aligned} &\Delta_h u^{(1)}_{N,k} + u^{(1)}_{N,k}(a - b u^{(1)}_{N,k}) = 0, \qquad k = 1, \ldots, N-1; \\ &u^{(1)}_{N,0} = u^{(1)}_{N,N} = 0. \end{aligned}$$

(Such solution exists uniquely due to (6.1-6), (6.1-2) and Theorems 6.1-1 and 6.1-2). Similarly, let $v^{(1)}_{N,k} > 0$, $k = 1, \ldots, N-1$ satisfy

$$\Delta_h v_{N,k}^{(1)} + v_{N,k}^{(1)}(e+fu_{N,k}^{(1)} - gv_{N,k}^{(1)}) = 0, \quad k - 1, \ldots, N-1;$$

(6.1-15)

$$v_{N,0}^{(1)} = v_{NN}^{(1)} = 0$$

(such solution exists uniquely due to (6.1-5) and Theorems 6.1-1 and 6.1-2).

Lemma 6.1-3. *For each* $k = 0, \ldots, N$, $v_{N,k}^{(1)} < c^{-1}[a-\lambda(N)]$.

Proof. By Theorems 6.1-1 and 6.1-2 we have $u_{N,k}^{(1)} \le a/b$. Hence by Theorems 6.1-1 and 6.1-2 again, we have $v_{N,k}^{(1)} \le \frac{1}{g}(e + f\frac{a}{b})$, $k = 0, \ldots, N$. However, hypothesis (6.1-6) and (6.1-2) implies that $a - \lambda(N) > \frac{c}{g}(e + f\frac{a}{b})$. This proves the lemma.

For $i = 2, 3, \ldots$, we now define $u_{N,k}^{(i)} > 0$, $v_{N,k}^{(i)} > 0$, for $k = 1, \ldots, N-1$, inductively as follows:

$$\Delta_h u_{N,k}^{(i)} + u_{N,k}^{(i)}(a-bu_{N,k}^{(i)} - cv_{N,k}^{(i-1)}) = 0$$

(6.1-16)

$$\Delta_h v_{N,k}^{(i)} + v_{N,k}^{(i)}(e+fu_{N,k}^{(i)} - gv_{N,k}^{(i)}) = 0$$

$u_{N,0}^{(i)} = u_{N,N}^{(i)} = v_{N,0}^{(i)} = v_{N,N}^{(i)} = 0$. They will be seen to be well defined uniquely below.

Lemma 6.1-4. *For each* $i = 1, 2, \ldots$ *we have* $0 < u_{N,k}^{(i)} \le u_{N,k}^{(1)}$, $0 < v_{N,k}^{(i)} \le v_{N,k}^{(1)}$ *for* $k = 1, \ldots, N-1$. *Such* $u_{N,k}^{(i)}$ *and* $v_{N,k}^{(i)}$ *are uniquely defined.*

Theorem 6.1-3. *For each nonnegative integer* i, *the following are true:*

$$u_{N,k}^{(2i+2)} \le u_{N,k}^{(2i+4)} \le u_{N,k}^{(2i+3)} \le u_{N,k}^{(2i+1)}$$

$$v_{N,k}^{(2i+2)} \le v_{N,k}^{(2i+4)} \le v_{N,k}^{(2i+3)} \le v_{N,k}^{(2i+1)}$$

for each $k = 0, \ldots, N$.

The above lemma and theorem are proved inductively by applying Theorems 6.1-1, 6.1-2 and Lemma 6.1-2 repeatedly for comparison. The details are exactly analogous to that of Lemma 5.2-4 and Theorem 5.2-2, and will be omitted here. Theorem 6.1-3 clearly implies that

$$0 \le u_{N,k}^{(2)} \le u_{N,k}^{(4)} \le u_{N,k}^{(6)} \le \ldots \le u_{N,k}^{(5)} \le u_{N,k}^{(3)} \le u_{N,k}^{(1)}, \text{ and}$$

$$0 \le v_{N,k}^{(2)} \le v_{N,k}^{(4)} \le v_{N,k}^{(6)} \le \ldots \le v_{N,k}^{(5)} \le v_{N,k}^{(3)} \le v_{N,k}^{(1)},$$

for $k = 0, \ldots, N$. In the next section, we will see that as $N \to \infty$, the piecewise linear extensions of $u_{N,k}^{(i)}$ and $v_{N,k}^{(i)}$ will tend to $u_i(x)$ and $v_i(x)$ which are solutions of corresponding continuous versions of (6.1-14) to (6.1-16), (cf. (5.2-8) to (5.2-10)).

6.2. Convergence to Solutions of Differential Equations and Computational Results

In this section, we first prove some convergence properties of solutions of finite difference equations as $N \to \infty$ (or $h \to 0$). We consider $\Delta_h w_{N,k} + f(z_{N,k}, w_{N,k}) = 0$, where $z_{N,k}$ is an independent function with some convergence properties. The important result is Theorem 6.2-1. This theorem is then used in Theorem 6.2-2 to prove that all grid functions constructed by the scheme in Section 6.1 will converge as $N \to \infty$. By means of this we can approximate solutions of (6.1-1).

Let $f\colon R^2 \to R$ be a continuously differentiable function satisfying $f(\cdot,0) = 0$. For each positive integer $N \ge 3$, let $z_{N,k}$, $k = 0, 1, \ldots, N$ denote a set of real numbers with the properties:

$$|h_N^{-1}(z_{N,k+1} - z_{N,k})| \le L, \quad \text{for each } k = 0, 1, \ldots, N-1, \tag{6.2-1}$$

where $h_N = \frac{\beta-\alpha}{N}$, and L is a positive constant independent of N;

$$z_{N,0} = z_{N,N} = 0; \text{ and} \tag{6.2-2}$$

the piecewise linear extensions $\hat{z}_N(x)$,

$$\hat{z}_N(x) = z_{N,k} + h_N^{-1}(x-x_{N,k})(z_{N,k+1}-z_{N,k}), \quad x_{N,k} \le x \le x_{N,k+1} \tag{6.2-3}$$

for $k = 0, \ldots, N-1$, where $x_{N,k} = \alpha + N^{-1}k(\beta-\alpha)$, satisfy $\hat{z}_N(x) \to z(x)$ uniformly for $\alpha \le x \le \beta$ as $N \to \infty$, (here $z(x)$ is some real function for $\alpha \le x \le \beta$).

For each positive integer $N \ge 3$, let $w_{N,k}$, $k = 0, \ldots, N$ be a set of real numbers with the properties:

$$(w_{N,k+1} - 2w_{N,k} + w_{N,k-1})h_N^{-2} + f(z_{N,k}, w_{N,k}) = 0 \quad \text{for } k = 1, \ldots, N-1; \tag{6.2-4}$$

$$w_{N,0} = w_{N,N} = 0; \tag{6.2-5}$$

$$\delta \le \max_{0\le k\le N} w_{N,k} \le M, \tag{6.2-6}$$

where M and δ are positive constants independent of N; and $0 \le \min_{0\le k\le N} w_{N,k}$.

Lemma 6.2-1. Let $z_{N,k}$ and $w_{N,k}$, $N \geq 3$, $k = 0, 1, \ldots, N$ be sets of real numbers satisfying properties (6.2-1) to (6.2-6). Let $m_{N,k} = h_N^{-1}(w_{N,k+1} - w_{N,k})$ for $k = 0, 1, \ldots, N-1$. Then there exists a number L_1 independent of k and N such that $|m_{N,k}| \leq L_1$.

Theorem 6.2-1. Let $z(x)$ be the continuous function described in (6.2-3) above. Suppose that the boundary value problem:

$$(6.2\text{-}7) \qquad \begin{cases} w''(x) + f(z(x), w(x)) = 0 & \text{for } \alpha < x < \beta \\ w(\alpha) = w(\beta) = 0 \end{cases}$$

has a unique nontrivial nonnegative solution $w(x) \geq 0$, $\alpha \leq x \leq \beta$. Then the piecewise linear extension $\hat{w}_N(x)$,

$$(6.2\text{-}8) \qquad \hat{w}_N(x) = w_{N,k} + h_N^{-1}(x - x_{N,k})(w_{N,k+1} - w_{N,k})$$

$x_{N,k} \leq x \leq x_{N,k+1}$, $k = 0, \ldots, N-1$, will satisfy:

$$\hat{w}_N(x) \to w(x), \text{ uniformly for } \alpha \leq x \leq \beta, \text{ as } N \to \infty.$$

(Here $w_{N,k}$, $z_{N,k}$ are numbers satisfying (6.2-4) - (6.2-6) and (6.2-1) - (6.2-3) above respectively). In particular, given any $\varepsilon > 0$, there exists $\bar{N}(\varepsilon)$, such that

$$|w_{N,k} - w(x_{N,k})| < \varepsilon \quad \text{for} \quad k = 0, 1, \ldots, N$$

if $N \geq \bar{N}(\varepsilon)$.

Proof of Lemma 6.2-1. Multiplying the kth equation in (6.2-4) by $w_{N,k}$ and summing k from 1 to N-1, we obtain

$$(6.2\text{-}9) \qquad \sum_{k=1}^{N-1} w_{N,k}h_N^{-1}(m_{N,k} - m_{N,k-1}) + \sum_{k=1}^{N-1} w_{N,k}f(z_{N,k}w_{N,k}) = 0.$$

However $\sum_{k=1}^{N-1} w_{N,k}\, h_N^{-1}(m_{N,k} - m_{N,k-1}) = h_N^{-1}[w_{N,N-1}m_{N,N-1} - \sum_{k=1}^{N-2} m_{N,k}(w_{N,k+1}-w_{N,k}) - w_{N,1}m_{N,0}] = -\sum_{k=0}^{N-1} m_{N,k}^2$, so (6.2-9) can be rewritten as

$$(6.2\text{-}10) \qquad \sum_{k=0}^{N-1} m_{N,k}^2 = \sum_{k=1}^{N-1} w_{N,k}f(z_{N,k}, w_{N,k}).$$

Let $m_{N,k_0} = \min_{0 \le k \le N-1} |m_{N,k}|$. (6.2-10) implies that

$$(6.2\text{-}11) \qquad Nm_{N,k_0}^2 \le (N-1)MC,$$

where M is described in (6.2-6) and C is the maximum of $|f(u,v)|$ as (u,v) ranges over a compact set containing all possible values of $(z_{N,k}, w_{N,k})$. (Note that (6.2-3) implies that $z_{N,k}$ is uniformly bounded for all N,k involved). Inequality (6.2-11) clearly shows that $|m_{N,k_0}| < \sqrt{MC}$. Let i be an integer with $k_0 < i < N$. We have

$$|m_{N,i} - m_{N,k_0}| = |\sum_{k=k_0+1}^{i} (m_{N,k}-m_{N,k-1})|$$

$$= |-h_N \sum_{k=k_0+1}^{i} f(z_{N,k}, w_{N,k})| \le h_N(i-k_0)C < (b-a)C.$$

Consequently $|m_{N,i}| \le |m_{N,k_0}| + (b-a)C < \sqrt{MC} + (b-a)C$. Similarly, we can obtain a uniform bound for $|m_{N,i}|$, with $0 \le i < k_0$. This proves the lemma.

Proof of Theorem 6.2-1: Hypothesis (6.2-4) can be rewritten as

$$h_N^{-1}(m_{N,k} - m_{N,k-1}) = -f(z_{N,k}, w_{N,k}),$$

for $k = 0, \ldots, N-1$. We extend $m_{N,k}$ in a piecewise linear way:

$$\hat{m}_N(x) = m_{N,k} - (x - x_{N,k})\, f(z_{N,k+1}, w_{N,k+1}), \tag{6.2-12}$$

for $x_{N,k} \le x \le x_{N,k+1}$, $k = 0, \ldots, N-2$, and

$$\hat{m}_N(x) = m_{N,N-1}, \tag{6.2-13}$$

for $x_{N,N-1} \le x \le x_{N,N} = \beta$. We will now show that the sequences of functions $\{\hat{w}_N(x)\}_{N=1}^{\infty}$ and $\{\hat{m}_N(x)\}_{N=1}^{\infty}$ are equicontinuous and uniformly bounded on $\alpha \le x \le \beta$.

Clearly, $\dfrac{d\hat{w}_N}{dx} = m_{N,k}$ for $x_{N,k} < x < x_{N,k+1}$, $k = 0, \ldots, N-1$. Hence $|\hat{w}_N'(x) - \hat{m}_N(x)| \le Ch_N$ a.e. (where C is the bound for f as described in the proof of last lemma). Together with the fact that $\hat{m}_N(x)$ are uniformly bounded for all N, $\alpha \le x \le \beta$ (due to the previous lemma), we conclude that $\{\hat{w}_N\}_{N=1}^{\infty}$ is equicontinuous. Hypothesis (6.2-6) clearly implies that $\{\hat{w}_N\}_{N=1}^{\infty}$ is uniformly bounded. Moreover, $\hat{w}_N(x) = \int_\alpha^x \hat{w}_N'(t)dt$, and hence for $\alpha \le x \le \beta$, we have

$$|\hat{w}_N(x) - \int_\alpha^x \hat{m}_N(t)dt| \le \int_\alpha^x Ch_N dt \le Ch_N(\beta-\alpha). \tag{6.2-14}$$

For the sequence $\{\hat{m}_N\}_{N=1}^{\infty}$, we observe that from (6.2-12) we have

$$\hat{m}_N'(x) = -f(z_{N,k+1}, w_{N,k+1}) \tag{6.2-15}$$

for $x_{N,k} < x < x_{N,k+1}$, $k = 0, \ldots, N-2$. However, $|\hat{w}_N(x) - w_{N,k+1}| = |(x-x_{N,k})m_{N,k} - h_N m_{N,k}| \le m_{N,k}h_N$; and hypothesis (6.2-1) implies that

$|\hat{z}_N(x)-z_{N,k+1}| = |h_N^{-1}(z_{N,k+1}-z_{N,k})[(x-x_{N,k})-h_N]| \le Lh_N.$

Consequently, for $x_{N,k} < x < x_{N,k+1}$, $k = 0, \ldots, N-2$ (6.2-15) gives

$$\text{(6.2-16)} \qquad |\hat{m}_N'(x) + f(\hat{z}_N(x), \hat{w}_N(x))| \le K_1Lh_N + K_2L_1h_N ,$$

where K_1, K_2 are respectively the bounds for $|\frac{\partial f}{\partial z}|$, $|\frac{\partial f}{\partial w}|$ over a compact set containing all possible range of $(\hat{z}_N(x), \hat{w}_N(x))$. For $x_{N,N-1} < x < x_{N,N} = \beta$, we have $\hat{m}_N'(x) = 0$; and $f(\hat{z}_N(x), \hat{w}_N(x)) = f(\hat{z}_N(x),0) + O(h_N) = O(h_N)$. Thus, together with (6.2-16), we have

$$\text{(6.2-17)} \qquad |\hat{m}_N'(x) + f(\hat{z}_N(x),\hat{w}_N(x))| \le O(h_N) \text{ a.e. for } \alpha \le x \le \beta.$$

Inequality (6.2-17) and the previous lemma imply that $\{\hat{m}_N(x)\}_{N=1}^{\infty}$ is equicontinuous and uniformly bounded. Moreover

$$\text{(6.2-18)} \qquad |\hat{m}_N(x) - \hat{m}_N(\alpha) + \int_{\alpha}^{x} f(\hat{z}_N(t), \hat{w}_N(t)dt| \le K_3(\beta-\alpha)h_N$$

for all $\alpha \le x \le \beta$, where K_3 is some positive constant.

By Ascoli's lemma, there exist subsequences $\{\hat{w}_{N_j}(x)\}_{j=1}^{\infty}$ and $\{\hat{m}_{N_j}(x)\}_{j=1}^{\infty}$ respectively of $\hat{w}_N$ and $\hat{m}_N$ such that $\hat{w}_{N_j}(x) \to \hat{w}(x)$ and $\hat{m}_{N_j}(x) \to \hat{m}(x)$ uniformly for $\alpha \le x \le \beta$, as $j \to \infty$. Here $\hat{w}(x)$ and $\hat{m}(x)$ are some continuous functions. From (6.2-14) and (6.2-18) we respectively conclude that

$$\text{(6.2-19)} \qquad \hat{w}(x) = \int_{\alpha}^{x} \hat{m}(t)\, dt, \quad \text{and}$$

$$\text{(6.2-20)} \qquad \hat{m}(x) - \hat{m}(\alpha) + \int_{\alpha}^{x} f(z(t), \hat{w}(t))dt = 0$$

for $\alpha \le x \le \beta$. From (6.2-19) and (6.2-20), we have

(6.2-21) $\hat{w}'(x) = \hat{m}(x)$, and

(6.2-22) $\hat{w}''(x) = \hat{m}'(x) = -f(z(x),\hat{w}(x))$

for $\alpha < x < \beta$. Moreover $\hat{w}(\alpha) = \lim_{j\to\infty} \hat{w}_{N_j}(\alpha) = w_{N_j,0} = 0$, and similarly $\hat{w}(\beta) = 0$. Consequently the function $\hat{w}(x)$ satisfies the boundary value problem (6.2-7). Hypothesis (6.2-6) implies that $\hat{w}(x)$ is a nontrivial nonnegative function. Therefore, uniqueness of such solution for (6.2-7) implies that $\hat{w}(x) = w(x)$, and the full sequences $\hat{w}_N(x) \to w(x)$ and $\hat{m}_N(x) \to w'(x)$ uniformly for $\alpha \le x \le \beta$, as $N \to \infty$.

Theorem 6.2-2. Suppose that (6.1-2) to (6.1-4) are satisfied, and for all large positive integer N, (6.1-5) and (6.1-6) are satisfied. For such N, and $i = 1, 2, \ldots$, let $u^{(i)}_{N,k}$, $v^{(i)}_{N,k}$ be functions as defined in (6.1-14) to (6.1-16), $k = 0, \ldots, N$. Let $u^{(i)}_N(x)$, $v^{(i)}_N(x)$ be piecewise linear extensions of $u^{(i)}_{N,k}$, $v^{(i)}_{N,k}(x)$ respectively:

$$u^{(i)}_N(x) = u^{(i)}_{N,k} + h_N^{-1}(x-x_{N,k})(u^{(i)}_{N,k+1} - u^{(i)}_{N,k})$$

$$v^{(i)}_N(x) = v^{(i)}_{N,k} + h_N^{-1}(x-x_{N,k})(v^{(i)}_{N,k+1} - v^{(i)}_{N,k})$$

$x_{N,k} \le x \le x_{N,k+1}$, $k = 0, \ldots, N-1$. Then $u^{(i)}_N(x) \to u^{(i)}(x)$ and $v^{(i)}_N(x) \to v^{(i)}(x)$ uniformly for $\alpha \le x \le \beta$ as $N \to \infty$. The functions $u^{(i)}(x)$, $v^{(i)}(x)$ are positive for $\alpha < x < \beta$, and satisfy:

$$\Delta u^{(1)} + u^{(1)}(a-bu^{(1)}) = 0,\ 0 < x < \beta,\ u^{(1)}(\alpha) = u^{(1)}(\beta) = 0;$$

$$\Delta v^{(i)} + v^{(i)}(e+fu^{(i)}-gv^{(i)}) = 0,\ \alpha < x < \beta,\ i = 1, 2, \ldots,$$

$$\Delta u^{(i)} + u^{(i)}(a-bu^{(i)}-cv^{(i-1)}) = 0,\ \alpha < x < \beta,\ i = 2, 3, \ldots,$$

$$u^{(i)}(\alpha) = u^{(i)}(\beta) = v^{(i)}(\alpha) = v^{(i)}(\beta) = 0.$$

Here, Δ denotes the second derivative. (Such functions $u^{(i)}(x)$, $v^{(i)}(x)$ correspond to $u_i(x)$, $v_i(x)$ in Section 5.2, and are unique nontrivial nonnegative solutions of the corresponding equations and boundary conditions).

Proof. If we let $f(z(x),w(x)) = w(x)[a-bw(x)]$, we can apply Theorem 6.2-1, since f is not dependent on z and (6.2-1) to (6.2-3) are vacuously true. Letting $w_{N,k} = u_{N,k}^{(1)}$, we conclude that $u_N^{(1)}(x) \to u^{(1)}(x)$ uniformly for $\alpha \le x \le \beta$ as $N \to \infty$. Further, by Lemma 6.2-1, $u_{N,k}^{(1)}$ will satisfy (6.2-1) to (6.2-3) with $z_{N,k}$, $\hat{z}_N$, z respectively replaced by $u_{N,k}^{(1)}$, $u_N^{(1)}$, $u^{(1)}$. Next, we let $f(z,w) = w(e+fz-gw)$, and $z_{N,k} = u_{N,k}^{(1)}$, $w_{N,k} = v_{N,k}^{(1)}$. By construction, (6.2-4) and (6.2-5) are satisfied; and the proof of Lemma 6.1-3 indicates that (6.2-6) is also satisfied with $w_{N,k} = v_{N,k}^{(1)}$. Theorem 6.2-1 therefore implies that $v_N^{(1)}(x) \to v^{(1)}(x)$ uniformly for $\alpha \le x \le \beta$ as $N \to \infty$. Moreover, Lemma 6.2-1 implies that $v_{N,k}^{(1)}$ will satisfy (6.2-1) to (6.2-3) with $z_{N,k}$, $\hat{z}_N$, z respectively replaced by $v_{N,k}^{(1)}$, $v_N^{(1)}$, $v^{(1)}$. Inductively, we can show in order that $u_N^{(2)}$, $v_N^{(2)}$, $u_N^{(3)}$, $v_N^{(3)}$, ... converge uniformly to $u^{(2)}$, $v^{(2)}$, $u^{(3)}$, $v^{(3)}$, ... respectively as $N \to \infty$.

In order to make this chapter self-contained, we now state precisely the properties of $u^{(i)}(x)$, $v^{(i)}(x)$ described in Theorem 6.2-2. Their proofs can be found in Section 5.1.

Theorem 6.2-3. $\lim_{n\to\infty} u^{(2n)}(x) = u_*(x) \le u^*(x) = \lim_{n\to\infty} u^{(2n+1)}(x)$, $\lim_{n\to\infty} v^{(2n)}(x) = v_*(x) \le v^*(x) = \lim_{n\to\infty} v^{(2n+1)}(x)$ for all $\alpha \le x \le \beta$. All solutions (u,v) of (6.1-1) with $u(x) > 0$, $v(x) > 0$ for $\alpha < x < \beta$ will satisfy $u_*(x) \le u(x) \le u^*(x)$, $v_*(x) \le v(x) \le v^*(x)$, for all $\alpha \le x \le \beta$. Such solution(s) (u,v) do exist. If $u_*(x) = u^*(x)$ or $v_*(x) = v^*(x)$, then $u_* \equiv u \equiv u^*$ and $v_* = v \equiv v^*$; that is, in this situation, such solution (u,v) is unique.

We close this section with some numerical results obtained by using the monotone scheme (6.1-14) to (6.1-16) for approximation of the solution of (6.1-1). In our example, we let $[\alpha, \beta] = [0, \pi]$ and discretize the problem by choosing $N = 64$, $h = \frac{\pi}{64}$. In all cases we plot some elements of the monotone decreasing grid sequences $u_{N,k}^{(2n-1)}$, $v_{N,k}^{(2n-1)}$, $n = 1, 2, \ldots$ and the monotone increasing grid sequences $u_{N,k}^{(2n)}$, $v_{N,k}^{(2n)}$. The "limiting" values (for all sequences) are obtained when $\max_{0 \le k \le N} |w_{N,k}^{(j+2)} - w_{N,k}^{(j)}| \le 10^{-4}$ where $w_{N,k}^{(j)} = u_{N,k}^{(2n-1)}$, $v_{N,k}^{(2n-1)}$, $u_{n,k}^{(2n)}$ or $v_{N,k}^{(2n)}$. The limiting values are plotted with full lines, and dashed lines are used for any other elements in the sequence.

<u>Example</u> 6.2-1. Figures 6.2-1, 6.2-2, 6.2-3, 6.2-4 show the corresponding sequences for $u_{N,k}^{(2n-1)}$, $v_{N,k}^{(2n-1)}$, $u_{N,k}^{(2n)}$, $v_{N,k}^{(2n)}$ respectively (with $n = 1, 2, 3$ in each figure). They describe the solution of (6.1-1) with $a = 5$, $b = 2$, $c = 1$ and $e = 3$, $f = 1$, $g = 2$. Conditions (6.1-2) to (6.1-4) are all satisfied with $cf < bg$ becoming $1 < 4$. The "limiting" value is attained for $n \approx 10$ and we have a unique solution for practical purposes. The proof of uniqueness in Section 5.1 may not include this case because cf is not small enough.

<u>Example</u> 6.2-2. Figures 6.2-5, 6.2-6, 6.2-7 and 6.2-8 show the corresponding sequences ($n = 1, 11, 19, 29, 40$) for (6.1-1) with $a = 80$, $b = 2$, $c = 1$ and $e = 1.5$, $f = 3.9$, $g = 2$. (6.1-2) to (6.1-4) are all satisfied, $cf = 3.9 < 4 = bg$ barely fulfilled. Note also that a is large relative to e. All these make computation here more difficult than Example 6.2-1. This time we need $n > 300$ in order to reach the "limiting" value. Computations suggest that we have uniqueness although we definitely do not have $cf \ll bg$.

<u>Example</u> 6.2-3. This is the same as Example 6.2-2 only with e changed to $e = -0.5$, violating (6.1-3). Although our proofs do not include this case, computations suggest that the conclusions concerning existence and uniqueness of positive nontrivial solution are the same as the previous example. This value of e is interesting because in many prey-predator models, the predator

is assumed to have negative intrinsic growth rate. The corresponding sequences (n = 1, 21, 37, 55, 79) are shown in Figures 6.2-9, 6.2-10, 6.2-11 and 6.2-12.

Remark: In each example, the vertical scale for the decreasing sequence for u is different from that for the increasing sequence (compare for example figures 6.2-1 and 6.2-3). Similar situation is true for v.

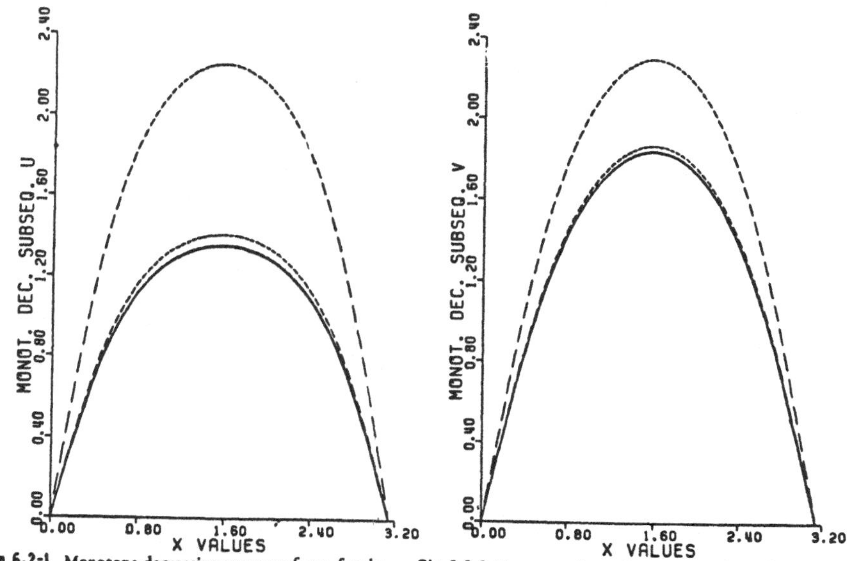

Fig 6.2-1 Monotone decreasing sequence for u, for the problem $u'' + u(5 - 2u - v) = 0$
$v'' + v(3 + u - 2v) = 0$

Fig 6.2-2 Monotone decreasing sequence for v, for the problem $u'' + u(5 - 2u - v) = 0$
$v'' + v(3 + u - 2v) = 0$

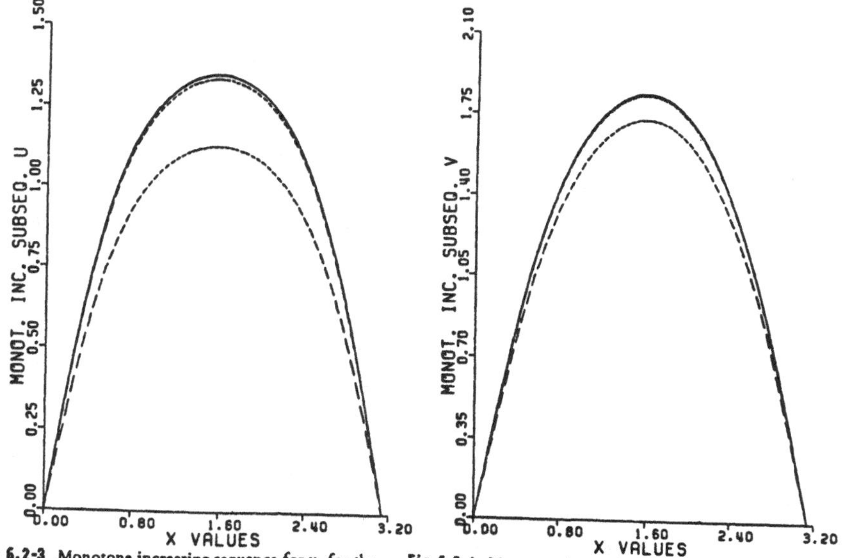

Fig 6.2-3 Monotone increasing sequence for u, for the problem $u'' + u(5 - 2u - v) = 0$
$v'' + v(3 + u - 2v) = 0$

Fig 6.2-4 Monotone increasing sequence v, for the problem $u'' + u(5 - 2u - v) = 0$
$v'' + v(3 + u - 2v) = 0$

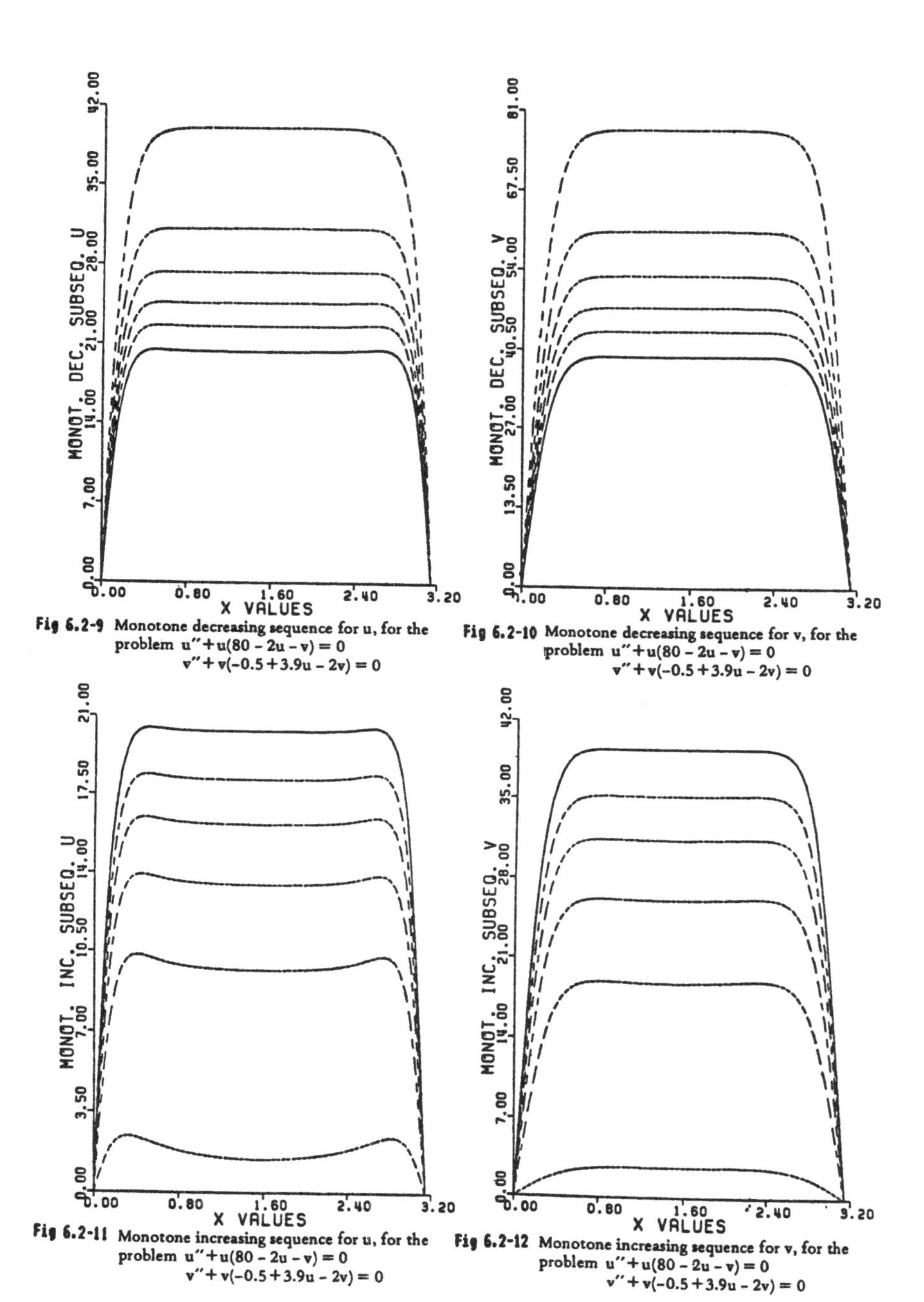

Fig 6.2-9 Monotone decreasing sequence for u, for the problem $u'' + u(80 - 2u - v) = 0$
$v'' + v(-0.5 + 3.9u - 2v) = 0$

Fig 6.2-10 Monotone decreasing sequence for v, for the problem $u'' + u(80 - 2u - v) = 0$
$v'' + v(-0.5 + 3.9u - 2v) = 0$

Fig 6.2-11 Monotone increasing sequence for u, for the problem $u'' + u(80 - 2u - v) = 0$
$v'' + v(-0.5 + 3.9u - 2v) = 0$

Fig 6.2-12 Monotone increasing sequence for v, for the problem $u'' + u(80 - 2u - v) = 0$
$v'' + v(-0.5 + 3.9u - 2v) = 0$

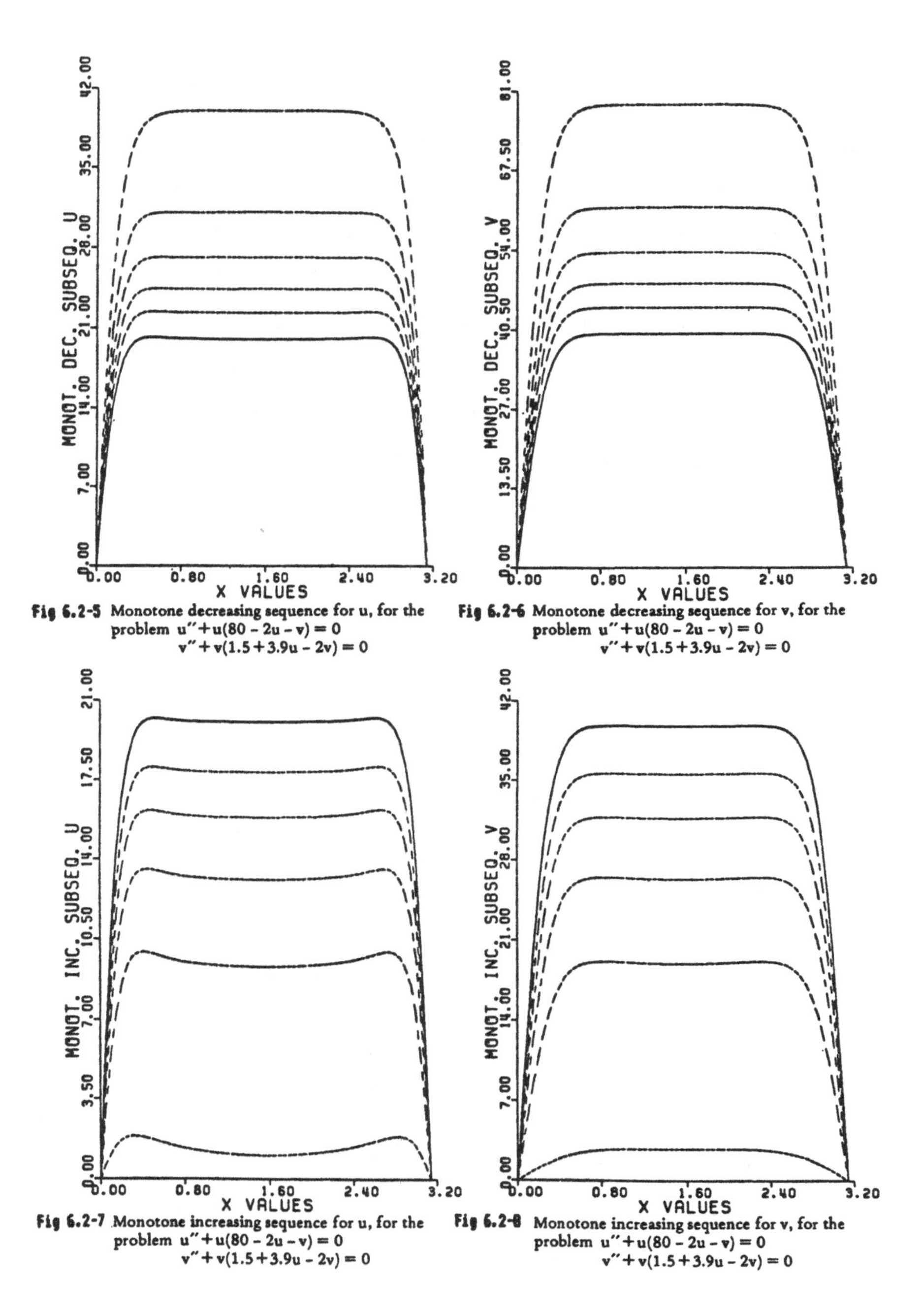

Fig 6.2-5 Monotone decreasing sequence for u, for the problem $u'' + u(80 - 2u - v) = 0$
$v'' + v(1.5 + 3.9u - 2v) = 0$

Fig 6.2-6 Monotone decreasing sequence for v, for the problem $u'' + u(80 - 2u - v) = 0$
$v'' + v(1.5 + 3.9u - 2v) = 0$

Fig 6.2-7 Monotone increasing sequence for u, for the problem $u'' + u(80 - 2u - v) = 0$
$v'' + v(1.5 + 3.9u - 2v) = 0$

Fig 6.2-8 Monotone increasing sequence for v, for the problem $u'' + u(80 - 2u - v) = 0$
$v'' + v(1.5 + 3.9u - 2v) = 0$

6.3. Accelerated Monotone Convergence

In the last section, for the computation of each $u_{N,k}^{(i)}$ or $v_{N,k}^{(i)}$, $i = 1, 2, \ldots$, we solve a scalar problem of the type (6.1-7). This scalar problem itself is solved by successive iterations as indicated in the proof of Theorem 6.1-1. When these computations are incorporated into the full iterative procedure for the system (6.1-14) to (6.1-16), the amount of computation is sometimes tremendous (see e.g., Example 6.2-2). In this section, we consider a method for accelerating the iterative process in solving the individual scalar problem (6.1-7). We first consider the accelerating procedure for the continuous problem, and then discretize to treat the discrete problem. For the continuous version of (6.1-7), we have

$$(6.3\text{-}1)\qquad \begin{aligned} &w''(x) + w(x)[p(x) - qw(x)] = 0, \quad \alpha < x < \beta \\ &w(\alpha) = w(\beta) = 0. \end{aligned}$$

Conditions (6.1-9) and (6.1-10) translate into

$$(6.3\text{-}2)\qquad q > 0 \text{ and } p(x) > \left(\frac{\pi}{\beta-\alpha}\right)^2 \text{ for } \alpha \le x \le \beta.$$

Here, we assume $p(x)$ is continuous on $[\alpha,\beta]$. For convenience, we write

$$f(w,x) = -w[p(x)-qw], \quad g(w,x) = \frac{\partial f}{\partial w}(w,x) = -p(x) + 2qw.$$

Define $w_1(x) \equiv K$, a constant such that $K \ge \frac{1}{q}\, p(x)$ for all $\alpha \le x \le \beta$ (Hence w_1 satisfies $w_1'' + w_1[p-qw_1] \le 0$ in (α,β)). Next, we define $\{w_n(x)\}$, $\alpha \le x \le \beta$, $n = 1, 2, \ldots$ recursively as the solution of

$$w_{n+1}'' = f(w_n(x),x) + (w_{n+1}-w_n(x))\, g(w_n(x),x), \quad \alpha < x < \beta$$

(6.3-3)

$$w_{n+1}(\alpha) = w_{n+1}(\beta) = 0.$$

Formula (6.3-3) is Newton's method of approximation applied to the differential equation (6.3-1). To see that the functions $w_n(x)$ are uniquely defined

positive functions in (α,β), we will use the following important property of f:

$$f(w,x) = \max_{-\infty<z<\infty} [f(z,x)+(w-z)g(z,x)] \tag{6.3-4}$$

for each $-\infty < w < \infty$, $\alpha < x < \beta$. Suppose $w_k(x)$ is uniquely defined by (6.3-3) with $w_k(x) \geq \varepsilon \sin(x-\alpha)\pi/(\beta-\alpha)$ $\alpha < x < \beta$, where ε is a small positive constant. If w_{k+1} exists, $u(x) = w_{k+1}(x)$ should satisfy

$$u'' = f(w_k(x),x) + (u-w_k(x))\, g(w_k(x),x) \overset{(def)}{\equiv} h_{k+1}(u,x), \quad \alpha < x < \beta$$

(6.3-5)

$$u(\alpha) = u(\beta) = 0.$$

For k = 1, we have $w_k = K$ satisfying:

$$w_1'' = 0 \leq -K[p(x)-qK] = f(K,x) = f(w_1,x) + (w_1-w_1)g(w_1,x) = h_2(w_1,x); \tag{6.3-6}$$

and for k > 1, we have by (6.3-3) and (6.3-4):

$$w_k''(x) = f(w_{k-1}(x),x) + (w_k-w_{k-1}(x))\, g(w_{k-1}(x),x) \tag{6.3-7}$$

$$\leq f(w_k(x),x) = h_{k+1}(w_k(x),x).$$

The two inequalities above indicate that $w_k(x)$ is an upper solution for problem (6.3-5). On the other hand, the function $v(x) = \varepsilon \sin(x-\alpha)\pi/(\beta-\alpha)$ satisfies

(6.3-8) $$v'' \geq f(v,x) \geq f(w_k(x),x) + (v-w_k(x))\, g(w_k(x),x) \equiv h_{k+1}(v,x)$$

for $\alpha < x < \beta$, $\varepsilon > 0$ sufficiently small. (Note that the last inequality above in again due to (6.3-4). The function $v(x)$ is therefore a lower solution for (6.3-5), and problem (6.3-5) must have a solution $u = z(x)$ with $v(x) \leq z(x) \leq w_k(x)$, $\alpha \leq x \leq \beta$ (cf. Section 5.1; we can start iterating from the upper solution and then take limits, while using the integral representation to deduce that the monotone limit is a classical solution). To see that $z(x)$ is unique, let $u = z_i(x)$, $i = 1, 2$ be two different solutions of (6.3-5) with $v \leq z_i \leq w_k$. The function $y = z_1 - z_2$ will satisfy

$$y'' = yg(w_k(x),x), \quad \alpha < x < \beta; \quad y(\alpha) = y(\beta) = 0.$$

That is, $\lambda = 1$ is an eigenvalue for the problem:

(6.3-9) $$\eta'' - 2qw_k(x)\eta + \lambda p(x)\eta = 0 \quad \text{for } \alpha < x < \beta, \quad \eta(\alpha) = \eta(\beta) = 0,$$

with $\eta = y(x)$ as the corresponding eigenfunction. Referring to (6.3-6), (6.3-7) and (6.3-8) again we observe that $w_k(x)$ and $v(x)$ are also upper and lower solutions for the problem:

$$u'' = f(u,x), \quad \alpha < x < \beta, \quad u(\alpha) = u(\beta) = 0.$$

Therefore (6.3-1) has a solution $w = \tilde{w}(x)$, with $v(x) \leq \tilde{w}(x) \leq w_k(x)$, $\alpha \leq x \leq \beta$. Clearly, $\eta = \tilde{w}(x)$ satisfies:

(6.3-10) $$\eta'' - q\tilde{w}(x)\eta + \lambda p(x)\eta = 0, \quad \alpha < x < \beta, \quad \eta(x) = \eta(\beta) = 0,$$

with $\lambda = 1$ and $\tilde{w}(x) > 0$ in (α,β). By Sturm-Liouville theory for ordinary differential equations, $\lambda = 1$ is the first eigenvalue for problem (6.3-10), since the corresponding eigenfunction does not change sign. By Sturm's theory again, sicne $2w_k(x) > \tilde{w}(x)$ in (α,β), the first eigenvalue for problem (6.3-9) must be strictly greater than 1. This is a contradiction, unless $y(x) \equiv 0$ and $\lambda = 1$ is not really an eigenvalue for (6.3-9). Consequently, $z(x)$ is uniquely defined, and by letting $w_{k+1}(x) = z(x)$, we have

(6.3-11) $$0 < v(x) \le w_{k+1}(x) \le w_k(x), \text{ for } \alpha < x < \beta.$$

By inducation (6.3-11) is true for $k = 1, 2, 3, \ldots$ The monotone decreasing sequence $\{w_n(x)\}$ must converge to a function $\hat{w}(x)$ $(\ge v(x))$ for $\alpha \le x \le \beta$ as $n \to \infty$. Using the integral representation for the solution of (6.3-3), one sees that $\hat{w}(x)$ is a solution of (6.3-1), with $\hat{w}(x) \ge v(x) > 0$, $\alpha < x < \beta$. (From the uniqueness of the positive solution in (α,β) for (6.3-1), cf. Lemma 5.2-2, we actually have $\hat{w}(x) = \tilde{w}(x)$).

We have proved the following:

Theorem 6.3-1. *Suppose that* $p(x)$ *is continuous on* $[\alpha,\beta]$ *with* $p(x) > (\pi/(\beta-\alpha))^2$ *for* $\alpha \le x \le \beta$; *and* $q > 0$. *Let* $w_1(x) \equiv K$ *where* K *is a positive constant satisfying* $K \ge \frac{1}{q} p(x)$ *for all* $x \in [\alpha,\beta]$, *and* $w_n(x)$, $n = 2, 3, \ldots$ *be defined recursively by* (6.3-3). *Then* $w_n(x)$ *converges monotonically to the unique positive* (for $\alpha < x < \beta$) *solution of the problem* (6.3-1).

The next theorem illustrates why our present method is called accelerated.

Theorem 6.3-2. *Under the conditions of* Theorem 6.3-1, *the sequence* $w_n(x)$, $n = 1, 2, \ldots$ *converges quadratically to the unique positive solution* $\hat{w}(x)$ *of the problem* (6.3-1) *in the sense that:*

(6.3-12) $$\max_{\alpha\le x\le\beta} |\hat{w}(x)-w_n(x)| \le L[\max_{\alpha\le x\le\beta} |\hat{w}(x)-w_{n-1}(x)|]^2, \quad n = 2, 3, \ldots$$

<u>for some</u> L <u>independent of</u> n.

<u>Proof</u>. From the equations which $\hat{w}$ and w_n satisfy, we find $(\hat{w}-w_n)'' = f(\hat{w},x) - f(w_{n-1}(x),x) - (w_n-w_{n-1}(x))\, g(w_{n-1}(x),x)$.

This can be rewritten as:

$$(\hat{w}-w_n)'' = f(\hat{w},x) - f(w_{n-1}(x),x) - (\hat{w}-w_{n-1}(x))\, g(w_{n-1}(x),x)$$

$$+ (\hat{w}-w_n(x))\, g(w_{n-1}(x),x).$$

Let $\eta = \hat{w} - w_n$. Then η satisfies for $n \ge 2$,

(6.3-13)
$$\eta'' - 2qw_{n-1}(x)\eta + p(x)\eta = f(\hat{w},x) - f(w_{n-1}(x),x) - (\hat{w}-w_{n-1}(x))\, g(w_{n-1}(x),x),$$
$$\eta(\alpha) = \eta(\beta) = 0.$$

We have shown at the end of the proof of Theorem 6.3-1 that the first eigenvalue λ for the problem (6.3-9), $k \ge 1$, must satisfy $\lambda > 1$. Consequently, from (6.3-13), the function $\hat{w} - w_n(x)$ is representable as:

(6.3-14) $$(\hat{w}-w_n)(x) = \int_\alpha^\beta G_{n-1}(x,\xi)\frac{1}{2}\frac{\partial^2 f}{\partial w^2}(\zeta_n(\xi),\xi)[\hat{w}(\xi)-w_{n-1}(\xi)]^2 d\xi$$

for $\alpha \le x \le \beta$, where G_{n-1} is the Green's function for the operator $(d^2/dx^2) - 2qw_{n-1} + p$ on the interval $[\alpha,\beta]$. (Here $\zeta_n(\xi)$ lies between $\hat{w}(\xi)$ and $w_{n-1}(\xi)$, and the mean value theorem is used). For our f, we readily verify that $\partial^2 f/\partial w^2 \equiv 2q$; hence (6.3-14) gives

(6.3-15)
$$\max_{\alpha\le x\le\beta} |\hat{w}(x)-w_n(x)| \le [\max_{\alpha\le x\le\beta} |\hat{w}(x)-w_{n-1}(x)|]^2 q \int_\alpha^\beta |G_{n-1}(x,\xi)|d\xi.$$

We now deduce that the functions $G_n(x,\xi)$ are uniformly bounded for $n = 1, 2, \dots, \alpha \le x \le \beta$. From the characterization of the Green's function, one readily sees that for $n = 2, 3, \dots$ if we let $u_n(x)$, $v_n(x)$ be solutions of

(6.3-16)
$$\eta'' + [p(x)-2qw_{n-1}(x)]\eta = 0, \quad \alpha < x < \beta$$

with $u_n(\alpha) = 0$, $u_n'(\alpha) = 1$ and $v_n(\beta) = 0$, $v_n'(\beta) = -1$, we have

(6.3-17)
$$G_{n-1}(x,\xi) = \begin{cases} (-v_n(\xi)/W_n(\xi))u_n(x) & \text{for } x < \xi \\ (-u_n(\xi)/W_n(\xi))v_n(x) & \text{for } \xi < x \end{cases}$$

Here

(6.3-18)
$$W_n(\xi) = u_n(\xi)v_n'(\xi) - v_n(\xi)u_n'(\xi)$$

which is the Wronskian for two independent solutions u_n, v_n. For convenience, let $u_0(x)$, $v_0(x)$ be solutions of

(6.3-19)
$$\eta'' + [p(x) - 2q\hat{w}(x)]\eta = 0, \quad \alpha < x < \beta$$

with $u_0(\alpha) = 0$, $u_0'(\alpha) = -1$ and $v_0(\beta) = 0$, $v_0'(\beta) = -1$. (Note that the eigenvalue problem $\eta'' - 2q\hat{w}(x)\eta + \lambda p(x)\eta = 0$, $\alpha < x < \beta$, $\eta(\alpha) = \eta(\beta) = 0$, must have its first eigenvalue $\lambda > 1$, since $2q\hat{w}(x) > q\hat{w}(x)$ and (6.3-10) has 1 as its first eigenvalue. Consequently, $v_0(\alpha) \ne 0$.) Let $W_0(x)$ be the Wronskian defined also by (6.3-18) with $n = 0$. From (6.3-16) and (6.3-19) which do not have η' term, we find that $W_n(\xi) = W_n(\alpha)$, for $\alpha \le \xi \le \beta$, $n = 2, 3, \dots$ and

$n = 0$. We will now see that $\lim_{n\to\infty} W_n(\alpha) = W_0(\alpha) = -v_0(\alpha)$. (The difficulty here is that we do not know that $\lim_{n\to\infty} w_n(x) = \hat{w}(x)$ uniformly in $[\alpha,\beta]$). From the uniform boundedness of $w_n(x)$ in $[\alpha,\beta]$, $n = 1, 2, \ldots$ and the equations (6.3-16), we must have $v_n(x)$, $v_n'(x)$ uniformly bounded for x in $[\alpha,\beta]$, $n = 2, 3, \ldots$ (cf. [98]). From (6.3-16) again, we have uniform boundedness of $v_n''(x)$ in $[\alpha,\beta]$. Consequently, we can extract a subsequence of $v_n(x)$ convergent uniformly in $[\alpha,\beta]$ to a function $\tilde{v}_0(x)$, furthermore a subsequence of $v_n'(x)$ converges uniformly to $\tilde{v}_0'(x)$. Expressing $v_n(x)$ and $v_n'(x)$ as an integral for an initial value problem at $x = \beta$ by means of (6.3-16), and taking limit for the subsequence under the integral by means of dominated convergence theorem, one sees that $\tilde{v}_0(x) = v_0(x)$ which is uniquely defined by the initial value problem at $x = \beta$. From the uniqueness of the initial value problem defining $v_0(x)$ again, we conclude that the full limit $\lim_{n\to\infty} v_n(\alpha) = v_0(\alpha)$. Therefore, we have $\lim_{n\to\infty} W_n(\alpha) = \lim_{n\to\infty} - v_n(\alpha) = -v_0(\alpha) \neq 0$, and $|W_n(\xi)| = |W_n(\alpha)| \geq c > 0$ for some positive constant c, $n = 2, 3, \ldots$ From (6.3-16) again, we have u_n, v_n uniformly bounded in $[\alpha,\beta]$; and using (6.3-17), we conclude that $G_n(x,\xi)$ are uniformly bounded for $n = 1, 2, \ldots, \alpha \leq x \leq \beta$, $\alpha \leq \xi \leq \beta$. From (6.3-15), we arrive at (6.3-12). This concludes the proof of Theorem 6.3-2.

In actual computations, we solve the finite difference version of (6.3-1), (i.e., (6.1-7)), by restricting (6.3-3) to the corresponding grid functions. That is, we solve

$$\begin{gathered} \Delta_h w_{N,k}^{(n+1)} + p_k w_{N,k}^{(n+1)} - 2q w_{N,k}^{(n)} w_{N,k}^{(n+1)} = -q(w_{N,k}^{(n)})^2 \\ w_{N,0}^{(n+1)} = w_{N,N}^{(n+1)} = 0, \; k = 1, \ldots, N-1 \end{gathered} \tag{6.3-20}$$

for $n = 1, 2, \ldots$, where $w_{N,k}^{(1)} = K$, $k = 1, \ldots, N-1$, is a known positive constant, upper solution of (6.1-7), with $w_{N,0}^{(1)} = w_{N,N}^{(1)} = 0$. One can proceed to prove the existence of a positive solution $w_{N,k}^{(n)}$, $n = 2, 3, \ldots$ for (6.3-20), where $w_{N,k}^{(n)}$ converges monotonically and quadratically as $n \to \infty$ to $w_{N,k}$, the solution of (6.1-7), as it is done in Theorem 6.3-1 and Theorem 6.3-2 for the continuous case under appropriate condition. We omit the details here.

We now apply the accelerated scheme (6.3-20) to solve each scalar equation in the monotone method of solving prey-predator systems (cf. (6.1-14) to (6.1-16)). Recall we let $u_{N,k}^{(1)} > 0$ $k = 1, \ldots, N-1$ satisfy

$$\Delta_h u_{N,k}^{(1)} + u_{N,k}^{(1)}(a-bu_{N,k}^{(1)}) = 0, \quad u_{N,0}^{(1)} = u_{N,N}^{(1)} = 0; \tag{6.3-21}$$

and similarly let $v_{N,k}^{(1)} > 0$, $k = 1, \ldots, N-1$ be the solution of

$$\Delta_h v_{N,k}^{(1)} + v_{N,k}^{(1)}(e + fu_{N,k}^{(1)} - gv_{N,k}^{(1)}) = 0, \quad v_{N,0}^{(1)} = v_{N,N}^{(1)} = 0. \tag{6.3-22}$$

For $i = 2, 3, \ldots$, define $u_{N,k}^{(i)} > 0$, $v_{N,k}^{(i)} > 0$ for $k = 1, \ldots, N-1$ inductively as follows:

$$\begin{aligned} &\Delta_h u_{N,k}^{(i)} + u_{N,k}^{(i)}(a - bu_{N,k}^{(i)} - cv_{N,k}^{(i-1)}) = 0 \\ &\Delta_h v_{N,k}^{(i)} + v_{N,k}^{(i)}(e + bu_{N,k}^{(i)} - gv_{N,k}^{(i)}) = 0 \\ &u_{N,0}^{(i)} = u_{N,N}^{(i)} = v_{N,0}^{(i)} = v_{N,N}^{(i)} = 0. \end{aligned} \tag{6.3-23}$$

When a, b, c, e, f, g are positive, and conditions (6.1-2), (6.1-5) and (6.1-6) are satisfied, we show in Section 6.1 that the sequences defined above will satisfy

$$0 \le u_{N,k}^{(2)} \le u_{N,k}^{(4)} \le u_{N,k}^{(6)} \le \ldots \le u_{N,k}^{(5)} \le u_{N,k}^{(3)} \le u_{N,k}^{(1)}, \text{ and}$$

$$0 \le v_{N,k}^{(2)} \le v_{N,k}^{(4)} \le v_{N,k}^{(6)} \le \ldots \le v_{N,k}^{(5)} \le v_{N,k}^{(3)} \le v_{N,k}^{(1)}$$

for $k = 0, \ldots, N$.

We let $[\alpha,\beta] = [0,\pi]$, $N = 64$, $h = \frac{1}{64}$. Experience in Example 6.2-2 shows that when the condition $cf < gb$ is barely fulfilled, and the coefficient a is large relative to e, the computations are difficult, in the sense that a large number of iterations are needed for the numerical sequences to satisfy a particular convergence criterior. We let $a = 80$, $b = 2$, $c = 1$ and $e = 1.5$, $f = 3.9$, $g = 2$ in Example 6.2-2, and 18003 iterations are needed to produce 1236 elements of the monotone sequences when computations are done in Section 6.2. We now solve exactly the same problem as Example 6.2-2, but use the accelerated scheme to solve the scalar equations (6.3-21) to (6.3-23). We stop the algorithm when the relative error in the maximum norm of two consecutive iterates in all four monotone sequences is less than 10^{-8}; i.e., when

$$(\|z_{N,k}^{(j+2)} - z_{N,k}^{(j)}\| / \|z_{N,k}^{(j+2)}\|) \leq 10^{-8}, \text{ with } \|z_{N,k}^{(j)}\| = \max_{0\leq k\leq N} |z_{N,k}^{(j)}|.$$

Here $z_{N,k}^{(j)} = u_{N,k}^{(2j-1)}$ or $v_{N,k}^{(2j)}$.

Now, the accelerated monotone scheme needs ony 3509 iterations to produce the same 1236 elements of the monotone sequences. The average number of iterations necessary to generate each element of the monotone sequences has been reduced from 15 to about 3. The monotone sequences produced are practically the same as those in Example 6.2-2 in section 6.2. For more details, the reader is referred to [147].

6.4. L_2 Convergence for Finite-Difference Solutions in Two Dimensional Domains

In this section, we extend the theories in Section 6.1 to two dimensional bounded domains. Let Ω be an open bounded connected domain in R^2, with its boundary $\delta\Omega$ in $H^{4+\alpha}$, $0 < \alpha < 1$ (cf. Section 1.3). We cover the plane with a square grid with mesh width h. The mesh points are the intersection of the grid lines. For any r, s = 0, ± 1, ± 2, ..., connect the mesh point (rh, sh) to the mesh points ((r±1)h, sh), (rh, (s±1)h), ((r+1)h, (s+1)h) and ((r-1)h, (s-1)h). Connect the adjacent mesh points above to form 6 triangles each with (rh, sh) as a vertex. The set Ω_h is defined to consist of those mesh points (rh, sh) in Ω such that all the 6 triangles described above with (rh, sh) as a vertex lie in $\Omega \cup \delta\Omega$. The set $\delta\Omega_h$ consists of those mesh points on $\delta\Omega$, and those in Ω not included in Ω_h. We assume that h is small, so that every two points in Ω_h can be connected by horizontal and vertical segments of length h joining only mesh points in Ω_h.

Let $\lambda = \lambda_1 > 0$ be the first eigenvalue for the problem $\Delta z + \lambda z = 0$ in Ω, $z = 0$ on $\delta\Omega$. We are concerned with the approximation of the solution of

(6.4-1) $$\Delta w + w[p(x,y)-qw] = 0 \text{ in } \Omega, \quad w = 0 \text{ on } \delta\Omega.$$

Here q is a positive constant, and p(x,y) is assumed to be in the class $H^{2+\alpha}(\bar{\Omega})$ with

(6.4-2) $$p(x,y) > \lambda_1 \quad \text{for all } (x,y) \in \bar{\Omega}$$

From Chapter 5, we know that (6.4-1) has a solution in $H^{2+\alpha}(\bar{\Omega})$, with $w(x,y) > 0$ in Ω. With the present additional smoothness assumption in $\delta\Omega$ and p(x,y), we can obtain that w is in $H^{4+\alpha}(\bar{\Omega})$. (cf. [89]). The discrete problem corresponding to (6.4-1) is

(6.4-3) $$\Delta_h v_h(P) + v_h(P)[p_h(P) - qv_h(P)] = 0 \quad \text{in } \Omega_h$$

$$v_h(P) = 0 \quad \text{on } \delta\Omega_h$$

where $\Delta_h v(rh, sh) = \frac{1}{h^2}[v((r+1)h, sh) + v((r-1)h, sh) + v(rh, (s+1)h) + v(rh, (s-1)h) - 4v(rh, sh)]$, for integers r, s, and $p_h(P)$ is the restriction of $p(x, y)$ to Ω_h. Let $\lambda_1^h > 0$ be the first eigenvalue for the discrete problem $\Delta_h z_h + \lambda z_h = 0$ in Ω_h, $z_h = 0$ in $\delta\Omega_h$. We will see that the condition

$$\min_{P \in \Omega_h} \{p_h(P) - \lambda_1^h\} > 0 \qquad (6.4\text{-}4)$$

is sufficient for obtaining a positive solution v_h of problem (6.4-3) (cf. Lemma 6.4-4). In Theorem 6.4-2, the positive solution will be shown to converge in the L_2 sense to $w_h(P)$, where $w_h(P)$ is the restriction of the positive solution $w(x, y)$ of (6.4-1) to Ω_h. (The positive solutions $v_h(P)$ and $w(x, y)$ are unique.) For the convergence proof, we will need a uniform lower bound for v_h as $h \to 0^+$, so that v_h does not tend to zero. For this purpose, we will see that the additional assumption

$$p_h(P) > \frac{4}{h^2}(1 - \cos\frac{h\pi}{\ell}), \quad P \in S_\ell \qquad (6.4\text{-}5)$$

is sufficient (cf. Lemma 6.4-3 and Remark 6.4-1). Here S_ℓ is a closed square of area ℓ^2 contained inside Ω. (Note that: $\frac{4}{h^2}(1-\cos\frac{h\pi}{\ell}) \to \frac{2\pi^2}{\ell^2}$ as $h \to 0$; and $\Delta(\sin\frac{\pi}{\ell}(x-a)\sin\frac{\pi}{\ell}(y-b)) = \frac{2\pi^2}{\ell^2}(\sin\frac{\pi}{\ell}(x-a)\sin\frac{\pi}{\ell}(y-b))$.

Recall that a grid function ϕ_h is an upper solution for problem (6.4-3) if it satisfies

$$(6.4\text{-}6) \qquad \begin{aligned} \Delta_h\phi_h(P) + \phi_h(P)[p_h(P) - q\phi_h(P)] &\le 0 \quad \text{in } \Omega_h \\ \phi_h(P) &\ge 0 \quad \text{on } \delta\Omega_h \end{aligned}$$

The grid function is a lower solution of problem (6.4-3) if both inequalities in (6.4-6) are reversed to $\geq$ and $\leq$ respectively. The following is immediate.

Lemma 6.4-1. *Suppose that*

$$q > 0 \ \text{and}\ p_h(P) > 0,\ P \in \Omega_h$$

Let C *be any constant such that* $C \geq \frac{1}{q} \max_{P\in\Omega_h} p_h(P)$. *Then the grid function* $\phi_h(P) \equiv C,\ P \in \Omega_h \cup \delta\Omega_h$ *is an upper solution for problem* (6.4-3).

For the rest of this section, we assume that Ω contains a closed square S_ℓ with sides of length $\ell >> 2h$. Let the vertices of S_ℓ be (a_1, a_2), $(a_1+\ell, a_2)$, $(a_1+\ell, a_2+\ell)$ and $(a_1, a_2+\ell)$. For $i = 1, 2$, let m_i^h and n_i^h be integers so that $a_i \leq m_i^h h < n_i^h h \leq a_i + \ell$ and $m_i^h h - a_i < h$, $(a_i+\ell) - n_i^h h < h$. For $P = (x,y) \in \Omega_h \cup \delta\Omega_h$, define the grid function

$$\theta_h(P) = \begin{cases} \sin \dfrac{(x-m_1^h h)\pi}{(n_1^h-m_1^h)h} \ \sin \dfrac{(y-m_2^h h)\pi}{(n_2^h-m_2^h)h} & \text{if } P \in S_\ell \cap \Omega_h \\ 0 & \text{if } P \notin S_\ell \cap \Omega_h \end{cases}$$

We make the assumption that in problem (6.4-1)

$$\text{(6.4-7)} \qquad p(x,y) > \frac{2\pi^2}{\ell^2} \ \text{for all } (x,y) \in S_\ell$$

Observe that $\frac{2}{h^2}(2 - \cos \frac{\pi}{n_1^h-m_1^h} - \cos \frac{\pi}{n_2^h-m_2^h}) \to \frac{2\pi^2}{\ell^2}$ as $h \to 0^+$. Thus, if we let

$$r_h = \min_{P\in S_\ell\cap\Omega_h} [p_h(P) - \frac{2}{h^2}(2 - \cos \frac{\pi}{n_1^h-m_1^h} - \cos \frac{\pi}{n_2^h-m_2^h})],$$

then $r_h > 0$ if h is small enough.

<u>Lemma</u> 6.4-2. <u>Assume inequality</u> (6.4-7). <u>Suppose that</u> $q > 0$ <u>and</u> $r_h > 0$. <u>Then for any</u> δ_h <u>such that</u> $0 < \delta_h \le (r_h/q)$, <u>the function</u> $S_h(P) \overset{(def)}{=} \delta_h\theta_h(P)$, $P \in \Omega_h \cup \delta\Omega_h$, <u>is a lower solution for problem</u> (6.4-3). <u>Moreover</u>, <u>if we let</u> $\hat{\delta} = (1/2q)\min\{p(x,y) - \frac{2\pi^2}{\ell^2} \mid (x,y) \in S_\ell\}$, <u>then</u> $\hat{\delta}\theta_h(P)$ <u>is a lower solution for problem</u> (6.4-3) <u>for all sufficiently small</u> h.

<u>Proof</u>. We notice that $S_h(P) = \theta_h(P) = 0$ for $P \notin S_\ell \cap \Omega_h$. Let S_ℓ^h and δS_ℓ^h, respectively, denotes the interior and boundary mesh points on the square with vertices $(m_1^h h, m_2^h h)$, $(n_1^h h, m_2^h h)$, $(n_1^h h, n_2^h h)$, $(m_1^h h, n_2^h h)$. For $P = (rh, sh)$ in S_ℓ^h, (i.e., $m_1^h < r < n_1^h$, $m_2^h < s < n_2^h$), we have

$$\Delta_h S_h(P) + S_h(P)[p_h(P) - qS_h(P)]$$

$$= \delta_h\Bigg[\sin\frac{(sh-m_2^h h)\pi}{(n_2^h-m_2^h)h}\Bigg\{\frac{\sin[(r-1)h-m_1^h h]\pi}{(n_1^h-m_1^h)h} - 2\sin\frac{(rh-m_1^h h)\pi}{(n_1^h-m_1^h)h}$$

$$+ \sin\frac{[(r+1)h-m_1^h h]\pi}{(n_1^h-m_1^h)h}\Bigg\}\frac{1}{h^2} + \sin\frac{(rh-m_1^h h)\pi}{(n_1^h-m_1^h)h}$$

$$\times\Bigg\{\sin\frac{[(s-1)h-m_2^h h]}{(n_2^h-m_2^h)h} - 2\sin\frac{(sh-m_2^h h)\pi}{(n_2^h-m_2^h)h}$$

$$+ \sin\frac{[(s+1)h-m_2^h h]\pi}{(n_2^h-m_2^h)h}\Bigg\}\frac{1}{h^2} + \theta_h(P)[p_h(P) - q\delta_h\theta_h(P)]\Bigg]$$

$$= \delta_h \sin\frac{(rh-m_1^h h)\pi}{(n_1^h-m_1^h)h}\sin\frac{(sh-m_2^h h)\pi}{(n_2^h-m_2^h)h}\left[\frac{2}{h^2}\left[\cos\frac{\pi}{(n_1^h-m_1^h)} - 1\right]\right.$$

$$+ \frac{2}{h^2}\left[\cos\frac{\pi}{(n_2^h - m_2^h)} - 1\right] + p_h(P) - \delta_h q\, \theta_h(P)\Bigg] .$$

The expression in [ ] above is ≥ 0 if $0 < \delta_h \leq (r_h/q)$. For P on δS_ℓ^h, say $P = (m_1^h h, sh)$, $m_2^h < s < n_2^h$, we have

$$\Delta_h S_h(P) + S_h(P)[p_h(P) - qS_h(P)] = \delta_h \sin\frac{\pi}{(n_1^h - m_1^h)} \sin\frac{(sh - m_2^h h)}{(n_2^h - m_2^h)h} > 0.$$

For $P \notin S_\ell^h \cup \delta S_\ell^h$, we clearly have $\Delta_h S_h(P) + S_h(P)[p_h(P) - qS_h(P)] = 0$. This proves the lemma.

Remark 6.4-1. If we let Γ be a square of length less than $\ell/2$ with the same center as S_ℓ, and denote $\hat{S}_h(P) \overset{(def)}{=} \hat{\delta}\theta_h(P)$, then $\hat{S}_h(P) > \hat{\delta}/2 > 0$ for all $P \in \Gamma \cap \Omega_h$, h is sufficiently small. Consequently, we have

$$\hat{S}_h(P) > \psi_h(P) \quad \text{for all } P \in \Gamma \cap \Omega_h,\ h \text{ sufficiently small},$$

where $\psi_h(P)$ can be considered as the restriction of some continuously differentiable function $\psi(x,y)$ defined on $\Omega \cup \delta\Omega$ with $\psi(x,y) \geq 0$ in $\Omega \cup \delta\Omega$ and $\psi(x,y) = \hat{\delta}/2$ in Γ. This $\psi(x,y)$ will be used as a uniform lower bound for the lower solution $\hat{S}_h(P)$, all h sufficiently small. Such bound will be used for comparison purposes in Theorem 6.4-2.

Lemma 6.4-3. *Assume* $q > 0$, $p(x,y) > 0$ *in* Ω *while satisfying inequality* (6.4-7) *in* S_ℓ, *and* $\hat{\delta}$ *is as defined in Lemma* 6.4-2. *Let* C_1 *be a constant larger than both* $\hat{\delta}$ *and* $\frac{1}{q} \max_{P \in \Omega_h} p_h(P)$. *Then for all* $h > 0$ *sufficiently small there exists a grid function* $v_h(P)$ *which is a solution of problem* (6.4-3) *such* that

$$\hat{S}_h(P) = \hat{\delta\theta}_h(P) \le v_h(P) \le C_1, \quad P \in \Omega_h \cup \delta\Omega_h. \tag{6.4-8}$$

Proof. Let Q be a nonnegative number satisfying $Q > 2qC_1 - \min_{P\in\Omega_h} p_h(P)$. Define the transformation T by $z_h(P) = (Ty_h)_h(P)$, $P \in \Omega_h \cup \delta\Omega_h$ if

$$(\Delta_h - Q)z_h(P) = -y_h(P)[p_h(P) - qy_h(P) + Q], \quad P \in \Omega_h, \text{ and}$$

$$z_h(P) = 0, \quad P \in \delta\Omega_h.$$

Further, if we define for $P \in \Omega_h \cup \delta\Omega_h$, $z_h^{(1)} = (T\phi_h)_h(P)$ where $\phi_h(P) \equiv C_1$, and $z_h^{(i+1)}(P) = (Tz_h^{(i)})_h(P)$, $i = 1, 2, \ldots$, we can show as in Theorem 6.1-1 that

$$\hat{S}_h(P) \equiv \hat{\delta\theta}_h(P) \le \ldots \le z_h^{(3)}(P) \le z_h^{(2)}(P) \le z_h^{(1)}(P) \le \phi(P) \equiv C_1$$

for $P \in \Omega_h \cup \delta\Omega_h$. As $i \to \infty$, $z_h^{(i)}(P) \to v_h(P)$, as stated. Note that $v_h(P)$ is uniquely defined here by the choice of the upper solution C_1 as the first iterate.

Let $\lambda_1^h > 0$ be as defined for (6.4-4) with $\omega_h(P)$ the corresponding normalized discrete eigenfunction. From the characterization of the first eigenvalue we can assume that $\omega_h(P) \ge 0$ for $P \in \Omega_h$ [cf. 77, pp. 336-337], because by replacing every value with its modulus, one will attain the same or smaller minimum for the characterization. Since $\Delta_h\omega_h(P) \le 0$ in Ω_h, i.e., $\omega_h(P)$ is greater than or equal to the average of its four neighbors, using the connectedness of interior mesh points, one sees that if $\omega_h(P) = 0$ at one $P \in \Omega_h$, then $\omega_h(P) = 0$. We therefore must have $\omega_h(P) > 0$ for all $P \in \Omega_h$.

Lemma 6.4-4. Assume that

$$q > 0, \min_{P\in\Omega_h} \{p_h(P)-\lambda_1^h\} \overset{(def)}{=} d_h > 0,$$

and let $\tilde{C}_1$ be a constant larger than $\frac{1}{q}\max_{P\in\Omega_h} p_h(P)$. Then for any δ_h such that $0 < \delta_h \le d_h/(q \max_{P\in\Omega_h} \omega_h(P))$, the grid function $\delta_h\omega_h(P)$, $P \in \Omega_h \cup \delta\Omega_h$ is a lower solution for problem (6.4-3). Furthermore, there exists a grid function $\tilde{v}_h(P)$ which is a solution of problem (6.4-3) such that $\delta_h\omega_h(P) \le \tilde{v}_h(P) \le \tilde{C}_1$, $P \in \Omega_h \cup \delta\Omega_h$.

Proof. For $P \in \Omega_h$, $\Delta_h\delta_h\omega_h(P) + \delta_h\omega_h[p_h-q\delta_h\omega_h] = \delta_h\omega_h(P)[p_h(P)-\lambda_1^h-q\delta_h\omega_h(P)]$ which is ≥ 0 if $\delta_h \le d_h/(q \max_{P\in\Omega_h} \omega_h(P))$. Thus $\delta_h\omega_h(P)$ is a lower solution. On the other hand, $\tilde{C}_1$ is an upper solution for problem (6.4-3), as in Lemma 6.4-1. (Note that $\max_{P\in\Omega_h} p_h(P) > d_h$ and thus $\tilde{C}_1 > \delta_h\omega_h(P)$, $P \in \Omega_h$.) Let T be the transformation defined in Lemma 6.4-3 and start the iterations using $\tilde{\phi}_h(P) \equiv \tilde{C}_1$, $\tilde{z}_h^{(1)} = (T\tilde{\phi}_h)_h(P)$ and $\tilde{z}_h^{(i+1)}(P) = (T\tilde{z}_h^{(i)})(P)$, $i = 1, 2, \ldots$ We obtain $\delta\omega_h(P) \le \ldots \le \tilde{z}_h^{(3)}(P) \le \tilde{z}_h^{(2)}(P) \le \tilde{z}_h^{(1)}(P) \le \tilde{\phi}_1(P) \equiv \tilde{C}_1$. As $i \to \infty$, $\tilde{z}_h^{(i)} \to \tilde{v}_h(P)$ as stated. Note that $\tilde{v}_h(P)$ is uniquely determined by the choice of the upper solution $\tilde{C}_1$ as the first iterate.

Theorem 6.4-1. Let q, p(x,y) and $\hat{\delta}$ be as described in Lemma 6.4-3. Suppose further that

$$\min_{P\in\Omega_h} \{p_h(P)-\lambda_1^h\} \overset{(def)}{=} d_h > 0$$

Let C_1 be larger than both $\hat{\delta}$ and $\frac{1}{q}\max_{P\in\Omega_h} p_h(P)$. Then for $h > 0$ sufficiently small, the solution $v_h(P)$ of problem (6.4-3) as constructed in Lemma 6.4-3 satisfies (6.4-8) as well as

$$(6.4\text{-}9) \qquad 0 < \delta_h \omega_h(P) \le v_h(P), \quad P \in \Omega_h.$$

Here δ_h is any constant such that $0 < \delta_h \le d_h/(q \max_{P \in \Omega_h} \omega_h(P))$.

Proof. In Lemma 6.4-4, choose $\tilde{C}_1$ to be the constant C_1 of this Theorem, then $\tilde{v}_h(P) \equiv v_h(P)$, $P \in \Omega_h \cup \partial\Omega_h$. (Since both are determined by iterating from the same upper solutions). Applying Lemma 6.4-3 and 6.4-4, we obtain (6.4-8) and (6.4-9) respectively.

The following lemma illustrates the uniqueness of positive solutions ($v_h(P) > 0$, $P \in \Omega_h$) of (6.4-3).

Lemma 6.4-5. Suppose that $q > 0$. Let $\bar{v}_h(P)$ and $\hat{v}_h(P)$ be solutions of problem (6.4-3) with the properties that $\bar{v}_h(P) > 0$ and $\hat{v}_h(P) > 0$ for all $P \in \Omega_h$. Then $\bar{v}_h(P) \equiv \hat{v}_h(P)$, $P \in \Omega_h$.

Proof. Suppose that $\bar{v}_h(P) \not\equiv \hat{v}_h(P)$. There exists $R \in \Omega_h$ such that, say, $\bar{v}_h(R) < \hat{v}_h(R)$. Let $S_1 = \{P \in \Omega_h : \bar{v}_h(P) < \hat{v}_h(P)\}$ and $S_2 = \{P \in \Omega_h \cup \partial\,\Omega_h : \bar{v}_h(P) \ge \hat{v}(P)\}$. Let $S_R = \{P \in S_1$: P is connected to R by horizontal and vertical line segments of length h with end points in $S_1\}$. Define $S_R^* = \{P^* \in S_2 : P^*$ is of distance h from some $P \in S_R\}$. Suppose that P is an interior point in Ω_h, we denote $\bar{v}_h(P)$ by $\bar{v}_{ij}$ and its four neighboring values by $\bar{v}_{i-1,j}$, $\bar{v}_{i+1,j}$, $\bar{v}_{i,j-1}$, $\bar{v}_{i,j+1}$; similarly, we denote $\hat{v}_{ij}$, etc. It is clear that

$$(6.4\text{-}10) \qquad \begin{aligned} h^2[\bar{v}_{ij}\Delta_h\hat{v}_{ij} - \hat{v}_{ij}\Delta_h\bar{v}_{ij}] &= \bar{v}_{ij}[\hat{v}_{i-1,j} + \hat{v}_{i+1,j} + \hat{v}_{i,j-1} + \hat{v}_{i,j+1}] \\ &\quad - \hat{v}_{ij}[\bar{v}_{i-1,j} + \bar{v}_{i+1,j} + \bar{v}_{i,j-1} + \bar{v}_{i,j+1}] \end{aligned}$$

From (6.4-10) we deduce that:

$$h^2 \sum_{P\in S_R} \bar{v}_h(P)\Delta_h\hat{v}_h(P) - \hat{v}_h(P)\Delta_h\bar{v}_h(P)$$

(6.4-11)

$$= \sum_{P^*\in S_R^*} \{ \sum_{P\in S_R} \bar{v}_h(P)\hat{v}_h(P^*) - \hat{v}_h(P)\bar{v}_h(P^*) \mid P \text{ is of distance } h \text{ from } P^* \}.$$

For $P \in S_R \subset \Omega_h$, we have

$$h^2[\bar{v}_h(P)\Delta_h\hat{v}_h(P) - \hat{v}_h(P)\Delta_h\bar{v}_h(P)]$$

$$= h^2\, \bar{v}_h(P)\hat{v}_h(P)q[\hat{v}(P)-\bar{v}(P)] > 0.$$

Thus the expression in (6.4-11) is > 0. However, for P, P^* considered in the right of (6.4-11), we have

$$\bar{v}_h(P)\hat{v}_h(P^*) - \hat{v}_h(P)\bar{v}_h(P^*) \le \bar{v}_h(P)\bar{v}_h(P^*) - \hat{v}_h(P)\bar{v}_h(P^*) \le 0$$

(since $P \in S_1$, $P^* \in S_2$). This implies the expression in (6.4-11) is ≤ 0, giving a contradiction. Hence $\bar{v}_h(P) \equiv \hat{v}_h(P)$.

Recall that we assume $p(x,y) > \lambda_1$ in (6.4-2), $p(x,y)$ in $H^{2+\alpha}(\bar{\Omega})$, and $\delta\Omega$ in $H^{4+\alpha}$. These hypotheses guarantee that there exist a solution $w(x,y)$ of (6.4-1) which is positive in Ω. Moreover, the positive solution is in $H^{4+\alpha}(\bar{\Omega})$ and is unique.

<u>Theorem</u> 6.4-2. <u>Let</u> $p(x,y)$ <u>satisfies</u> (6.4-2), λ_1^h <u>be as described in Theorem</u> 6.4-1, <u>and the closed square</u> $S_\ell \subset \Omega$ <u>be as described after Lemma</u> 6.4-1. <u>Assume that</u>

(6.4-12) $$\min_{P\in\Omega_h} \{p_h(P)-\lambda_1^h\} > 0, \text{ and}$$

(6.4-13) $$p_h(P) > \frac{4}{h^2}(1-\cos\frac{h\pi}{\ell})$$

<u>for all</u> P <u>at the mesh points on</u> S_ℓ <u>and all</u> h <u>small enough</u>, $q > 0$. <u>Then the unique positive solution</u> $w(x,y)$ <u>in</u> Ω <u>and</u> $v_h(P)$ <u>in</u> Ω_h <u>satisfy</u>

$$\|w_h(P) - v_h(P)\|_2 \to 0 \text{ as } h \to 0^+,\ P \in \Omega_h \cup \delta\Omega_h.$$

Here $\|e_h(P)\|_2 = h\{\sum_{P\in\Omega_h} [e_h(P)]^2\}^{1/2}$.

<u>Remark</u> 6.4-2. Suppose that $p(x,y) > \frac{2\pi^2}{\ell^2}$ for all $(x,y) \in S_\ell$, then $p_h(P)$ will satisfy inequality (6.3-13) for all h small enough. From Remark 6.4-1 and Lemma 6.4-3, (6.3-13) is used to establish the existence of a continuously differentiable function $\psi(x,y) \geq 0$ in $\Omega \cup \delta\Omega$, with $\psi(x,y) = \frac{\hat{\delta}}{2} > 0$ in a square Γ inside S_ℓ so that $v_h(P) \geq \psi_h(P)$, $P \in \Omega_h$, for h small enough ($\psi_h(P)$ is the restriction of $\psi(x,y)$ to the mesh points). Hypothesis (6.4-12), Lemma 6.4-4 and Theorem 6.4-1 insure that $v_h(P)$ is positive in Ω_h, and is uniquely defined.

<u>Proof</u>. For $P \in \Omega_h \cup \delta\Omega_h$, define the grid error function $e_h(P) = w_h(P) - v_h(P)$. From the boundedness of the first four partial derivatives of w in Ω, we have

(6.4-14)
$$\Delta_h e_h(P) + [p_h(P)-q(w_h(P)+v_h(P))]e_h(P) = O(h^2) \text{ in } \Omega_h,$$
$$e_h(P) = O(h) \qquad \text{on } \delta\Omega_h,$$

(cf. [77]). Let $y_h(P)$ be the grid function which satisfies the linear problem

(6.4-15) $$\Delta_h y_h(P) + [p_h(P) - q(w_h(P)+v_h(P))-E]y_h(P) = 0 \text{ in } \Omega_h,$$

$$y_h(P) = e_h(P) \qquad \text{on } \delta\Omega_h,$$

where $E > 0$ is a large constant so that $p_h(P) - q(w_h(P) + v_h(P)) - E < 0$ for all P in Ω_h, all small $h > 0$. Since $e_h(P)$ is of the order $O(h)$ on $\delta\Omega_h$, the maximum principle for the finite difference equation (6.4-15) implies that $y_h(P) = O(h)$ in Ω_h. Finally, let $z_h(P) = e_h(P) - y_h(P)$, we deduce from (6.4-14), (6.4-15) and $y_h = O(h)$ that $z_h(P)$ satisfies:

$$\text{(6.4-16)} \quad \begin{aligned} &\Delta z_h(P) + [p_h(P)-q(w_h(P) + v_h(P))]z_h(P) = O(h^2) - Qy_h(P) = O(h) \quad \text{in } \Omega_h, \\ &z_h(P) = 0 \qquad \text{on } \delta\Omega_h. \end{aligned}$$

In matrix form, (6.4-16) can be expressed as

$$\text{(6.4-17)} \qquad L_h \vec{z}_h(P) = O(h), \ P \in \Omega_h,$$

where L_h is a symmetric matrix with least eigenvalue μ_1^h for the problem $(L_h+\mu)\vec{z} = 0$. (Here, only the values of $z(P)$, $P \in \Omega_h$ are used in the vector $\vec{z}(P)$). We will show that $\mu_1^h \geq c > 0$ for all h under consideration. Let $\psi(x,y)$ be as described in Remark 6.4-2 above. For comparison purpose, we consider the continuous eigenvalue problem:

$$\text{(6.4-18)} \quad \begin{aligned} &\Delta u + [p(x,y)-q(w(x,y)+\psi(x,y))-K]u + \theta u = 0 \text{ in } \Omega, \\ &u = 0 \qquad \text{on } \partial\Omega, \end{aligned}$$

where $K > 0$ is a large enough constant so that $p-q(w+\psi)-K < 0$ in $\Omega \cup \partial\Omega$. Let the frist eigenvalue of (6.4-18) be $\theta = \mu_1+K$. Due to the relationship $\psi_h(P) \leq v_h(P)$, since $\psi_h(P)$ is the restriction of $\psi(x,y)$ to the grid, one can compare the eigenvalues of (6.4-17) and (6.4-18) with those of

$$\Delta_h u_h(P)+[p_h(P)-q(w_h(P)+\psi_h(P))-K]u_h(P)+\theta u_h(P) = 0 \text{ in } \Omega_h,$$

(6.4-19)

$$u_h(P) = 0 \qquad \text{on } \partial\Omega_h.$$

Denoting the first eigenvalue of (6.4-19) by $\theta = \tilde{\mu}_1^h+K$, we have

(6.4-20) $$\mu_1^h \geq \tilde{\mu}_1^h$$

because $\psi_h(P) \leq v_h(P)$. We will next show that for the considered values of h we have,

(6.4-21) $$\tilde{\mu}_1^h \geq \mu_1-Rh \text{ for some } R > 0 \text{ independent of } h.$$

For convenience, define $Q(x,y) = -p(x,y)+q(w(x,y)+\psi(x,y))+K$ for (x,y) in Ω. For any continuous function with piecewise continuous partial derivatives in Ω and vanishing on $\partial\Omega$, define the norms:

$$I(u,u) = \int_\Omega u^2\, dxdy,$$

(6.4-22)

$$A(u,u) = \int_\Omega [(\nabla u)^2+Q(x,y)u^2]dxdy.$$

For any grid function z on the plane, defined as zero on $\partial\Omega_h$ and outside Ω_h, define the norms:

$$I_h(z,z) = \sum z(rh,sh)^2h^2 ,$$

(6.4-23) $$A_h(z,z)= \sum\{[\frac{z((r+1)h,sh)-z(rh,sh)}{h}]^2 + [\frac{z(rh,(s+1)h)-z(rh,sh)}{h}]^2\}h^2$$

$$+ \sum Q(rh, sh)z(rh, sh)^2h^2,$$

where the summations for r,s abover are over all integers. The eigenvalues $\tilde{\mu}_1^h$ and μ_1 will be compared by means of their corresponding Rayleigh quotients (cf. [227]).

For any $r, s = 0, \pm 1, \pm 2, \ldots$, connect the mesh point (rh, sh) to the mesh points $((r\pm 1)h, sh)$, $(rh, (s\pm 1)h)$, $((r+1)h, (s+1)h)$ and $((r-1)h, (s-1)h)$. For those (r, s) such that all such triangles having (rh, sh) as a vertex lie in $\Omega \cup \partial\Omega$, define $v_{rs}(x, y)$ to be the piecewise linear function which is one at (rh, sh) and 0 at the other mesh points. For other (r, s) such that (rh, sh) is the vertex of a triangle that goes outside $\Omega \cup \partial\Omega$, define $v_{rs}(x, y) = 0$. For any grid function z which is zero on $\partial\Omega_h$ and outside Ω_h, define

$$M_hz = \sum z(rh, sh)v_{rs}(x, y),$$

where the sum is over $r, s = 0, \pm 1, \pm 2, \ldots$. Thus M_hz is a continuous piecewise linear function whose value at (rh, sh) is $z(rh, sh)$, if (rh, sh) is a vertex such that all neighboring triangles as described above lie in $\Omega \cup \partial\Omega$. M_hz is zero at the boundary $\partial\Omega$, and its derivatives have discontinuities only at the lines $x = ih$, $y = jh$ and $x-y = kh$. For any grid function z of the form described, we have:

$$A(M_hz, M_hz) = A_h(z, z) + \int_\Omega Q(M_hz)^2dxdy - \sum Q(rh, sh)z(rh, sh)^2h^2, \tag{6.4-24}$$

using an identity of G. Polya [187]:

$$\sum\{[z((r+1)h, sh)-z(rh, sh)]^2 + [z(rh, (s+1)h)-z(rh, sh)]^2 = \int_\Omega (\nabla M_hz)^2dxdy.$$

From the identity (cf. [227], p. 135):

$$\int_{\Omega} (M_h z)^2 dxdy = \frac{1}{12} h^2 \sum \{12\ z(rh,sh)^2 - [z((r+1)h,sh) - z(rh,sh)]^2$$

(6.4-25)

$$-[z(rh,(s+1)h) - z(rh,sh)]^2 - [z(r+1)h,(s+1)h) - z(rh,sh)]^2\};$$

and by the mean value theorem, we have:

(6.4-26)
$$\int_{\Omega} Q(x,y)(M_h z)^2 dxdy - \sum Q(rh,sh) z(rh,sh)^2 h^2$$

$$\leq -\frac{1}{12} h^2\ Q(x^*,y^*) \sum \{[z((r+1)h,sh) - z(rh,sh)]^2 + [z(rh,(s+1)h) - z(rh,sh)]^2$$

$$+ [z((r+1)h,(s+1)h) - z(rh,sh)]^2 + E,$$

where $|E| \leq Bh \sum z(rh,sh)^2 h^2$ and B is $\sqrt{2}$ times a bound for $|\nabla Q|$ independnet of z. The point $(x^*,y^*) \in \Omega$ and $Q(x^*,y^*) > 0$. From equations (6.4-24) and (6.4-26), we obtain

(6.4-27)
$$A_h(z,z) \geq A(M_h z, M_h z) - BhI_h(z,z).$$

In order to compare the Rayleigh quotients for problems (6.4-18) and (6.4-19), we need to estimate $I_h(z,z)/A(M_h z, M_h z)$. We have

$$A(M_h z, M_h z) = \int_{\Omega} [(\nabla M_h z)^2 + Q(M_h z)^2] dxdy$$

$$\geq \int_{\Omega} (\nabla M_h z)^2 dxdy + \min_{\Omega} Q \int_{\Omega} (M_h z)^2 dxdy$$

(6.4-28)
$$\geq \int_{\Omega} (\nabla M_h z)^2 dxdy + \min_{\Omega} Q\{\sum z(rh,sh)^2 h^2$$

$$- \frac{1}{4} h^2 \sum([z((r+1)h, sh)-z(rh, sh)]^2 + [z(rh, (s+1)h)-z(rh, sh)]^2)\}$$

$$= \int_\Omega (\nabla M_h z)^2 dxdy + \min_\Omega Q\{\sum z^2 h^2 - \frac{1}{4} h^2 \int_\Omega (\nabla M_h z)^2 dxdy$$

$$\geq \min_\Omega Q \sum z(rh, sh)^2 h^2$$

for h sufficiently small. Here we have used (6.4-25), the triangle inequality and the identity of G. Polya after (6.4-25). From (6.4-28), we have the estimate:

$$\left| \frac{I_h(z,z)}{A(M_h z, M_h z)} \right| \leq \frac{I_h(z,z)}{(\min_\Omega Q) I_h(z,z)} = \frac{1}{\min_\Omega Q} . \tag{6.4-29}$$

For problem (6.4-19), let u_h^* be the eigenvector for which $\frac{1}{\tilde{\mu}_1^h + K} = \frac{I_h(u_h^*, u_h^*)}{A_h(u_h^*, u_h^*)}$.

Using equation (6.4-25) and the triangle inequality,

$$\begin{aligned} I(M_h u_h^*, M_h u_h^*) &\geq I_h(u_h^*, u_h^*) - \frac{1}{4} h^2 \sum\{[u_h^*((r+1)h, sh)-u_h^*(rh, sh)]^2 \\ &\qquad +[u_h^*(rh, (s+1)h)-u_h^*(rh, sh)]^2\} \\ &\geq I_h(u_h^*, u_h^*) - \frac{1}{4} h^2 A_h(u_h^*, u_h^*) \\ &\geq \left[\frac{I_h(u_h^*, u_h^*)}{A_h(u_h^*, u_h^*)} - \frac{1}{4} h^2\right] [A(M_h u_h^*, M_h u_h^*)-BhI_h(u_h^*, u_h^*)], \end{aligned} \tag{6.4-30}$$

where inequality (6.4-27) is used for the last inequality. Consequently, we obtain from (6.4-30) and (6.4-29):

$$\frac{I(M_h u_h^*, M_h u_h^*)}{A(M_h u_h^*, M_h u_h^*)} \ge \left[\frac{I_h(u_h^*, u_h^*)}{A_h(u_h^*, u_h^*)} - \frac{1}{4} h^2\right] [1-Bh(\min_\Omega Q)^{-1}].$$

Using the Rayleigh quotient characterization for problem (6.4-18), we deduce that

$$(6.4\text{-}31) \quad \frac{1}{\mu_1+K} \ge \frac{I(M_h u_h^*, M_h u_h^*)}{A(M_h u_h^*, M_h u_h^*)} \ge \left[\frac{1}{\tilde{\mu}_1^h+K} - \frac{1}{4} h^2\right] [1-Bh(\min_\Omega Q)^{-1}] = \frac{1}{\tilde{\mu}_1^h+K} [1+O(h)].$$

From inequality (6.4-31), we obtain condition (6.4-21) and then we use (6.4-20) to assert

$$(6.4\text{-}32) \quad \mu_1^h \ge \tilde{\mu}_1^h \ge \frac{\mu_1}{2} > 0$$

for h sufficiently small.

The problem (6.4-16) or (6.4-17) is therefore invertible. Since L_h is symmetric, we obtain from expressions (6.4-32) and (6.4-17) that

$$(6.4\text{-}33) \quad \|e_h(P)-y_h(P)\|_2 = \|z_h(P)\|_2 \le \|L_h^{-1}\|_2 \, O(h) \le \frac{2}{\mu_1} O(h)$$

as $h \to 0^+$. Finally, since $y_h(P) = O(h)$ in Ω_h, we conclude that

$$(6.4\text{-}34) \quad \|w_h(P)-v_h(P)\|_2 = \|e_h(P)\|_2 \le \|e_h(P)-y_h(P)\|_2 + \|y_h(P)\|_2 = O(h) \to 0$$

as $h \to 0^+$. This proves the theorem.

Remark 6.4-3. Another version of identity (6.4-25) can be stated as follows: Let Δ be the triangle joining the vertices (rh, sh), $((r+1)h, sh)$ and $(rh, (s-1)h)$ on the plane, and let $M_h z$ be defined on Δ as a linear function whose values at the vertices above are respectively $z(rh, sh)$, $z((r+1)h, sh)$ and

$z(rh,(s-1)h)$. Then

$$\int_{\Delta}(M_h z)^2 dxdy = \frac{1}{6} h^2[z(rh,sh)^2+z((r+1)h,sh)^2+z(rh,(s-1)h)^2]$$

$$(6.4\text{-}35) \qquad - \frac{1}{24} h^2\{[z(rh,sh)-z((r+1)h,sh)]^2 + [z((r+1)h,sh)-z(rh,(s-1)h)]^2$$

$$+ [(z(rh,(s-1)h)-z(rh,sh)]^2\}.$$

For the domain Ω, each mesh point in Ω_h is used 6 times and each side of a triangle is used twice. Summing up, we obtain (6.4-25) from (6.4-35).

For the close of this section, we show some computational results by applying the monotone scheme described in this section for problem (6.4-1). We let Ω be a "L-shaped" domain such that by choosing $h = 0.05$, Ω_h consists of all mesh points in the union of the product of the intervals $(0,0.5) \times (0,0.5)$ and $[0.5,1) \times (0,0.2)$. $\delta\Omega_h$ are those mesh points on the boundary of this L-shaped polygon. We discretize the problem with this choice of $h = 0.05$. The limiting values for all sequences are obtained when

$$(6.4\text{-}36) \qquad \frac{\max_{P\in\Omega_h} |v_h^{(j+1)}(P)-v_h^{(j)}(P)|}{\max_{P\in\Omega_h} |v_h^{(j)}(P)|} \le 10^{-8}.$$

We first consider problem (6.4-1), (6.4-3), with $q = 1$ and $p(x,y) = 100$. In this case, the conditions in Theorem 6.4-2 are all satisfied since λ_1^h, the first eigenvalue for the problem $\Delta_h z_h + \lambda^h z_h = 0$ in Ω_h, $z_h = 0$ on $\delta\Omega_h$, is smaller than the first eigenvalue for the same problem restricted to the square $(0,0.5) \times (0,0.5)$ whose value is ≈ 80. Figures 6.4-1 to 6.4-4 correspond to the monotone decreasing sequence $v_h^{(j)}(P)$ with $j = 1, 10, 20, 96$ respectively. We start the iterations with $v_h^{(0)}(P) \equiv 100$ and the "limiting" value is attained for $j \approx 96$.

Figures 6.4-5 to 6.4-8 show some elemnts of the monotone decreasing sequence $v_h^j(P)$, $j = 1, 10, 30, 764$, respectively for problem (6.4-1), (6.4-3) with $q = 1$ and $p(x,y) = 10,000 \exp\{-40[(x-0.25)^2 + (y-0.25)^2]\} + 1$. We notice that $p(x,y) > \frac{2\pi^2}{\ell^2}$ in S_ℓ, where S_ℓ is the square centered at (0.25, 0.25) and $\ell = 0.05$, but $p(x,y)$ is not larger than the first eigenvalue for the problem $\Delta_h z_h + \lambda^h z_h = 0$ in Ω_h, $z_h = 0$ on $\delta\Omega_h$ for every (x,y) in Ω_h (cf. Theorem 6.4-1 and Lemma 6.4-5). However, computations with our scheme indicate the existence of a unique positive discrete solution even in this case. In this example, the stopping criterion (6.4-36) is reached when $j \approx 764$, and the iterations start with $v_h^{(0)}(P) \equiv 10,001$.

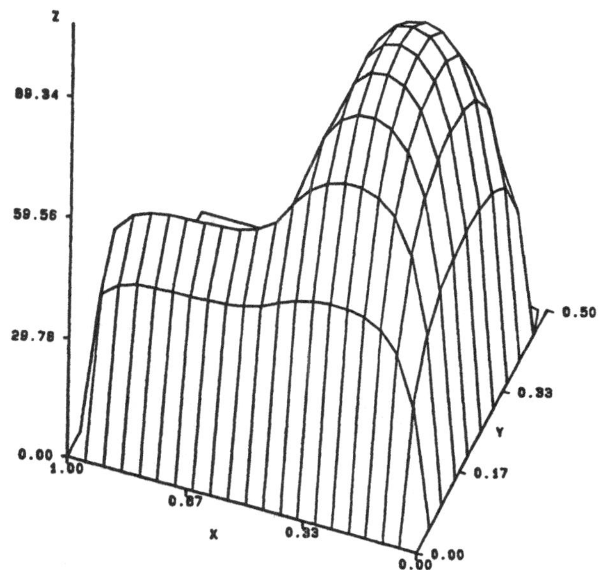

Fig 6.4-1 Monotone decreasing sequence $v_h^{(j)}(P)$, $j = 1$ for the problem $\Delta v + v[100 - v] = 0$ in Ω, $v = 0$ on $\partial\Omega$

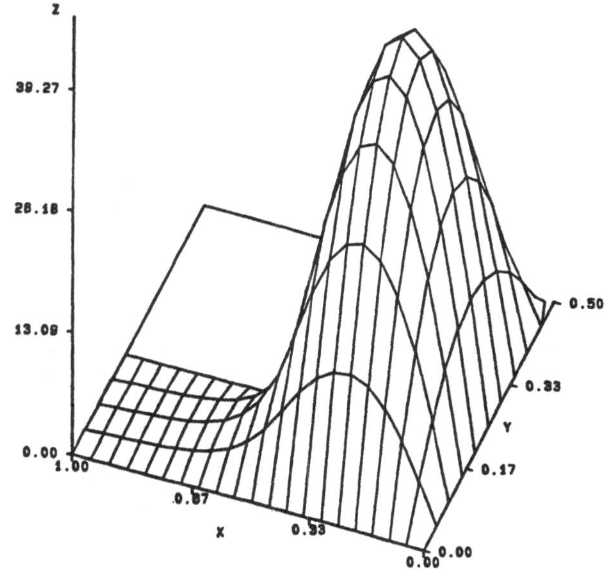

Fig 6.4-2 Monotone decreasing sequence $v_h^{(j)}(P)$, $j = 10$ for the problem $\Delta v + v[100 - v] = 0$ in Ω, $v = 0$ on $\partial\Omega$.

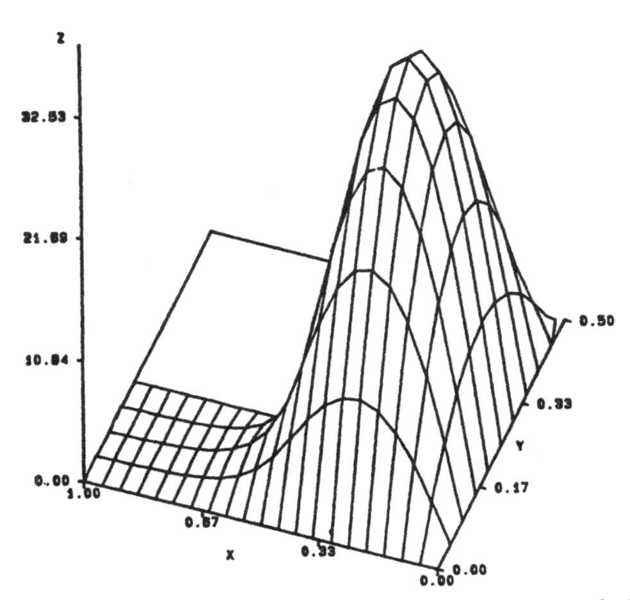

Fig 6.4-3 Monotone decreasing sequence $v_k^{(j)}(P)$, $j = 20$ for the problem $\Delta v + v[100 - v] = 0$ in Ω, $v = 0$ on $\partial\Omega$.

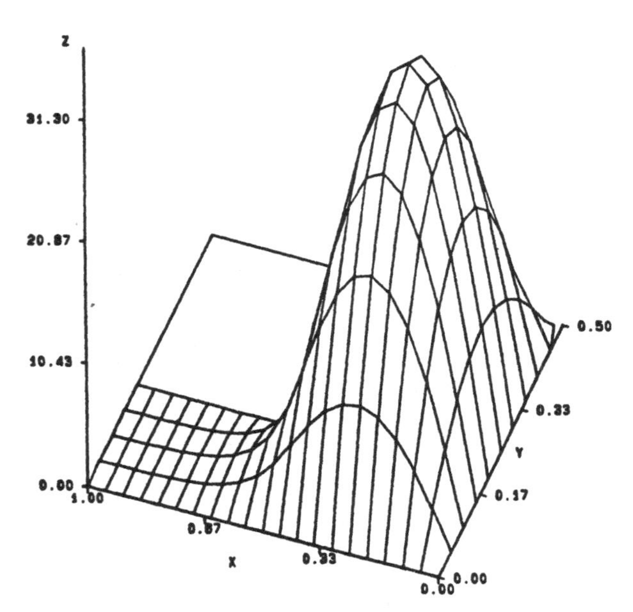

Fig 6.4-4 Monotone decreasing sequence $v_k^{(j)}(P)$, $j = 96$ for the problem $\Delta v + v[100 - v] = 0$ in Ω, $v = 0$ on $\partial\Omega$.

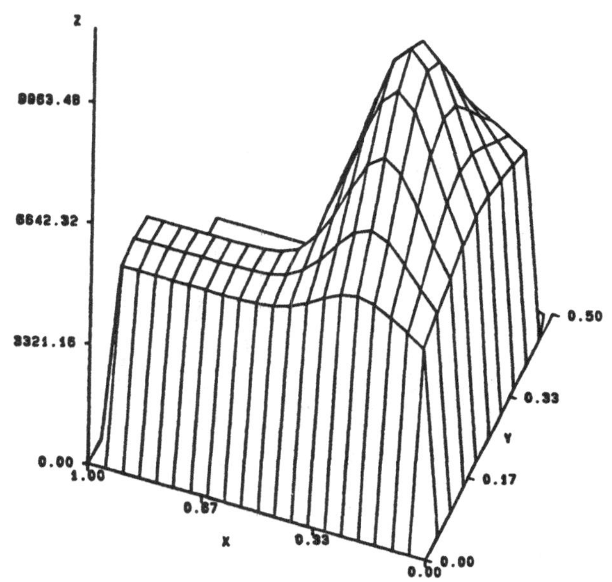

Fig 6.4-5 Monotone decreasing sequence $v_2^{(j)}(P)$, $j = 1$ for the problem $\Delta v + v[10{,}000 \times \exp\{-40[(x-0.25)^2 + (y-0.25)^2]\} + 1 - v] = 0$ in Ω, $v = 0$ on $\partial\Omega$.

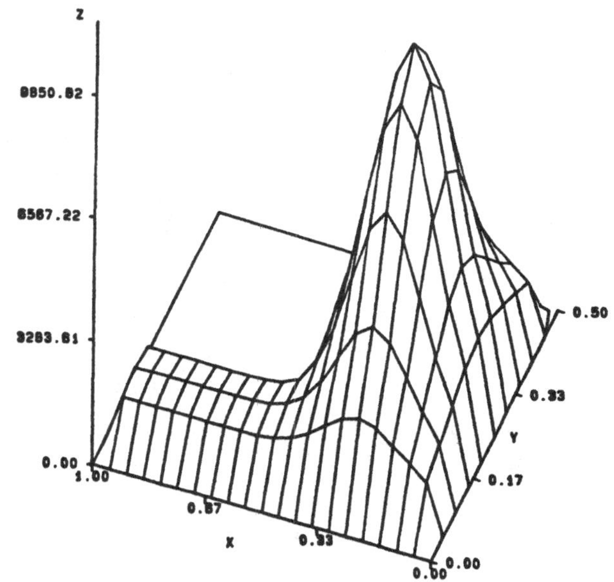

Fig 6.4-6 Monotone decreasing sequence $v_2^{(j)}(P)$, $j = 10$ for the problem $\Delta v + v[10{,}000 \times \exp\{-40[(x-0.25)^2 + (y-0.25)^2]\} + 1 - v] = 0$ in Ω, $v = 0$ on $\partial\Omega$.

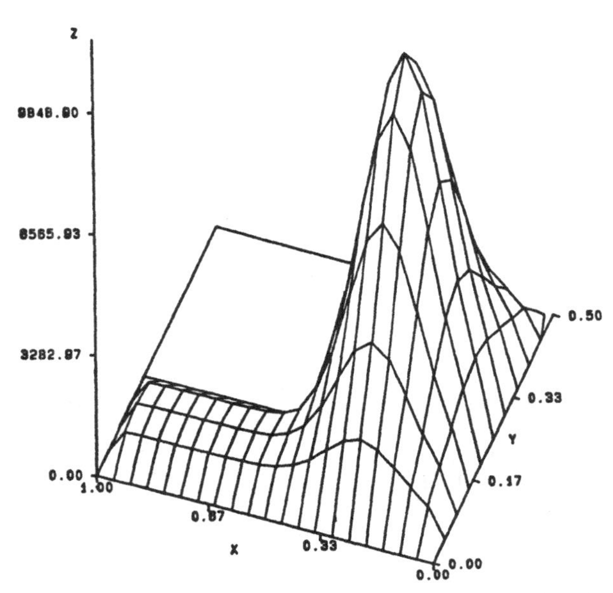

Fig 6.4-7 Monotone decreasing sequence $v_h^{(j)}(P)$, $j=30$ for the problem $\Delta v + v[10{,}000 \times \exp\{-40[(x-0.25)^2+(y-0.25)^2]\} + 1 - v] = 0$ in Ω, $v=0$ on $\partial\Omega$.

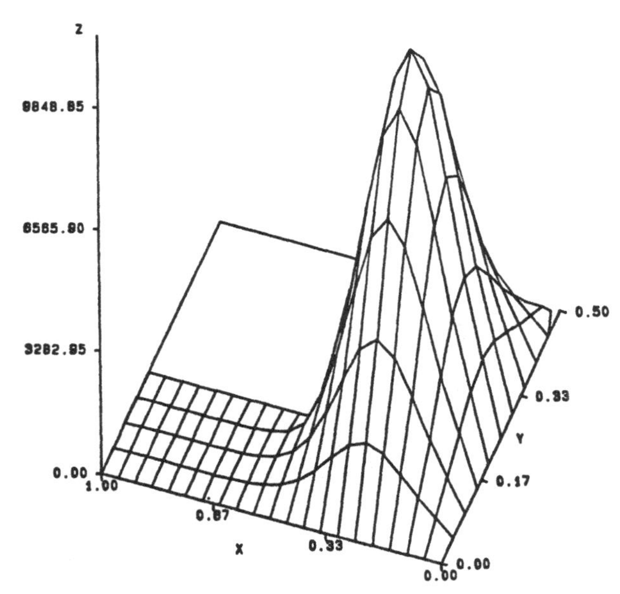

Fig 6.4-8 Monotone decreasing sequence $v_h^{(j)}(P)$, $j=764$ for the problem $\Delta v + v[10{,}000 \times \exp\{-40[(x-0.25)^2+(y-0.25)^2]\} + 1 - v] = 0$ in Ω, $v=0$ on $\partial\Omega$.

Notes

Finite difference solutions for various nonlinear elliptic and parabolic equations had been investigated by Douglas [67], Parter [183], Bramble and Hubbard [27], [28], Adams and Ames [2] and others. Discretization errors had also been considered in these articles. The special form of nonlinearity considered in this chatper, however, does not exactly satisfy the conditions of these earlier mentioned papérs. Interesting numerical considerations of other nonlinear parabolic systems can also be found in Turner and Ames [221] and Hoff [106]. These articles are not primarily concerned with the study of steady states, as in this chapter. The materials presented in Sections 6.1 and 6.2 are adapted from Lazer, Leung and Murio [131], Leung and Murio [146]. The monotonic or oscillating scheme of iterations for systems originated from materials presented in Chapter 5. Other related work can be found in Huy, McKenna and Walter [112] and others. Newton's acceleration method as in Section 6.3 for other equations was done in Bellman [22] and Kalaba [115]. For the Volterra-Lotka system considered in Section 6.3, we follow the presentation in Leung and Murio [147]. In Section 6.4, the discretization error is deduced through inverting an operator which is dependent on the solution of the nonlinear equation. The invertibility of this operator is obtained by means of comparing eigenvalues for related equations. We follow the presentation in Leung and Murio [148]. More discussions of the method can be found in Weinberger [226], [227]. As to references for an introduction and other widely used theories in finite difference solutions for elliptic equations, see Forsythe and Wasow [77], Ames [6], and Birhoff and Lynch [24].

CHAPTER VII

Large Systems under Neumann Boundary Conditions, Bifurcations

7.1 Introduction

In Sections 7.2 and 7.3, we will give a careful treatment of large parabolic systems of Volterra-Lotka prey-predator type under zero Neumann boundary condition. We follow the methods in [191], [192], [193] and [194], combining graph-theoretic technique with the use of Lyapunov functions. The earlier use of Lyapunov functions to study such reaction-diffusion systems began with [197], [229], [134], [190], and others. The recent results presented here give very general and elegant insight into the problem. Such systems had also been investigated by many others by invariant rectangles and comparison methods as indicated in the notes at the end of this chapter. Results using these other techniques had been summarized in other books, e.g. [211], [31], and are thus not included here. More precisely we will consider Volterra-Lotka system of the form:

$$\frac{\partial u_i}{\partial t} = \Delta u_i + u_i[e_i + \sum_{i=1}^{n} p_{ij}u_j] , \quad i=1,\ldots,n$$

where e_i and p_{ij} are constants. The reader should also refer to Section 4.6, which indicates the possibility of spatially dependent steady states for similar systems. Such distinctly different results of course can only occur to systems with interacting coefficients which do not satisfy the conditions in the Main Theorem 7.3-6 below.

In Section 7.4, we go back to the Dirichlet boundary value problem. We use functional-analytic techniques of [206] and [201] to study the bifurcations of steady states of Volterra-Lotka type elliptic systems.

We follow the results obtained in [32] and [25]. Although some resu overlap with those in earlier chapters, the method illustrates a us ful alternative which is sometimes more readily applicable for even other similar systems. It further gives insight to the bifurcation of solutions as some parameters vary globally.

Our space domain $\Omega \subset R^N$ is open, connected and bounded, with its boundary $\delta\Omega \in H^{2+\alpha}$ as described in Section 1.3.

7.2 Lyapunov Functions for Volterra-Lotka Systems

Before our detailed study of the Volterra-Lotka system (problem (7.2-5) to (7.2-7)), we first obtain some basic properties concerning the following more general parabolic systems under zero Neumann boundary condition:

$$(7.2\text{-}1)\quad \begin{cases} \dfrac{\partial u_i}{\partial t} = \Delta u_i + g_i(u_1,\ldots,u_n), & i=1,\ldots,n \text{ in } \Omega\times(0,T], \\ \dfrac{\partial u_i}{\partial \eta} = 0, & i=1,\ldots,n \text{ on } \delta\Omega\times(0,T], \\ u_i(x,0) = u_i^0(x), & i=1,\ldots,n \text{ on } \overline{\Omega}. \end{cases}$$

For simplicity, we are using only the Laplacian operator Δ in (7.2-1). It can be readily replaced by a more general elliptic operator of the form $L = \sum_{r,s=1}^{N} \frac{\partial}{\partial x_r}(a_{rs}(x)\frac{\partial}{\partial x_s})$, with each $a_{rs}(x)$ in $H^{2+\alpha}(\overline{\Omega})$, $0<\alpha<1$ and $\sum_{r,s=1}^{N} a_{rs}(x)\xi_r\xi_s \geq \beta|\xi|^2$ for all $\xi \in R^N$ some $\beta>0$, as given in [191]. Because of the use of the Lyapunov function later in the section, the operator L is restricted to be the same for each component for simplicity. The outward unit normal at the boundary $\delta\Omega$ is denoted by η. For an interval I, the symbol $C^{2,1}(\overline{\Omega}\times I)$ denotes the set of functions defined on $(x,t) \in \overline{\Omega}\times I$ with all partial derivatives with respect to x_i up to second order and first derivative with respect to t continuous in $\overline{\Omega}$xI. The set $\{(y_1,\ldots,y_n) : y_i>0, i=1,\ldots,n\}$ will be denoted by R^n_+, and its boundary denoted by δR^n_+. For any integer k>0, $C^k(\overline{\Omega})$ denotes the set of functions on $\overline{\Omega}$ with all partial derivatives up to order k continuous in $\overline{\Omega}$.

Theorem 7.2-1. *Suppose that $g_i:R^n\to R$ are locally Lipschitz continuous for each i=1,...,n, and each component $u_i^0(x)$ is of class $H^{2+\alpha}(\overline{\Omega})$ with zero normal derivative on $\delta\Omega$. Then the initial boundary value problem (7.2-1) has a solution in $C^{2,1}(\overline{\Omega}x[0,\delta])$ for some $\delta>0$. Furthermore, all spatial partial derivatives of u up to second order are Hölder continuous in x, and $u(\cdot,t)$ maps $[0,\delta]$ continuously into $H^{2+\alpha}(\overline{\Omega})$. If there is an a priori estimate $|u(x,t)| \le K$ (K independent of T), then u(x,t) exists for $0 \le t < \infty$, and is bounded in $H^{2+\alpha}(\overline{\Omega})$ (i.e., $|u(\cdot,t)|_{\Omega}^{(2+\alpha)} \le K_1$ for all $t \in [0,\infty)$).*

For a reference of the proofs, see e.g. [191] and [189], One can also compare with an analogous theorem 2.1-1 in Chapter 2. The following two lemmas will be needed for Theorem 7.2-2 which shows that solutions of the Volterra-Lotka system (7.2-5) to (7.2-7) will tend to spatial independence as $t\to+\infty$, under appropriate condition on p_{ij}.

Lemma 7.2-1. *Suppose that the scalar function $\varphi \in C^{2,1}(\overline{\Omega}x[0,\infty))$ satisfies*

$$\frac{\partial\varphi}{\partial t}(x,t) = \Delta\varphi(x,t) \quad \text{for } (x,t) \text{ in } \Omega x(0,\infty)$$

$$\frac{\partial\varphi}{\partial\eta}(x,t) = 0 \qquad \text{for } (x,t) \text{ on } \delta\Omega x(0,\infty)$$

and $\varphi(x,0) \in H^{2+\alpha}(\overline{\Omega})$. *Then*

$$(7.2\text{-}2) \qquad \lim_{t\to\infty} |\varphi(\cdot,t)-c|_{\Omega}^{(2)} = 0, \ \underline{\text{where}} \ c = \frac{1}{|\Omega|}\int_{\Omega}\varphi(x,0)dx,$$

and $|\Omega|$ = measure of Ω.

Proof. Let $h(t) = \int_{\Omega}\varphi(x,t)dx$, one has $\dot{h}(t) = \int_{\Omega}\Delta\varphi(x,t)dx = \int_{\delta\Omega}\frac{\partial\varphi}{\partial\eta}ds = 0$ and thus $h(t) \equiv \int_{\Omega}\varphi(x,0)dx = c|\Omega|$. Using the comparison Theorem 1.2-4 (with $f=0$, $\alpha=1$, $\beta=0$ and v, w as φ and constant functions) we deduce that $\varphi(x,t)$ is bounded. Theorem 7.2-1 implies that $|\varphi(\cdot,t)|_{\Omega}^{(2+\alpha)} \le K_1$ for $t\in[0,\infty)$. Thus, for any sequence tending to ∞, there exists a subsequence $\{t_m\}$ such that

$$(7.2\text{-}3) \qquad \lim_{m\to\infty}\varphi(\cdot,t_m) = z\in C^2(\overline{\Omega})$$

where the convergence is in C^2. On the other hand, let $p(t) = \int_{\Omega}\varphi^2(x,t)dx$, one has

$$(7.2\text{-}4) \qquad \dot{p}(t) = \int_{\Omega}2\varphi\Delta\varphi dx = -\int_{\Omega}2(\nabla\varphi)^2dx \le 0$$

Consequently $\lim_{t\to\infty}p(t) = \overline{\alpha} \ge 0$, and there exists a sequence $\{t_k\}$ tending to ∞ such that $\dot{p}(t_k)\to 0$ (otherwise, one has $p(t_k)\to-\infty$ which is a contradiction). However, from the first part of (7.2-4) and (7.2-3), we can choose a subsequence $\{t_m\}$ of $\{t_k\}$ so that

$$\lim_{m\to\infty}\dot{p}(t_m) = -2\int_{\Omega}(\nabla z)^2dx.$$

Therefore, one must have $\int_{\Omega}(\nabla z)^2dx = 0$, and $z(x)$ = const. on $\overline{\Omega}$. To find this constant, one evaluates

$$c|\Omega| = h(t_m) \to \int_{\Omega}z(x)dx = \text{const.}\ |\Omega|$$

and consequently $z(x) \equiv c$ in $\overline{\Omega}$.

Now, apply Theorem 1.2-4 again (with $f=0$, $\alpha=1$, $\beta=0$, $v=\varphi$, $w=$ constant $= \max\{\varphi(x,t_0) \mid x \in \bar{\Omega}\}$ to the region $\Omega\times(t_0,\infty)$, we can show that $\varphi(x,t) \leq \max\{\varphi(x,t_0) \mid x \in \bar{\Omega}\}$ for $t \geq t_0$, and hence $M(t) \overset{(def)}{=} \max\{\varphi(x,t) \mid x \in \bar{\Omega}\}$ is decreasing in t. Using (7.2-3) and $z \equiv c$, we thus have $M(t) \to c$ as $t \to \infty$. Similarly, we can deduce $m(t) \overset{(def)}{=} \min\{\varphi(x,t) \mid x \in \bar{\Omega}\} \to c$ and is increasing, as $t \to \infty$. This gives $\varphi(x,t) \to c$ uniformly in $x \in \bar{\Omega}$ as $t \to \infty$. Finally, from Theorem 7.2-1, we have $|\varphi(\cdot,t)|_{\Omega}^{(2+\alpha)} \leq K_1$ for all $t \in [0,\infty)$, and thus for any sequence (t_k) tending to infinity there exists a subsequence (τ_k) such that $\varphi(\cdot,\tau_k) \to c$ in $C^2(\bar{\Omega})$. Consequently, we have $\varphi(\cdot,t) \to c$ in $C^2(\bar{\Omega})$ as $t \to \infty$.

Lemma 7.2-2. *Let $\tilde{u}$, $\hat{u}$ be solutions of the initial boundary value problem (7.2-1) in the class $C^{2,1}(\bar{\Omega}\times[0,T])$ (with the initial functions for $\tilde{u}$, $\hat{u}$ changed to $\tilde{u}^0(x)$ and $\hat{u}^0(x)$ respectively). Suppose that*

$$|\tilde{u}^0(x) - \hat{u}^0(x)| \leq \delta \text{ for all } x \in \bar{\Omega},$$

and $g: R^n \to R^n$ satisfies the Lipschitz condition

$$|g(z) - g(z')| \leq M|z-z'|$$

for all z, z' in R^n. (Here, $g = (g_1,\ldots,g_n)$). Then

$$|\tilde{u}(x,t)-\hat{u}(x,t)| \leq \delta e^{Mt} \text{ for all } (x,t) \in \bar{\Omega}\times[0,T].$$

Proof. Let $v = |\tilde{u}-\hat{u}|^2$. Direct calculation gives

$$\begin{aligned}\frac{\partial v}{\partial t} &= \Delta v - 2\sum_{i=1}^{n} |\nabla(\tilde{u}_i-\hat{u}_i)|^2 + 2\sum_{i=1}^{n} (\tilde{u}_i-\hat{u}_i)(g_i(\tilde{u}) - g_i(\hat{u})) \\ &\leq \Delta v + 2Mv \quad \text{in } \Omega\times(0,T].\end{aligned}$$

Let $w = \delta^2 e^{2Mt}$. One readily sees that $\Delta w + 2Mw - \frac{\partial w}{\partial t} = 0$ $\leq \Delta v + 2Mv - \frac{\partial v}{\partial t}$ in $\Omega x(0,T]$, and $\frac{\partial v}{\partial \eta} = \frac{\partial w}{\partial \eta} = 0$ on $\delta\Omega x(0,T]$. Apply Theorem 1.2-4, we conclude that $v(x,t) \leq w(x,t)$ in $\bar{\Omega}x[0,T]$. This proves the lemma.

In order to investigate the behavior for large t of the solutions of the diffusive Volterra-Lotka system under zero Neumann boundary condition:

$$(7.2\text{-}5) \qquad \frac{\partial u_i}{\partial t} = \Delta u_i + u_i[e_i + \sum_{j=1}^{n} p_{ij}u_j], \quad i=1,\dots,n \quad \text{in } \Omega x(0,\infty)$$

$$(7.2\text{-}6) \qquad \frac{\partial u_i}{\partial \eta} = 0 \qquad , \quad i=1,\dots,n \quad \text{on } \delta\Omega x(0,\infty)$$

$$(7.2\text{-}7) \qquad u_i(x,0) = u_i^0(x) > 0 \text{ (with } \frac{\partial u_i^0}{\partial \eta} = 0 \text{ on } \delta\Omega) \text{ for } i=1,\dots,n \text{ on } \bar{\Omega}.$$

(e_i, p_{ij} are constants), we will construct an appropriate Lyapunov function for the system. This can be done successfully if we assume that:

$$(7.2\text{-}8) \quad \begin{cases} \text{There exists } q=(q_1,\dots,q_n), \text{ with each } q_i>0 \text{ so that} \\ e_i + \sum_{j=1}^{n} p_{ij}q_j = 0, \ i=1,\dots,n; \text{ and} \end{cases}$$

$$(7.2\text{-}9) \quad \begin{cases} \text{There exist } a_i>0, \ i=1,\dots,n \text{ such that } (a_i p_{ij}) \leq 0, \\ \text{i.e. } \sum_{i,j=1}^{n} a_i p_{ij} w_i w_j \leq 0 \text{ for all } (w_1,\dots,w_n) \in R^n \end{cases}$$

(Note that in (7.2-8) q may not be unique.) An nxn matrix (p_{ij}) with only property (7.2-9) will be called "admissible." We will assume (7.2-8) (7.2-9) in the rest of this section unless otherwise stated. Define $V(u_1,\dots,u_n)$ for $u_i>0$, $i=1,\dots,n$ by

(7.2-10) $$V(u) = \sum_{i=1}^{n} \left[a_i(u_i-q_i) - a_i q_i \ln\frac{u_i}{q_i} \right]$$

where a_i, q_i are described in (7.2-8) and (7.2-9). Observe that V has the properties:

(7.2-11)
$$V(q)=0,\ V(u)>0 \text{ for } u \in R_+^n,\ u \neq q,$$
$$V(u) \to \infty \text{ as } |u| \to \infty \text{ or } u \to \delta R_+^n .$$

If we denote $f(u) \overset{(def)}{=} (u_1[e_1+\sum_{j=1}^{n} p_{1j}u_j], \ldots, u_n[e_n+\sum_{j=1}^{n} p_{nj}u_j])$, one has

(7.2-12) $$[\text{grad } V(u)] \cdot f(u) = \sum_{i=1}^{n} a_i(1-\frac{q_i}{u_i})u_i(e_i+\sum_{j=1}^{n} p_{ij}u_j)$$
$$= \sum_{i=1}^{n} a_i(u_i-q_i) \sum_{j=1}^{n} p_{ij}(u_j-q_j)$$

By (7.2-9) we have grad $V \cdot f \leq 0$, and thus $V(u)$ will serve the purpose of a Lyapunov function for (7.2-5) (i.e. if there are no Δu_i terms $V(u(t))$ will decrease along solutions).

By Theorem 7.2-1, and the fact that $u_i^0(x)>0$ in $\bar{\Omega}$, we can assume the existence of $u_i(x,t)>0$ for small $t>0$, $x \in \bar{\Omega}$. (Here we assume that $u_i^0(x)$ in $H^{2+\alpha}(\bar{\Omega})$.) Define

(7.2-13) $$y_i(x,t) = \ln[u_i(x,t)/q_i]$$

for convenience. One clearly has $\frac{\partial u_i}{\partial t} = u_i \frac{\partial y_i}{\partial t}$ and $\frac{\partial^2 u_i}{\partial x_k^2} = u_i \frac{\partial^2 y_i}{\partial x_k^2} + u_i \left[\frac{\partial y_i}{\partial x_k}\right]^2$. Cancelling u_i, equation (7.2-5) can be rewritten as

(7.2-14) $$\frac{\partial y_i}{\partial t} = \Delta y_i + |\text{grad } y_i|^2 + \sum_{j=1}^{n} p_{ij}(u_j-q_j),\ i=1,\ldots,n,$$

Define $U(x,t) = V(u(x,t))$, one has

(7.2-15) $$\frac{\partial U}{\partial t} = [\text{grad } V(u(x,t))]\cdot\frac{\partial u}{\partial t} = \sum_{i=1}^{n} a_i(u_i-q_i)\frac{\partial y_i}{\partial t}$$

(7.2-16) $$\Delta U = \sum_{i=1}^{n}\left\{a_i(u_i-q_i)\left[\Delta y_i + |\text{grad } y_i|^2\right] + a_i u_i |\text{grad } y_i|^2\right\}$$

Combining (7.2-14) with (7.2-16), we obtain

(7.2-17) $$\frac{\partial U}{\partial t} = \Delta U - \sum_{i=1}^{n} a_i u_i |\text{grad } y_i|^2 + \sum_{i,j=1}^{n} a_i p_{ij}(u_i-q_i)(u_j-q_j)$$

From (7.2-9) we therefore obtain $\Delta U - \frac{\partial U}{\partial t} \geq 0$ in Ω as long as solution exists. Let $M \geq \{\max U(x,0) | x \in \bar{\Omega}\}$, and define $w(x,t) \equiv M$, we have

$$\Delta U - \frac{\partial U}{\partial t} \geq \Delta w - \frac{\partial w}{\partial t}$$

and $\frac{\partial U}{\partial \eta} = \frac{\partial w}{\partial \eta} = 0$ on $\delta\Omega$. From Theorem 1.2-4, we conclude that $U(x,t) \leq M$ for $x \in \bar{\Omega}$, $t \in [0,T]$ as long as solution exists to time T. From the properties of the function $V(u)$ as $|u|\to\infty$ or $u\to\delta R_+^n$, it is clear that

(7.2-18) $$0 < \alpha \leq u_i(x,t) \leq \beta, \quad (x,t) \in \bar{\Omega}x[0,T], \quad i=1,\ldots,n,$$

for some constants α,β which depends only on M (that is only on the function $u^0(x)$). From Theorem 7.2-1, the solution $u(x,t)$ for (7.2-5) to (7.2-7) exists for $0\leq t<\infty$.

The next lemma shows that $U = V(u(x,t))$ tends to a constant as $t\to\infty$, for all $x \in \bar{\Omega}$.

Lemma 7.2-3. *Let* $\tilde{h}(t) \overset{(def)}{=} \frac{1}{|\Omega|}\int_\Omega U(x,t)dx$. *It must be a decreasing function of* t, *and* $\lim_{t\to\infty} \tilde{h}(t) = a \geq 0$. *Moreover,*

$\lim_{t\to\infty} V(u(\cdot,t)) \equiv \lim_{t\to\infty} U(\cdot,t) = a$ in $C^2(\bar{\Omega})$.

Proof. Differentiating with respect to t, $\dot{\tilde{h}}(t) = \frac{1}{|\Omega|}\int_\Omega \frac{\partial U}{\partial t}(x,t)dx$ $\leq \frac{1}{|\Omega|}\int_\Omega \Delta U dx = \frac{1}{|\Omega|}\int_{\delta\Omega} \frac{\partial U}{\partial \eta} ds = 0$. Moreover $U(x,t)$ is nonnegative, the function $\tilde{h}(t)$ must be decreasing and $\lim_{t\to\infty} \tilde{h}(t) = a \geq 0$. Let $\varepsilon>0$ be arbitrary and t_0 be such that $\tilde{h}(t_0) < a + \varepsilon$. Define $w(x,t)$ as the solution of the problem:

$$\Delta w - \frac{\partial w}{\partial t} = 0 \quad \text{in } \Omega \times (t_0,\infty),$$

$$\frac{\partial w}{\partial \eta} = 0 \text{ on } \delta\Omega \times (t_0,\infty), \quad w(x,t_0) = U(x,t_0) \text{ in } \bar{\Omega}.$$

Since $\Delta U - \frac{\partial U}{\partial t} \geq 0$, Theorem 1.2-4 implies that $U(x,t) \leq w(x,t)$ for $(x,t) \in \bar{\Omega}\times[t_0,\infty)$. However, lemma 7.2-1 implies that $\lim_{t\to\infty} w(x,t) = \frac{1}{|\Omega|}\int_\Omega U(x,t_0)dx = \tilde{h}(t_0)$ uniformly in $\bar{\Omega}$. Consequently, $U(x,t) \leq a + 2\varepsilon$ for large t or $\limsup_{t\to\infty} U(x,t) \leq a$. Now suppose $\lim_{k\to\infty} u(\cdot,t_k) = z$ (in C^2) for some $t_k\to\infty$. Clearly, we have

$$\lim_{k\to\infty} U(\cdot,t_k) = V(z) \quad \text{in } C^2(\bar{\Omega}),$$

and $V(z) \leq a$. However, we also have $\frac{1}{|\Omega|}\int_\Omega V(z(x))dx =$ $\frac{1}{|\Omega|}\int_\Omega \lim_{k\to\infty} U(x,t_k)dx = \lim_{k\to\infty} \tilde{h}(t_k) = a$. Consequently, we must have $V(z(x)) \equiv a$ in $\bar{\Omega}$. For any sequence (s_k) tending to infinity, one can choose a subsequence $(\tilde{s}_k)$ such that $U(\tilde{s}_k,\cdot)\to a$ in C^2. This implies that $U(\cdot,t)\to a$ in C^2 as $t\to\infty$.

Remark: Note that the number a depends on initial function $u^0(x)$. Moreover $V(u(\cdot,t)) = U(\cdot,t)\to a$ does not imply that $u(\cdot,t)$ tends to one constant as $t\to\infty$.

Let $u(x,t)$ be a solution for (7.2-5) to (7.2-7) for $0 \le t < \infty$. The limit set of $u(\cdot,t)$, $\Lambda^+(u)$, is defined as:

$$\Lambda^+(u) = \{z \in C^2(\bar{\Omega}) : \lim_{k\to\infty} |u(\cdot,t_k) - z(\cdot)|_{\Omega}^{(2)} = 0 \text{ for some}$$

sequence $t_k \to \infty\}$.

The next theorem shows that all function(s) in $\Lambda^+(u)$ are constant(s).

Theorem 7.2-2. Assume (7.2-8), (7.2-9) and $u_i^0(x) > 0$ are in $H^{2+\alpha}(\bar{\Omega})$, with $\frac{\partial u_i^0}{\partial \eta} = 0$ on $\delta\Omega$, $i = 1, \ldots, n$. Then the initial boundary value problem (7.2-5) to (7.2-7) has a unique solution $u(x,t)$ in $C^{2,1}(\bar{\Omega}\times[0,\infty))$. The function $u(\cdot,t) : [0,\infty) \to H^{2+\alpha}(\bar{\Omega})$ is continuous and $\sup_{t\ge 0} |u(\cdot,t)|_{\Omega}^{(2+\alpha)} < \infty$. The limit set $\Lambda^+(u)$ is nonempty, and all its element(s) are constant function(s).

Proof. From earlier theorem, lemmas and discussions in this section, the only thing left to prove is that the element(s) in $\Lambda^+(u)$ are constant function(s) only. Let $\bar{z} = \lim_{j\to\infty} u(\cdot,t_j)$ in $C^2(\bar{\Omega})$ for some $t_j \to \infty$. We may assume that $\bar{z} \in H^{2+\alpha'}(\bar{\Omega})$ for some $0 < \alpha' < \alpha$. Let $z(x,t)$ be the solution of (7.2-5) to (7.2-7) with initial value u^0 replaced by $\bar{z}$. Lemma 7.2-2 applied to $u(x,t_j+t)$ and $z(x,t)$, and equation (7.2-5) implies that

$$|u(x,t_j+t) - z(x,t)| \le \max_{\bar{\Omega}} |u(x,t_j) - \bar{z}(x)| e^{Mt} \tag{7.2-19}$$

for $(x,t) \in \bar{\Omega}\times[0,\infty)$, where M is an appropriate Lipschitz constant. (Note that as in the argument for $u(x,t)$, we can utilize $V(z(x,t))$ to deduce positive upper and lower bounds for $z(x,t)$ as in (7.2-18), and thus obtaining the Lipschitz constant M.)

From (7.2-19), we deduce that $\lim_{j\to\infty} u(x,t_j+t) = z(x,t)$ uniformly in x, for fixed t. Consequently, we obtain

$$\lim_{j\to\infty} V(u(x,t_j+t) = V(z(x,t)),\ x \in \overline{\Omega},$$

From Lemma 7.2-3, we therefore obtain $V(z(x,t)) = \text{const} = a$ for $(x,t) \in \overline{\Omega}x[0,\infty)$. If we let $W(x,t) = V(z(x,t))$ and $y_i = \ln[z_i(x,t)/q_i]$, we see that W is a solution (7.2-17) with U replaced by W. Now $W \equiv$ const and $(a_i p_{ij}) \leq 0$ imply that $|\text{grad } y_i| \equiv 0$. Consequently $z(x,t)$, and in particular $\overline{z}(x) = z(x,0)$, is independent of x. This proves the theorem.

Remark: The function $z(\cdot,t)$ above may still depend on t. It is also true that $\frac{\partial u}{\partial x_i}(x,t)$ and $\frac{\partial^2 u}{\partial x_i \partial x_j}(x,t) \to 0$ as $t\to\infty$ uniformly for $x \in \overline{\Omega}$, for arbitrary $1 \leq i,j \leq d$.

Otherwise, there exist $\varepsilon > 0$ and $t_k \to \infty$ so that $|\frac{\partial u}{\partial x_i}(x_k,t_k)| \geq \varepsilon$, for example. One can then select subsequence (τ_k) so that $u(\cdot,\tau_k) \to$ constant in $C^2(\overline{\Omega})$, since every element in $\Lambda^+(u)$ is a constant. In particular $\frac{\partial u}{\partial x_i}(x,\tau_k) \to 0$ uniformly in x, giving rise to a contradiction.

We now consider a few examples for the initial-boundary value problem (7.2-5) to (7.2-7):

Example 7.2-1. Consider the prey-predator system with crowding effect:

$$\frac{\partial u_1}{\partial t} = \Delta u_1 + u_1[1 - u_1 - 2u_2]$$

$$\frac{\partial u_2}{\partial t} = \Delta u_2 + u_2[-1 + 2u_1 - u_2]$$

Here $(q_1,q_2) = (3/5,1/5)$. Letting $a_1 = a_2 = 1$, $\sum_{i,j=1}^{2} a_i p_{ij} w_i w_j = -w_1^2 - w_2^2 \leq 0$. Formula (7.2-10) becomes $V(u_1,u_2) = (u_1 - 3/5) - 3/5 \ln \frac{5u_1}{3} + (u_2 - 1/5) - 1/5 \ln 5u_2$. From the proof in Theorem 7.2-2, we found that $W(x,t) = V(z(x,t))$ satisfies (7.2-17), with u_i replaced by z_i. Consequently $W \equiv$ constant, and the fact that $\sum_{i,j=1}^{n} a_i p_{ij}(z_i - q_i)(z_j - q_j) = 0$ implies that $z_i = q_i$ $(i=1,2)$ leads to the conclusion that $z(x,t) \equiv (q_1,q_2)$. Therefore, any $\overline{z} \in \Lambda^+(u)$ must satisfy $\overline{z} = (q_1,q_2)$. We conclude That $u(\cdot,t) \to (q_1,q_2) = (3/5,1/5)$ as $t \to \infty$ in $C^2(\overline{\Omega})$.

Example 7.2-2. Consider the following special case for 2 predators with 1 prey:

$$
\begin{aligned}
&\frac{\partial u_1}{\partial t} = \Delta u_1 + u_1[2 - u_1 - u_2 - 2u_3] \\
(7.2\text{-}21)\quad &\frac{\partial u_2}{\partial t} = \Delta u_2 + u_2[-1 + u_1] \\
&\frac{\partial u_3}{\partial t} = \Delta u_3 + u_3[-1 + u_1]
\end{aligned}
$$

There are many positive steady states (q_1,q_2,q_3) on the interior of the line segment joining $(1,0,1/2)$ and $(1,1,0)$. Choosing $a_1 = a_2 = 1$, $a_3 = 2$, we obtain $\sum_{i,j=1}^{3} a_i p_{ij} w_i w_j = -w_1^2 \leq 0$. Considering the equation (7.2-17) satisfied by $V(z(x,t)) \equiv$ constant, the nature of $(a_i p_{ij})$ implies that $z_1(x,t) \equiv q_1 = 1$. In view of Theorem 7.2-1, one expects that $u(x,t)$ tends to some point on the line segment described above in a spatially independent manner. The limit point should depend on initial conditions.

Example 7.2-3.

$$\frac{\partial u_1}{\partial t} = \Delta u_1 + u_1[1 - u_2]$$
(7.2-22)
$$\frac{\partial u_2}{\partial t} = \Delta u_2 + u_2[-1 + u_1]$$

This is the prey-predator case with no crowding (self-inhibiting) effect. There is the positive equilibrium $(q_1,q_2) = (1,1)$; and letting $a_1 = a_2 = 1$, one has $\sum_{i,j=1}^{2} a_i p_{ij} w_i w_j \equiv 0$. The formula (7.2-10) takes the form $V(u_1,u_2) = \sum_{i=1}^{2} [(u_i-1) - \ln u_i]$. The set $V(u_1,u_2) = a > 0$ is a closed curve in the first open quadrant. Thus the fact that $\lim_{t\to\infty} V(u(x,t)) = a$ and Theorem 7.2-2 suggest that $u(x,t)$ eventually winds around the curve $V(u_1,u_2) = a$ in a spatially independent manner (if $a>0$).

The theorems and lemmas above will not be able to completely justify all the behavior suggested by the three examples above. The following theorem will clarify the behavior in examples 7.2-1 and 7.2-2. Further, the discussions in the next section provide a simple efficient procedure to analyze large systems, and classify them into cases typified by the three above examples.

Lemma 7.2-4. *Under the assumptions of Theorem 7.2-2, the limit set $\Lambda^+(u)$ is a compact subset of $C^2(\bar{\Omega})$. Furthermore $\Lambda^+(u)$ is invariant with respect to the ordinary differential equations*

$\frac{dz_i}{dt} = z_i[e_i + \sum_{j=1}^{n} p_{ij}z_j]$, $i=1,\ldots,n$. (*That is, if* $z(t)$ *is a solution of the ordinary differential equations above for* $t \in [-T_1,T_2]$, $T_1,T_2 > 0$, *with* $z(0) \in \Lambda^+(u)$, *then* $z(t) \in \Lambda^+(u)$ for $t \in [-T_1,T_2]$).

Proof. (Outline) Recall that elements of $\Lambda^+(u)$ are spatially constants. One utilizes inequalities of the form:

$$(7.2\text{-}23) \qquad |u(x,t_k+t) - z(t)| \leq \max_{\bar{\Omega}}|u(x,t_k) - z(0)|e^{Mt}$$

for $0 \leq t \leq T_2$ and an appropriate Lipschitz constant M, where $u(\cdot,t_k) \to z(0)$ uniformly in $\bar{\Omega}$ as $t_k \to \infty$. Inequality (7.2-23) is due to Lemma 7.2-2. Letting $t_k \to \infty$, we conclude that $z(t) \in \Lambda^+(u)$ for $0 < t \leq T_2$. Similar argument is used for $-T_1 \leq t \leq 0$. The details can be found in [191].

The previous lemma and following theorem show that some properties of the limit set $\Lambda^+(u)$ is the same as that for the ODE case.

Theorem 7.2-3. *Under the assumptions of Theorem* 7.2-2, *let* $u(x,t)$ *be the solution to* (7.2-5) *to* (7.2-7) *with initial condition* u^0. *Define for* $z_i > 0$, $i=1,\ldots,n$,

$$(7.2\text{-}24) \qquad \dot{V}(z) = \sum_{i,j=1}^{n} a_i p_{ij}(z_i - q_i)(z_j - q_j), \quad z = (z_1,\ldots,z_n).$$

Then $\dot{V}(z) = 0$ *if* $z \in \Lambda^+(u)$. *Consequently*, $\Lambda^+(u)$ *in contained in the largest invariant (with respect to the ODE given in lemma* 7.2-4) *subset of the set* $\{z \in R^n_+ : \dot{V}(z) = 0\}$.

Proof. (Outline) From property (7.2-9), we have $\dot{V}(z) \leq 0$.

Suppose $\dot{V}(\bar{z}) = -2\bar{\alpha} < 0$ at a point $\bar{z} \in \Lambda^+(u)$, let $u(x,t_k) \to \bar{z}$ as $t_k \to \infty$. Recall that $U(x,t) \overset{(def)}{=} V(u(x,t))$ satisfies (7.2-17), and define $r(t) = \max\{U(x,t) | x \in \bar{\Omega}\}$. By differential inequalities, one shows that $r(t)$ is decreasing. When $u(\cdot,t)$ is close to $\bar{z}$, the last term on the right of (7.2-17) is $<-\bar{\alpha}$. Applying differential inequalities related to (7.2-17) for regions $\bar{\Omega}\times[t_k, t_k+\delta]$, δ some small positive constant, one shows that $r(t_k+\gamma) \leq r(t_k) - \bar{\alpha}\delta$. (Note that $w(t) \overset{(def)}{=} r(t_k) - \bar{\alpha}(t-t_k)$ satisfies $\frac{\partial w}{\partial t} = \Delta w - \bar{\alpha}$.) Since the last inequality is true for all large k, we obtain $\lim_{t\to\infty} r(t) = -\infty$, contradicting the fact that U is nonnegative. This shows that $\dot{V}(\bar{z})$ cannot be negative in $\Lambda^+(u)$. More details can be found in [191].

Remark. In view of (7.2-12), we see that $\dot{V}$ is the derivative of V along a trajectory of the ODE given in lemma 7.2-4.

In example 7.2-1, $\dot{V}(z) = 0$ implies that $(z_1,z_2) = (3/5,1/5)$. Consequently, Theorem 7.2-3 says that $\Lambda^+(u)$ consists of the only point (3/5,1/5). In example 7.2-2, $\dot{V}(z) = 0$ implies that $z_1 = 1$. In the largest invariant subset described in Theorem 7.2-3, we must have $\frac{dz_1}{dt} = 0$; and thus $1 - z_2 - 2z_3 = 0$. Consequently $\Lambda^+(u)$ consists of the interior of the line segment joining (1,0,1/2) and (1,1,0). In example 7.2-3, $\dot{V}$ is identically 0 everywhere. Thus Theorem 7.2-3 does not give much information, and one has to refer back to Theorem 7.2-2 for better insight.

7.3 Stably Admissibility, Graph Theory

By Theorem 7.2-3, the limit set, Λ^+, of solutions of (7.2-5) to (7.2-7) is characterized by

$$\sum_{i,j=1}^{n} a_i p_{ij}(z_i-q_i)(z_j-q_j) = 0 \quad \text{and}$$

$$\frac{dz_i}{dt} = z_i[e_i + \sum_{j=1}^{n} p_{ij}z_j], \quad i=1,\ldots,n$$

Letting $w_i = z_i-q_i$ and $y_i = \ln \frac{z_i}{q_i}$, $i=1,\ldots,n$, the above equations become

$$\sum_{i,j=1}^{n} a_i p_{ij} w_i w_j = 0, \quad \text{and} \tag{7.3-1}$$

$$\frac{dy_i}{dt} = \sum_{j=1}^{n} p_{ij} w_j\,, \quad i=1,\ldots,n \tag{7.3-2}$$

In this section, we will assume further assumptions on (p_{ij}) motivated by ecological considerations. We will also develop some graphical techniques to utilize (7.3-1) and (7.3-2) to analyze the limit set Λ^+ efficiently. In ecological models, one cannot expect (p_{ij}) be known exactly, consequently, the "admissibility" assumption (7.2-9) for (p_{ij}) can only be meaningful if it is also true for slight "perturbations" of (p_{ij}). Furthermore, if certain $p_{ij} = 0$ (or $p_{ij} \neq 0$), a "perturbation" $\tilde{p}_{ij}$ of p_{ij} should not allow $\tilde{p}_{ij} \neq 0$ (or $\tilde{p}_{ij} = 0$), because this means additional dependence (or lack of dependence) of the ith species on the jth species, and is thus a drastic change of structure. We therefore define the concept "stably admissible" as follows:

Definition. An nxn matrix (p_{ij}) is called stably admissible if there is an $\varepsilon>0$ such that whenever an nxn matrix $(\tilde{p}_{ij})$ satisfies (i) $\max_{1\le i,j\le n} |\tilde{p}_{ij}-p_{ij}| < \varepsilon$ and (ii) $\tilde{p}_{ij} = 0 \Leftrightarrow p_{ij} = 0$, then $(\tilde{p}_{ij})$ is admissible.

For a stably admissible matrix, one can choose $a_i > 0$ which satisfy a useful property additional to that of (7.2-9).

Theorem 7.3-1. Suppose (p_{ij}) is an nxn stably admissible matrix, then there exist $a_i>0$, $i=1,\dots,n$ so that

(7.3-3) $$(a_i p_{ij}) \le 0, \quad \text{and}$$

(7.3-4) $$\text{Whenever } \sum_{i,j=1}^{n} a_i p_{ij} w_i w_j = 0, \text{ then } p_{ii} w_i = 0$$

each $i=1,\dots,n$.

Proof. Let $\tilde{p}_{ii} = (1-\delta)p_{ii}$, $\tilde{p}_{ij} = p_{ij}$ for $i\neq j$. For $\delta>0$ sufficiently small, the matrix $(\tilde{p}_{ij})$ is admissible. Let $a_i>0$, $i=1,\dots,n$, be such that $(a_i\tilde{p}_{ij}) \le 0$. We have

(7.3-5) $$\sum_{i,j=1}^{n} a_i \tilde{p}_{ij} w_i w_j = \sum_{i,j=1}^{n} a_i p_{ij} w_i w_j - \delta \sum_{i=1}^{n} p_{ii} w_i^2 \le 0$$

for all $(w_1,\dots,w_n) \in R^n$. On the other hand the admissibility of (p_{ij}) implies that $p_{ii}\le 0$, $i=1,\dots,n$. Consequently (7.3-5) implies that

(7.3-6) $$\sum_{i,j=1}^{n} a_i p_{ij} w_i w_j \le \delta \sum_{i=1}^{n} p_{ii} w_i^2 \le 0, \text{ for all } (w_1,\dots,w_n) \in R^n.$$

Moreover, if $\sum_{i,j=1}^{n} a_i p_{ij} w_i w_j = 0$, then each $p_{ii} w_i^2 = 0$, i.e. either $p_{ii}=0$ or $w_i=0$, $i=1,\dots,n$. This proves (7.3-4).

In example 7.3-1, $p_{ii} = -1$, $p_{12} = -2$, $p_{21} = 2$. With a slight perturbation $(\tilde{p}_{ij})$, one can always choose $a_i > 0$ so that $a_1\tilde{p}_{12} + a_2\tilde{p}_{21} = 0$ and $\sum_{i,j=1}^{2} a_i\tilde{p}_{ij}w_iw_j = -m_1w_1^2 - m_2w_2^2$ where m_1, m_2 are constants close to 1. Clearly (p_{ij}) is stably admissible. In example 7.2-2, we have $p_{22} = p_{33} = 0$, and $p_{23} = p_{32} = 0$. Let $(\tilde{p}_{ij})$ be a slight perturbation in the sense described above (thus $\tilde{p}_{22} = \tilde{p}_{33} = \tilde{p}_{23} = \tilde{p}_{32} = 0$). Note we should also have $\tilde{p}_{12}\tilde{p}_{21} < 0$ and $\tilde{p}_{13}\tilde{p}_{31} < 0$, therefore choosing $a_1 = 1$, $a_2 = -a_1\tilde{p}_{12}/\tilde{p}_{21} > 0$, $a_3 = -a_1\tilde{p}_{13}/\tilde{p}_{31} > 0$, we have $\sum_{i,j=1}^{2} a_i\tilde{p}_{ij}w_iw_j = \tilde{p}_{11}w_1^2$ where $\tilde{p}_{11}$ is close to -1. Again, we clearly must have (p_{ij}) stably admissible. In example 7.2-3, $p_{11} = p_{22} = 0$, $p_{12} = -1$, $p_{21} = 1$. For a slight perturbation, $\tilde{p}_{11} = \tilde{p}_{22} = 0$ and $\tilde{p}_{12}\tilde{p}_{21} < 0$. One can find $a_i > 0$ so that $\sum_{i,j=1}^{2} a_ip_{ij}w_iw_j \equiv 0$. The matrix (p_{ij}) is therefore again stably admissible.

For a matrix $p = (p_{ij})$, $1 \leq i,j \leq n$, in order to analyze the equations (7.3-1) and (7.3-2) we consider the graph G(p) as follows. A graph G(p) consists of n "vertices" (or nodes) 1,2,...,n. For $i \neq j$, the vertex i is "adjacent" to the vertex j if and only if $p_{ij} \neq 0$ or $p_{ji} \neq 0$. Adjacent vertices are said to be connected by an "edge." The graph G(p) has a black dot ● at the vertex i if $p_{ii} < 0$, and an open circle o at vertex i if $p_{ii} = 0$. An edge which connects two adjacent black dots is called a "strong link," otherwise the "link" between the adjacent vertices is "weak." A "path" from vertex j to k is defined by a chain of adjacent vertices $n_1 = j$, $n_2, \ldots, n_m = k$. If j=k and there is no repeated edge in the chain from vertex j to k, we say the path forms a "loop." The entire configuration of vertices, with their •, o, edges, path, etc. is the graph, G(p), of the matrix $p = (p_{ij})$. If the graph consists of two or more separate pieces, the differential equation (7.2-5) can be uncoupled into two or more independent subsystems. Consequently, we will always assume, without loss of generality, that the graph is connected. As an example, let

$$p = \begin{bmatrix} c_1 & 0 & 0 & c_2 \\ c_3 & 0 & 0 & 0 \\ 0 & c_4 & c_5 & 0 \\ 0 & c_6 & 0 & c_7 \end{bmatrix}$$

where c_i, $i=1,\ldots,7$ are nonzero constants, $c_i<0$ for $i=1,5,7$. The graph G(p) can be pictured as:

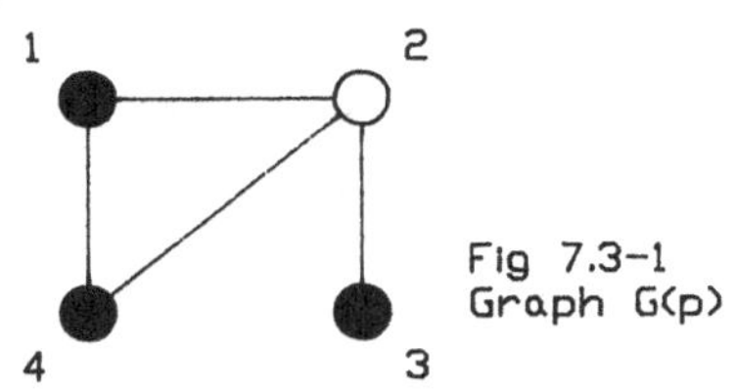

Fig 7.3-1
Graph G(p)

Remark: If p is admissible, then $p_{ii} \le 0$. In such situation, a vertex of G(p) is therefore either a • or o.

In the example above, the edge connecting vertex 1 to 4 is a strong link, and all others are weak links. The path connecting vertices 1,2,4,1 forms a loop.

Since for a pertubation $\tilde{p}$ of p, we assume $\tilde{p}_{ij} = 0$ if and only if $p_{ij} = 0$, consequently the graphs $G(\tilde{p})$ and G(p) are the same, provided $\max_{1\leq i,j\leq n} |\tilde{p}_{ij} - p_{ij}|$ is small enough. We next give a simple sufficient condition on the structure of G(p), so that $p = (p_{ij})$ is stably admissible.

Theorem 7.3-2. *Suppose the* nxn matrix $p = (p_{ij})$ *satisfies the prey-predator conditions*:

(7.3-7) $p_{ij}p_{ji} < 0$ *whenever* $(i-j)p_{ij} \neq 0$, *and* $p_{ii} \leq 0$ *for* $1\leq i, j \leq n$. *Furthermore, assume that the graph* G(p) *does not have any loop* (*such graph is said to be a "tree"*). *Then* p *is stably admissible*.

Proof. If n=2, choose $a_1 = 1$, $a_2 = -p_{12}/p_{21} > 0$, then

$$a_i p_{ij} + a_j p_{ji} = 0, \text{ for all } i \neq j. \tag{7.3-8}$$

We have $\sum_{i,j=1}^{2} a_i p_{ij} w_i w_j = a_1 p_{11} w_1^2 + a_2 p_{22} w_2^2 \leq 0$, because of (7.3-7). The choice of a_i can be done similarly for a small perturbation $\tilde{p}$, therefore the theorem is true when $n = 2$ (Note that if $n = 2$ there can be no loop). Suppose the theorem is true if $n = k$. Consider the case for $n = k+1$ and p satisfies all assumptions in the theorem. Then there must be a vertex m which is connected with exactly one adjacent vertex, otherwise G(p) would have a loop. For $i\neq m$, we can choose $a_i > 0$ so that $\sum_{i,j\neq m} a_i p_{ij} w_i w_j \leq 0$ for all $(w_1, \ldots w_{m-1}, w_{m+1}, \ldots, w_n) \in R^{n-1}$, by induction hypotheses. Choose a_m so that $a_m p_{mr} + a_r p_{rm} = 0$ where r is the vertex which is adjacent to m. Hypothesis (7.3-7) implies that $a_m > 0$, and moreover

$\sum_{i,j=1}^{n} a_i p_{ij} w_i w_j = \sum_{i,j\neq m} a_i p_{ij} w_i w_j + a_m p_{mm} w_m^2 \leq 0$. Thus p is admissible. The same arguments can be used for a small enough perturbation $\tilde{p}$, therefore p is stably admissible.

Before we analyze more sufficient conditions for stable admissibility, we next consider a simple necessary condition.

<u>Theorem</u> 7.3-3. <u>Let</u> p <u>be an</u> nxn <u>matrix which is stably admissible</u>. <u>Then every loop in the graph</u> G(p) <u>must contain a strong link</u>.

<u>Proof</u>. Let $a_i > 0$ be chosen so that $(a_i p_{ij}) \leq 0$. Suppose that there is a weak link connecting vertices r,s, then $p_{rr} p_{ss} = 0$. Choose $w_k = 0$ if $k \neq r$ or s, then $\sum_{j=1}^{n} a_i p_{ij} w_i w_j$ becomes

$$(a_r p_{rs} + a_s p_{sr}) w_r w_s + a_r p_{rr} w_r^2 \text{ or } (a_r p_{rs} + a_s p_{sr}) w_r w_s + a_s p_{ss} w_s^2$$

The quadratic expressions above cannot be ≤ 0 for all (w_r, w_s) unless we have

$$(7.3\text{-}9) \qquad a_r p_{rs} + a_s p_{sr} = 0$$

Suppose that the edges connecting the successive distinct vertices $s_1, \ldots, s_k$ forms a loop with every edge being a weak link. Then equations of the form (7.3-9) applying to each weak link s_i, s_{i+1} determine the ratios $a_{s_{i+1}}/a_{s_i}$ $= -p_{s_i s_{i+1}}/p_{s_{i+1} s_i}$, i=1,...,k-1, and the product determines a_{s_k}/a_{s_1}. However the edge connecting s_k to s_1 also determines

$a_{s_k}/a_{s_1} = -p_{s_1 s_k}/p_{s_k s_1}$. Equating the formulas for a_{s_k}/a_{s_1}, we obtain that "balanced" relation:

$$(7.3\text{-}10) \quad |p_{s_1 s_2} p_{s_2 s_3} \cdots p_{s_{k-1} s_k} p_{s_k s_1}| = |p_{s_2 s_1} p_{s_3 s_2} \cdots p_{s_k s_{k-1}} p_{s_1 s_k}|$$

which has to hold for each loop containing only weak links so that p is admissible. Let $\tilde{p}$ be a small perturbation of p so the $G(\tilde{p})$ is the same as $G(p)$. We may assume $\tilde{p}$ is admissible since p is stably admissible. Applying the above arguments to the same loop connecting $s_1, \ldots, s_k$, we obtain the balanced relation (7.3-10) with all p's replaced by $\tilde{p}$'s, so that $\tilde{p}$ remains admissible. However, one can always choose $\tilde{p}$ with $|p-\tilde{p}|$ arbitrarily small so that the balanced relation fails. Consequently p cannot be stably admissible, unless the loop has a strong link.

We next state a useful sufficient criteria for stable admissibility. Suppose that p satisfies condition (7.3-7) and that every loop of G(p) has at least one strong link. Let L be a set of strong links in the graph of G(p) so that the removal of L from G(p) simplifies the graph to a tree (i.e. connected graph with no loop). Suppose L contains a strong link joining vertex i to j, we say L_{ij} is in L (Do not include L_{ji}, if L_{ij} is already included, since they represent the same edge). List all the links L_{ij} in the set L, with each link only once. Define n(i) to be the number of times the index i occurs as subscript in the list. For example, if L_{12}, L_{34}, L_{68}, L_{53}, L_{15}, L_{25} are all the links in L, then $n(1) = 1$,

$n(2) = 2$, $n(3) = 2$, $n(5) = 3$, $n(6) = 1$, $n(8) = 1$. Since the graph becomes a connected tree after removal of L_{ij}'s, after the removal of L, there is a unique path connecting vertex i to j in the simplified graph, with successive distinct vertices $i, s_1, s_2, \ldots, s_\ell, j$, $\ell \geq 1$. Thus $p_{is_1}p_{s_1s_2}\cdots p_{s_\ell j}p_{ji} \neq 0$. Let

$$(7.3\text{-}11) \qquad A_{ij} = \frac{|p_{is_1}p_{s_1s_2}\cdots p_{s_\ell j}p_{ji}|}{|p_{s_1i}p_{s_2s_1}\cdots p_{js_\ell}p_{ij}|} + \frac{|p_{s_1i}p_{s_2s_1}\cdots p_{js_\ell}p_{ij}|}{|p_{is_1}p_{s_1s_2}\cdots p_{s_\ell j}p_{ji}|} - 2$$

$$(7.3\text{-}12) \qquad B_{ij} = \frac{p_{ii}p_{jj}}{p_{ij}p_{ji}}$$

if L_{ij} is in the set L.

Theorem 7.3-4. *Suppose that an nxn matrix p satisfies conditions (7.3-7). Let L_{ij} be strong links whose removal simplify G(p) to a connected graph with no loop (i.e. tree) as described above. Let $\{L_{ij}\}$ denote this set of strong links so that any of its proper subset will not suffice to generate a tree upon removal. Suppose that*

$$(7.3\text{-}13) \qquad A_{ij} < 4\,\frac{B_{ij}}{n(i)n(j)}$$

for every pair of subscripts (i,j) in the set $\{L_{ij}\}$, then the matrix p associated with the original graph is stably admissible.

We only discuss the proof of a simple case. Suppose that G(p) consists of one loop joining vertices 1,2,3,...,m,1 with a strong link joining 1 and m. Choose $a_i>0$ so that $a_ip_{ii+1} + a_{i+1}p_{i+1,i} = 0$, for $i=1,2,\ldots,m-1$. Then

$\sum_{i,j=1}^{m} a_i p_{ij} w_i w_j$ becomes $\sum_{i=2}^{m-1} a_i p_{ii} w_i^2 + a_1 p_{11} w_1^2 + (a_1 p_{1m} + a_m p_{m1}) w_1 w_m$ $+ a_m p_{mm} w_m^2$. Clearly a sufficient condition for stable admissibility is

$$(7.3\text{-}14) \qquad (a_1 p_{1m} + a_m p_{m1})^2 < 4 a_1 a_m p_{11} p_{mm}$$

From the choice of a_i above, one has

$$\frac{a_{i+1}}{a_i} = \left|\frac{p_{ii+1}}{p_{i+1,i}}\right| \qquad \text{for } i=1,\ldots,m-1 .$$

Comparing with (7.3-11), we see that in our case $i, s_1, s_2, \ldots s_\ell, j$ becomes $1,2,3,\ldots,m-1,m$. Simple calculation shows that (7.3-14) is equivalent to $A_{1m} < 4B_{1m}$, as in (7.13). The proof for the general case can be found in [194].

Remark: We have only stated sufficient conditions for stably admissibility for the prey-predator case in Theorem 7.3-2 and 7.3-4. However, the theory we develop below does not assume the prey-predator condition (7.3-7), while we will assume p is stably admissible.

If $p = (p_{ij})$ is stably admissible, choose $a_i > 0$ as described in Theorem 7.3-1 and use them to define V(u) in (7.2-10). To find the largest invariant subset (with respect to the O.D.E. given in lemma 7.2-4) contained in $\{z \in R_+^n : \dot{V}(z) = 0\}$, one is led by (7.3-1) and (7.3-2) to finding all solutions of:

$$(7.3\text{-}15) \qquad p_{ii} w_i = 0 \qquad i+1,\ldots,n ,$$

(7.3-16) $$\frac{dy_i}{dt} = \sum_{j=1}^{n} p_{ij}w_j \qquad i=1,\ldots,n\,,$$

We will now use G(p) as a tool to describe the solutions of these two equations. We keep track of the deductions by relabelling G(p) to a 'reduced graph' R(p).

Recall $w_i = z_i - q_i$, $y_i = \ln \frac{z_i}{q_i}$. Consequently, we have

$$z_i(t) \equiv q_i \iff w_i(t) \equiv 0 \iff y_i(t) \equiv 0\,,$$

$$\frac{dz_i}{dt}(t) \equiv 0 \iff \frac{dw_i}{dt}(t) \equiv 0 \iff \frac{dy_i}{dt}(t) \equiv 0\,.$$

If we can deduce that for a solution of (7.3-15) and (7.3-16), that $w_i(t) \equiv 0$, relabel (if necessary) the vertex i in the graph by a ●. Similarly, if we can deduce that such a solution must have $\frac{dy_i}{dt}(t) \equiv 0$, (i.e. $w_i(t) \equiv$ constant) but cannot conclude $w_i \equiv 0$, relabel the vertex i in the graph by a ⊕. (Note that if $p_{ii}<0$, we already label i by ●, and (7.3-15) also imply $w_i(t) \equiv 0$, hence a ●).

We need one additional theorem concerning the property of stably admissible matrices before the graph reduction process.

<u>Theorem</u> 7.3-5. <u>Suppose</u> p <u>is stably admissible and</u> i,j <u>are adjacent vertices of</u> G(p). <u>Then</u> $p_{ij}p_{ji} < p_{ii}p_{jj}$.

<u>Proof</u>. Let $a_i>0$ be chosen so that $(a_ip_{ij}) \leq 0$. Let $w_k=0$ for $k \neq i$ or j. One has $a_ip_{ii}w_i^2 + (a_ip_{ij} + a_jp_{ji})w_iw_j + a_jp_{jj}w_j^2 \leq 0$ for all w_i, w_j, therefore one must have

(7.3-17) $$(a_ip_{ij} + a_jp_{ji})^2 \leq 4a_ia_jp_{ii}p_{jj}.$$

If $p_{ii}p_{jj} = 0$, one can only satisfy (7.3-17) if $p_{ij}p_{ji} < 0$ and the lemma is proved. Suppose that $p_{ii}p_{jj} \neq 0$, one must have $p_{ii}p_{jj} > 0$. Then it suffices to show that p_{ij} and p_{ji} cannot be of the same sign. Dividing (7.3-17) by $a_i a_j$, we obtain

$$\left[\frac{\sqrt{a_i}}{\sqrt{a_j}}\, p_{ij} + \frac{\sqrt{a_j}}{\sqrt{a_i}}\, p_{ji}\right]^2 \leq 4p_{ii}p_{jj} \tag{7.3-18}$$

as a necessary condition. If $p_{ii}p_{jj} > 0$, one readily finds that the expression on the left above is $\geq 4p_{ij}p_{ji}$, with equality attained when $\frac{\sqrt{a_i}}{\sqrt{a_j}} = \sqrt{\frac{p_{ji}}{p_{ij}}}$. Thus a necessary condition for p to be admissible is $p_{ij}p_{ji} \leq p_{ii}p_{jj}$ (otherwise, (7.3-18) can never be satisfied for any choice of a_i, a_j). However, if $p_{ij}p_{ji} = p_{ii}p_{jj}$, we can choose a small perturbation $\tilde{p}$ with $\tilde{p}_{ij}\tilde{p}_{ji} > \tilde{p}_{ii}\tilde{p}_{jj}$. The above discussion shows that such $\tilde{p}$ cannot be admissible. Thus p cannot be stably admissible, unless $p_{ij}p_{ji} \leq 0 < p_{ii}p_{jj}$ (if $p_{ii}p_{jj} \neq 0$). This completes the proof.

Suppose i is labelled ● or ⊕, and there are all ● at vertices adjacent to i except for a single vertex j adjacent to i. Equation (7.3-16) for the vertex i becomes

$$0 = p_{ij}w_j \tag{7.3-19}$$

because $\frac{dy_i}{dt} = 0$, and the other black dots imply those w's are zero on the right. Since j is not yet a black dot, we have $p_{jj} = 0$. From Theorem 7.3-5, if p is stably admissible, we have $p_{ij}p_{ji} < 0$, hence $p_{ij} \neq 0$ and (7.3-19) shows that $w_j \equiv 0$. We therefore obtain the following reduction rule (R1) for G(p).

We also list two more reduction rules for G(p), p stably admissible. The explanations are given below.

(R1) Suppose there is ● or ⊕ at i, and ● at all vertices adjacent to i except for a single vertex j adjacent to i. Then we put ● at j.

(R2) Suppose there is ● or ⊕ at i, and ● or ⊕ at all vertices adjacent to i except for a single vertex j adjacent to i. The we put ⊕ at j.

(R3) Suppose there is O at i, and ● or ⊕ at each vertex adjacent to i. Then we put ⊕ at i.

For (R2), the assumptions imply that equation (7.3-16) for the vertex i becomes $0 = \text{constant} + p_{ij}w_j$. Theorem 7.3-5 again implies $p_{ij} \neq 0$, and thus $w_j(t) \equiv \text{constant}$. Hence we put ⊕ at vertex j. For (R3), the assumptions imply that equation (7.3-16) for the vertex i becomes $\frac{dy_i}{dt} = \text{constant}$. However, from the previous section we know that y_i is bounded. Consequently y_i is a bounded linear function of t, hence $y_i(t) \equiv$ constant. Thus we put ⊕ at vertex i.

The reduced graph R(p) is the graph obtained from G(p) after repeated application of the rules (R1), (R2) and (R3) until no further change is possible. We say R(p) is of type (●) if every vertex of R(p) is a black dot. R(p) is of type (●,⊕) if every vertex is either a ● or ⊕. Finally, R(p) is of type (●,⊕,O) if at least one of its vertices is O. These classifications exhaust all cases possible. We now come to the main Theorem of this section.

Theorem 7.3-6. (Main) Assume (7.2-8) and $u_i^0(x) > 0$ are in $H^{2+\alpha}(\bar{\Omega})$, with $\frac{\partial u_i^0}{\partial \eta} = 0$ on $\delta\Omega$, $i=1,\ldots,n$. Further, let $p = (p_{ij})$ be stably admissible and $R(p)$ be its reduced graph.

(a) If $R(p)$ is of type (●), then p is nonsingular, the steady state solution $q = (q_1,\ldots,q_n)$ is unique, and all solutions of (7.2-5) to (7.2-7) with positive initial functions as described satisfy $u(x,t) \to q$ in $C^2(\bar{\Omega})$ as $t \to +\infty$. (i.e. "global asymptotic stability" in R_+^n)

(b) If $R(p)$ is of type (●,⊕), then p is singular, the steady state solutions are not unique, and every solution with positive initial function as described tend to a constant steady state as $t \to +\infty$ in $C^2(\bar{\Omega})$. The limiting constant depends on initial function $(u_1^0(x),\ldots,u_n^0(x))$.

(c) If $R(p)$ is of type (●,⊕,O), then there exists a stably admissible matrix $\tilde{p}$ with $G(\tilde{p}) = G(p)$, such that (7.2-5) and (7.2-6), with p_{ij} changed to $\tilde{p}_{ij}$, has a (spatially constant) t-periodic nonconstant solution.

Proof. (Outline) (a) From Theorem 7.2-3, $\Lambda^+(u)$ is contained in the largest invariant (with repsect to the ODE given in lemma 7.2-4) subset of the set $\{z \in R_+^n : \dot{V}(z) = 0\}$. From the deduction above, $R(p)$ is of type (●) means that such invariant subset is precisely the only point $(w_1,\ldots,w_n) = (0,\ldots,0)$, i.e. $z = q$. Hence $\Lambda^+ = \{q\}$, and all solutions described will tend to q as stated by Theorem 7.2-2.

(b) It seems that $R(p)$ may belong to more than one type if one applies the reduction in different order of procedure. However, it is shown in [192], that this is not possible, and the type of $R(p)$ is uniquely determined by $G(p)$. It is also shown in [192], lemma 10, that if $R(p)$ is of type (0,0), then p is singular. Consequently, if we let q^* be any fixed constant positive steady state solution of (7.2-5) to (7.2-7), then any other positive steady state solutions are of the form $q = q^* + r$ where $r = (r_1,\dots,r_n)$ is a solution of $\sum_{j=1}^{n} p_{ij}r_j = 0$, $i=1,\dots,n$. (Recall, from Theorem 7.2-2, steady state solutions in $H^{2+\alpha}(\bar{\Omega})$ must be spatially constant.) Let $\lim_{k\to\infty} u(\cdot,t_k) = \tilde{q} \in \Lambda^+(u)$, where $t_k\to\infty$. Define the Lyapunov function V on R_n^+ as in (7.2-10), by using the coordinates of $\tilde{q}_i$ for q_i. Since $u(\cdot,t_k)\to\tilde{q}$, we must have $V(u(\cdot,t_k)) \to 0$. Moreover, using (7.2-17), one can prove by means of differential inequality as in Theorem 7.2-3 that $\max\{V(u(x,t))\,|\,x \in \bar{\Omega}\}$ is decreasing, and therefore $V(u(x,t)) \to 0$ uniformly in $\bar{\Omega}$ as $t\to\infty$. This implies that $u(x,t) \to \tilde{q}$ uniformly in $\bar{\Omega}$ as $t\to\infty$. Since $u(\cdot,t_k) \to \tilde{q}$ in $C^2(\bar{\Omega})$ by Theorem 7.2-2, we can obtain finally that $u(x,t) \to \tilde{q}$ in $C^2(\bar{\Omega})$ as $t\to\infty$.

(c) The proof is given in Theorem 1 of [192], or Theorem 3 of [191] and will be omitted here.

Remark. In case (b) of the above theorem, one can further show that every positive constant steady state solution $\tilde{q}$ of (7.2-5) to (7.2-7) is stable i.e. for any $\varepsilon>0$, there exists $\delta>0$ such that any solution with $|u(x,0) - \tilde{q}|<\delta$ will satisfy $|u(x,t) - \tilde{q}|<\varepsilon$ for all $t\geq 0$, $x \in \bar{\Omega}$. For the proof, first define the Lyapunov function V on R_n^+ as in (7.2-10) by using the coordinates of

$\tilde{q}_i$ for q_i. Let $\varepsilon>0$ be given, choose $\alpha>0$ so that $V(z)<\alpha$ implies $|z-\tilde{q}|<\varepsilon$. Then choose $\delta>0$ so that $|z-\tilde{q}|<\delta$ implies $V(z)<\alpha$. Suppose $|u(x,0)-\tilde{q}|<\delta$, for all $x\in\bar{\Omega}$, then $V(u(x,0))<\alpha$ for all $x\in\bar{\Omega}$. Since the function $\max\{V(u(x,t)) \mid x\in\bar{\Omega}\}$ is decreasing in t, we have $V(u(x,t))<\alpha$ for all $t>0$, and hence $|u(x,t)-\tilde{q}|<\varepsilon$ for all $t\geq 0$, $x\in\bar{\Omega}$. This shows that $\tilde{q}$ is stable (but not asymptotically stable).

<u>Remark</u>: Generalization of the theories in this section has been carried out in [191], [192] for the system:

$$\frac{\partial u_i}{\partial t} = Lu_i + N(u)\, f_i(u_i)\,[e_i + \sum_{j=1}^{n} p_{ij} g_j(u_j)], \quad i=1,\ldots,n,$$

with the same boundary condition (7.2-6). The functions $N: R_+^n \to R$ and f_i, g_i: $R^+ \to R$ are locally Lipschitz continuous and positive, g_i is strictly increasing and $\int_0^1 \frac{ds}{f_i(s)} = \int_1^\infty \frac{ds}{f_i(s)} = \infty$. Condition (7.2-8) is replaced by the existence of $q=(q_1,\ldots q_n)$, $q_i>0$ so that $e_i + \sum_{j=1}^{n} p_{ij} g_j(q_j) = 0$, $i=1,\ldots,n$. Parts (a) and (b) of the Main Theorem 7.3-6 remain valid.

Closing this section, we study a few examples in the application of rules and the Main Theorem 7.3-6.

<u>Example</u> 7.3-1. Suppose p has graph G(p) of the form:

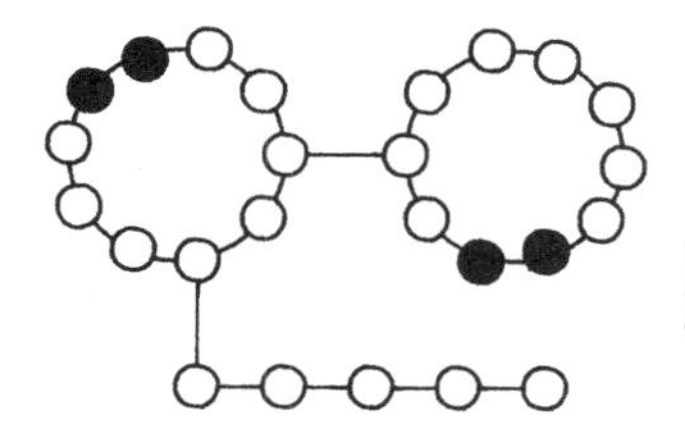

Fig. 7.3-2
Graph G(p) for example 7.3-1

Use the two black dots on the circle on the left as vertex i and apply rule (R1) successively. Similarly, applying the same procedure to the circle on the right, we obtain an intermediate step:

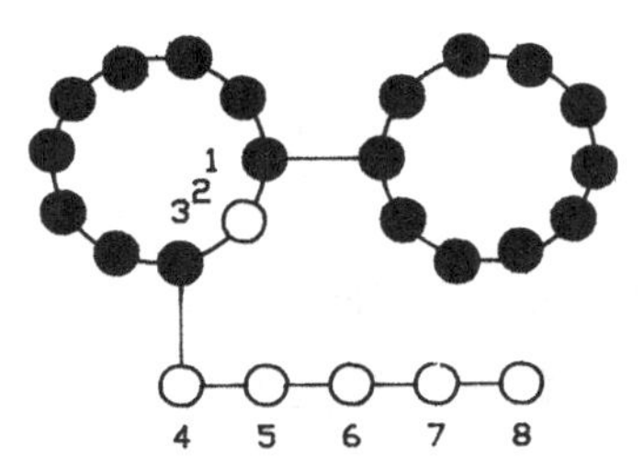

Fig. 7.3-3
An intermediate step in graph reduction

For convenience, we attach some identifications to some vertices above. Use vertex 1 as i in (R1), we obtain ● at 2. Then, use vertex 3 as i in (R1), we obtain ● at 4. Similarly, we successively use (R1) to obtain black dots at 5, 6, 7 and 8. Hence R(p) is of type (●). Furthermore, if p is stably admissible, then part (a) of the Main Theorem 7.3-6 applies. (Note that G(p) has two strong links, and Theorem 7.3-4 can be used to deduce sufficient conditions for stable admissibility.)

<u>Example</u> 7.3-2. Suppose G(p) has the form:

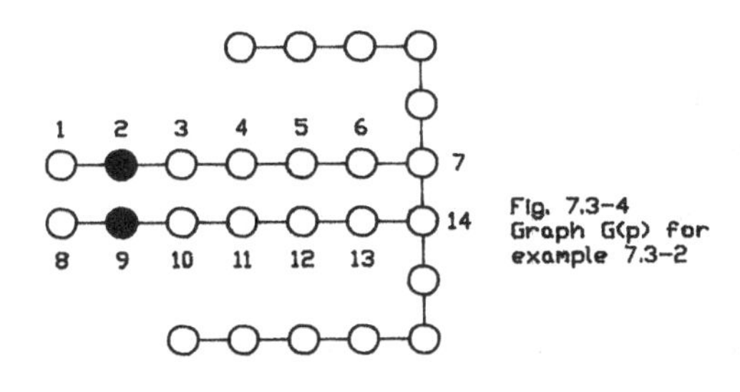

Fig. 7.3-4
Graph G(p) for example 7.3-2

Use (R3) with 1 and 8 as i, we obtain ⊕ at 1 and 8. Use (R2) with 2 and 9 as i, we obtain ⊕ at 3 and 10. Use (R1) with 3 and 10 as i, we obtain ● at 4 and 11. Successively applying (R2), (R1) and (R2) as before, we obtain a first stage in Fig. 7.3-5.

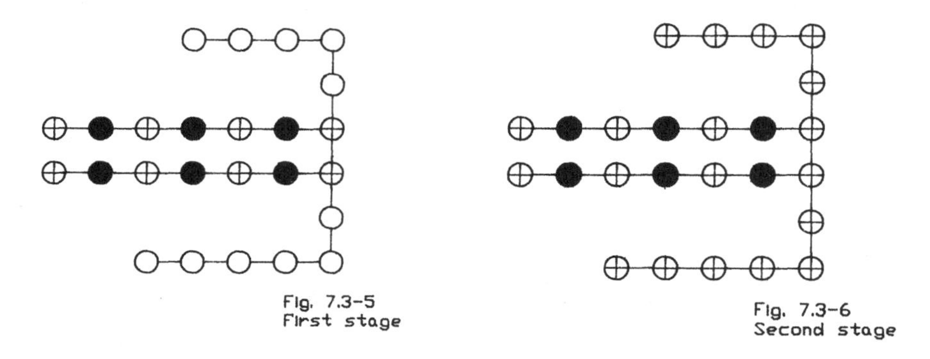

Fig. 7.3-5
First stage

Fig. 7.3-6
Second stage

Successively applying (R2), we then obtain a second stage in Fig. 7.3-6. Use (R1) with the vertices at the free end at the top and bottom as i, we obtain ● at 15 and 17. Similarly, we obtain ● at 16, 18, 7 and 19. Finally, we use (R1) further to obtain ● at 5, 3 and 1. Hence we arrive at the final form R(p) in Fig. 7.3-7.

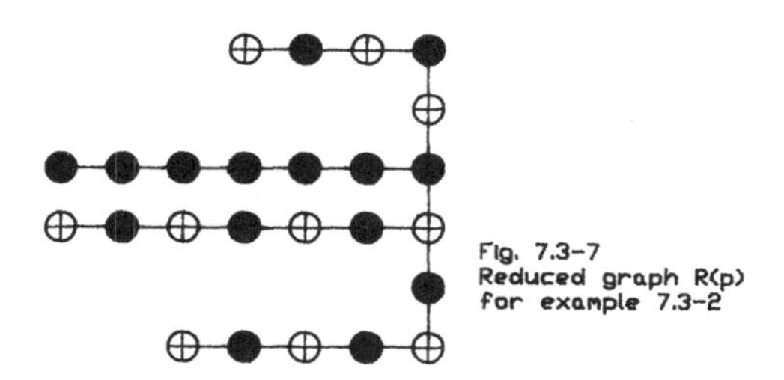

Fig. 7.3-7
Reduced graph R(p)
for example 7.3-2

Clearly R(p) is of type (●,⊕). Moreover, G(p) is a "tree." Thus if $p = (p_{ij})$ satisfies prey-predator conditions (7.3-7):

$p_{ij}p_{ji} < 0$ whenever $(i-j)p_{ij} \neq 0$ and $p_{ij} \leq 0$, then p is stably admissible by Theorem 7.3-2. Consequently, part (b) of the Main Theorem 7.3-6 applies.

Example 7.3-3. If we simply add one more vertex to the bottom line in the last example to start G(p) in the form,

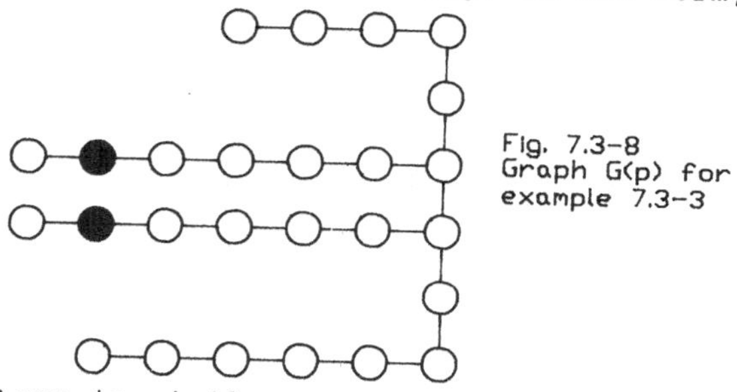

Fig. 7.3-8 Graph G(p) for example 7.3-3

the second stage is similar to that in the last example. However, further reduction will bring R(p) into type (●).

Remark: The type of R(p) in the examples above determines whether p is singular (if case (b)) or nonsingular (if case (a)). This procedure is much more convenient than calculating determinants.

7.4 Global Bifurcations of Steady-States in Prey-Predator Systems

We go back to the study of Dirichlet boundary value problem for systems of prey-predator reaction-diffusion equations in this section. Such problems have been discussed in Chapters 2 and 5. However, we now use a different method of analysis. We employ the technique of bifurcation theory in functional analysis, via degree theory. Although the results overlap with those in earlier chapters, the method here illustrates a useful alternative which is sometimes more readily applicable for other related problems. Furthermore, it gives insight to the bifurcation of solutions as some parameters vary globally. More specifically, we consider the homogeneous Dirichlet problem:

$$\text{(7.4-1)} \qquad \begin{aligned} &\sigma_1 \Delta u + u(a - bu - cv) = 0 \\ &\sigma_2 \Delta v + v(e + fu - gv) = 0 \end{aligned} \qquad \text{in } \Omega$$

$$u = v = 0 \qquad \text{on } \delta\Omega$$

where σ_1, σ_2, b, c, f and g are positive parameters and a, e are any constants. The functions u and v represent the concentrations of prey and predator respectively in the domain Ω. We will analyze the bifurcation of nonnegative solutions as the parameter e varies globally while the others are fixed. When the other parameters vary, similar treatments can be made, but will however be omitted here. In this section, we follow the investigations presented in [25] and [26]; the problem was first proposed in[145]. We will assume that Ω is a bounded domain in R^N, with boundary $\delta\Omega \in H^{2+\alpha}$, $0 < \alpha < 1$ as in Section 1.3.

For any q(x) in $H^{\alpha}(\overline{\Omega})$ (without any restriction on sign), $\sigma > 0$, the linear eigenvalue problem:

$$\text{(7.4-2)} \qquad -\sigma\Delta u + q(x)u = \mu u \quad \text{on } \Omega,\ u=0 \text{ on } \delta\Omega$$

has an infinite sequence of eigenvalues for μ which are bounded below. This can be readily obtained from information described in Section 1.3. It is also known that the first eigenvalue, denoted as $\mu_1(\sigma,q)$, is simple (see e.g. [227]); i.e. all solutions of (7.4-2) with $\mu = \mu_1(\sigma,q)$ are multiples of a particular eigenfunction. From Section 1.3, we also know that this eigenfunction does not change sign on Ω and its normal derivatives on the boundary $\delta\Omega$ never vanish.

Consider now the boundary value problem:

$$(7.4\text{-}3) \qquad -\sigma\Delta u + q(x)u = u(a - bu) \quad \text{in } \Omega,\ u=0 \text{ on } \delta\Omega,$$

where σ, $q(x)$, a and b are as described. Suppose that $a \leq \mu_1(\sigma,q)$. Let $\rho(x)>0$ be an eigenfunction for (7.4-2) with $\mu = \mu_1(\sigma,q)$. Using the family of upper solutions $\varepsilon\rho(x)$, $\varepsilon>0$ for (7.4-3) and the sweeping principle Theorem 1.4-3, one readily deduces that $u=0$ is the only nonnegative solution of (7.4-3) if $a \leq \mu_1(\sigma,q)$. On the other hand, suppose $a > \mu_1(\sigma,q)$. We use large constant as upper solution and small multiple of $\rho(x)$ as lower solution for (7.4-3) to deduce the existence of a solution which is positive in Ω by the method in Section 5.1. Furthermore, from Lemma 5.2-2, such positive solution to (7.4-3) is unique, when $a > \mu_1(\sigma,q)$.

The approach for the study of the system here involves decoupling the two equations in (7.4-1). We write the first equation in (7.4-1) in the form:

$$(7.4\text{-}4) \qquad -\sigma_1\Delta u + cvu = u(a - bu) \text{ in } \Omega,\ u=0 \text{ on } \delta\Omega$$

which can be regarded as a special case of (7.4-3) with $\sigma=\sigma_1$, $q(x) = cv(x)$. Thus, if $a \leq \mu_1(\sigma_1, cv)$, then (7.4-4) has no positive solution; while if $a > \mu_1(\sigma_1, cv)$, then (7.4-4) has a unique solution which is positive in Ω. Let v be an arbitrary function in $C^1(\bar{\Omega})$, we define u(v) as a function in $\bar{\Omega}$ by

$$(7.4\text{-}5) \qquad u(v) = \begin{cases} 0 \quad \text{if } a \le \mu_1(\sigma_1, cv) \\ \text{unique positive solution of (7.4-4) if} \\ a > \mu_1(\sigma_1, cv) \end{cases}$$

From Chapter 5, we see that u(v) is in $H^{2+\alpha}(\bar{\Omega})$. Clearly, if v satisfies the single equation:

$$(7.4\text{-}6) \qquad -\sigma_2 \Delta v = v(e - gv + fu(v)) \text{ in } \Omega, \; v=0 \text{ on } \Omega,$$

then the pair (u(v),v) will be a solution of (7.4-1). To analyze (7.4-6), we need to consider the properties of the the mapping $v \to u(v)$ as described below.

Lemma 7.4-1 (i) The mapping: $v \to u(v)$ defined by (7.4-5) considered as a function from $C^1(\bar{\Omega})$ to $C^1(\bar{\Omega})$ is continuous; (ii) if $v_1(x) \ge v_2(x)$ for all $x \in \bar{\Omega}$, then $u(v_1)(x) \le u(v_2)(x)$ for all $x \in \bar{\Omega}$.

Proof. (i) Let $v_n \to v$ in $C^1(\bar{\Omega})$, we now prove $u(v_n) \to u(v)$ in $C^1(\bar{\Omega})$. The first eigenvalues are characterized by:

$$(7.4\text{-}7) \qquad \mu_1(\sigma_1, cv) = \inf\{\int_\Omega \sigma_1 |\nabla u|^2 + cvu^2 \, dx : u \in W_0^{1,2}(\Omega),$$
$$\int_\Omega u^2 \, dx = 1\}$$

where $W_0^{1,2}(\Omega)$ is defined in the appendix (cf. Section 1.3 and [79]). When v is replaced by v_n, $\mu_1(v_1, cv_n)$ is characterized by the right of (7.4-7) with v replaced by v_n. From such characterizations, it follows that if v_n converges to v uniformly in $\overline{\Omega}$ then $\mu_1(\sigma_1, cv_n)$ converges to $\mu_1(\sigma_1, cv)$ as $n \to \infty$.

First, consider the case when $u(v)$ is non-zero. Then, we must have $a > \mu_1(\sigma_1, cv)$; this implies that for sufficiently large n we have $a > \mu_1(\sigma_1, cv_n)$ and hence $u(v_n)$ is non-zero. Let ϕ_n denote the positive eigenfunction of $-\sigma_1\Delta + cv_n$ corresponding to the eigenvalue $\mu_1(\sigma_1, cv_n)$ such that $\sup\{\phi_n(x) : x \in \Omega\} = 1$. The problem

$$-\sigma_1\Delta u + cv_n u = u(a - bu) \quad \text{in } \Omega, \ u=0 \text{ on } \delta\Omega$$

has large positive constants as upper solutions and $b^{-1}(a - \mu_1(\sigma_1, cv_n))\phi_n$ as lower solutions. From the theory in Chapter 5, $u(v_n)$ is between the upper and lower solutions; and thus there must exist $\delta > 0$ such that $u(v_n) \geq \delta b^{-1}\phi_n$ for sufficinetly large n. Consequently, for sufficiently large n, there exists $x_n \in \Omega$ such that $u(v_n)(x_n) \geq \delta b^{-1}$. Therefore no subsequence of $\{u(v_n)\}$ can converge to the zero function. Next, we assume that $u(v_n)$ does not converge to $u(v)$ in $C^1(\overline{\Omega})$; we will then deduce a contradiction. We find a subsequence of $\{u(v_n)\}$, denoted again by $\{u(v_n)\}$ for convenience, lying outside a certain C^1 neighborhood of $u(v)$. Since $\{v_n\}$ is uniformly bounded, there exists $k > 0$ such that $k(a - bk - cv_n(x)) < 0$ for all n and $x \in \Omega$. Let $U_n = \{x \in \Omega : u(v_n)(x) > k\}$. For all $x \in U_n$, we have $-\Delta u(v_n)(x) = u(v_n)(x)[a - bu(v_n)(x) - cv_n(x)] < 0$. Thus $u(v_n)$ must attain its maximum on $\overline{U}_n$ at a point on δU_n. However $u(v_n)(x) = k$ on δU_n and so $u(v_n)(x) \leq k$ for all $x \in \Omega$. Consequently, the sequence $\{u(v_n)\}$ is uniformly bounded, and $u(v_n)[a - bu(v_n) - cv_n]$ is uniformly bounded in $L^p(\Omega)$ for any $p \geq 1$. From the equations satisfied by $u(v_n)$ and Theorem A3. in the appendix, we obtain uniform $W^{2,p}(\Omega)$ bound for $\{u(v_n)\}$. Using Sobolev's embedding theorem, we thus find that $\{u(v_n)\}$ has uniform bound in $H^{\alpha}(\bar{\Omega})$; consequently, we can use Theorem 1.3-3 and the equations satisfied by $u(v_n)$ to deduce that $\{u(v_n)\}$ is uniformly bounded in $H^{2+\alpha}(\bar{\Omega})$. We can therefore choose a subsequence of $\{u(v_n)\}$ converging in $C^2(\bar{\Omega})$ to a function w. Now $w \neq u(v)$ and $w \not\equiv 0$ from the arguments above. However, taking limit with the equations satisfied by the subsequence of $u(v_n)$, we see that w must be a non-negative solution of

$$(7.4\text{-}8) \qquad -\sigma_1 \Delta w = w(a - bw - cv) \quad \text{in } \Omega, \ w=0 \text{ on } \delta\Omega.$$

This means that w must be equal to the unique solution $u(v)$, giving rise to a contradiction. Thus we have proved that $u(v_n) \to u(v)$ in $C^1(\bar{\Omega})$ in the case when $u(v) \neq 0$.

Next, suppose that $u(v)$ is the zero function. Then the zero function is the unique non-negative solution of (7.4-4). Assume that $\{u(v_n)\}$ does not converge to the function $u(v) \equiv 0$ in $C^1(\bar{\Omega})$, then we can find a subsequence, denoted again by $\{u(v_n)\}$ for convenience, which lies outside a certain C^1 neighborhood of $u(v) \equiv 0$. By using the same arguments as in the last paragraph, we can show that $\{u(v_n)\}$ contains a subsequence which converges in $C^2(\bar{\Omega})$ to a function w say. Taking limit with the equations for this subsequence of $\{u(v_n)\}$, we see that w must be a non-negative solution of (7.4-8); however, w cannot be the zero function by selection of subsequence. Consequently, $\{u(v_n)\}$ must converge to the zero function in

$C^1(\bar{\Omega})$. This completes the proof of part (i).

(ii) Suppose that $v_1(x) \geq v_2(x)$ for all $x \in \bar{\Omega}$. There exists a constant $k > 0$ such that $a - bk - cv_i(x) < 0$ for each $x \in \Omega$, $i=1,2$. For each $i=1,2$ the function $u=k$ is an upper solution of the problem: $\sigma_1 \Delta u + u(a - bu - cv_i) = 0$ in Ω, $u=0$ on $\delta\Omega$. If we choose $M > 0$ such that $u \to -u(a - bu - cv_i(x)) - Mu$ are increasing functions for $i=1,2$ and all $x \in \Omega$, we obtain as in Section 5.1 and Lemma 5.2-2 that $u(v_i)$ is the limit of the decreasing sequence $u_n^{(i)}$ which is defined inductively by $u_0^{(i)} \equiv k$ in $\bar{\Omega}$, and

$$\sigma_1 \Delta u_{n+1}^{(i)} - Mu_{n+1}^{(i)} = -u_n^{(i)}(a - bu_n^{(i)} - cv^{(i)}) - Mu_n^{(i)} \quad \text{in } \Omega,$$

$$u_{n+1}^{(i)} = 0 \text{ on } \delta\Omega.$$

Using $v_1 \geq v_2$, we obtain from the maximum principle, Theorem 1.1-2, that $u_1^{(1)} \leq u_1^{(2)}$ and, by induction, that $u_n^{(1)} \leq u_n^{(2)}$ in $\bar{\Omega}$ for all n. Consequently, we conclude that $u(v_1) \leq u(v_2)$ in $\bar{\Omega}$.

Returning to the original problem (7.4-1), we first consider the simplest situation when the growth rate a of the prey is small (cf Theorem 2.2-1).

<u>Theorem</u> 7.4-1. <u>Suppose that</u> $a \leq \mu_1(\sigma_1,0)$ <u>and</u> (u,v) <u>is a non-negative solution of</u> (7.4-1). <u>Then the following are true</u>:

(i) $u \equiv 0$ <u>in</u> $\bar{\Omega}$.

(ii) <u>If</u> $e \leq \mu_1(\sigma_2,0)$, <u>then we also have</u> $v \equiv 0$ <u>in</u> $\bar{\Omega}$; <u>if</u> $e > \mu_1(\sigma_2,0)$, <u>then either</u> $v \equiv 0$ <u>in</u> $\bar{\Omega}$ <u>or</u> v <u>is the unique positive solution of</u>

$$(7.4\text{-}9) \qquad \sigma_2 \Delta v + v(e - gv) = 0 \ \underline{\text{in}}\ \Omega, \ v=0 \ \underline{\text{on}}\ \delta\Omega .$$

(Note that solutions will have each component in $H^{2+\alpha}(\bar{\Omega})$.

<u>Proof</u>. Multiplying the first equation of (7.4-1) by u, and integrating over Ω, we obtain

$$(7.4\text{-}10) \qquad -\sigma_1 \int_\Omega u\Delta u \, dx \le a\int_\Omega u^2 \, dx - b\int_\Omega u^3 \, dx .$$

On the other hand, the characterization of first eigenvalue gives

$$(7.4\text{-}11) \qquad \mu_1(\sigma_1, 0)\int_\Omega u^2 dx \le \int_\Omega \sigma_1 |\nabla u|^2 dx = -\sigma_1 \int_\Omega u\Delta u \, dx$$

Inequalities (7.4-10) and (7.4-11) imply that $\int_\Omega u^2 dx \le \frac{a}{\mu_1(\sigma_1, 0)} \int_\Omega u^2 dx$, and thus u must be $\equiv 0$ in $\bar{\Omega}$. Since $u \equiv 0$, the function v must satisfy (7.4-9). Consequently, assertion (ii) follows from the discussion in the beginning part of this section concerning the solution of a single equation.

To study more interesting solutions, we now suppose that

$$(7.4\text{-}12) \qquad a > \mu_1(\sigma_1, 0).$$

Problem (7.4-1) clearly has the two nonnegative solutions (0,0) and (u(0),0) for all values of e. We will consider the global bifurcations as e varies in the decoupled equation (7.4-6). This leads to the bifurcation from the line of solution (u(0),0) to solutions of (7.4-1) with both components positive in Ω.

Let L be the operator defined by

$Lv = -\sigma_2 \Delta v - fu(0)v.$

Without loss of generality, we may assume that $\mu_1(\sigma_2, -fu(0)) \neq 0$. Otherwise, we replace L by L+k for an appropriate constant k (cf. Section 1.3). For each h in $C^1(\bar{\Omega})$, let Kh denote the unique solution of the problem: $Lu = h$ in Ω, $u=0$ on $\delta\Omega$. From Theorem 1.3-3 and the fact that the identification map from $H^{2+\alpha}(\bar{\Omega})$ into $C^1(\bar{\Omega})$ is compact (cf Section 1.3), we may consider the map $K : C^1(\bar{\Omega}) \to C^1(\bar{\Omega})$ is a compact linear operator. Let $F : C^1(\bar{\Omega}) \to C^1(\bar{\Omega})$ be defined by

$$F(v) = -gv^2 + f[u(v) - u(0)]v$$

By Lemma 7.4-1, F is continuous and $\|F(v)\| = o(\|v\|)$ as $v \to 0$ in $C^1(\bar{\Omega})$, where $\| \ \|$ denotes the norm in $C^1(\bar{\Omega})$. We now write (7.4-6) in the form

$$(7.4\text{-}13) \qquad v - eKv - KF(v) = 0$$

Since $\|KF(v)\| = o(\|v\|)$ as $v \to 0$ in $C^1(\bar{\Omega})$, we can apply the global bifurcation results Theorem A2-3 and A2-4, as the parameter e varies. (Here the role of e, v, K, KF will be respectively replaced by λ, u, L and h in Theorem A2-3.) Moreover, if we let $L_o = I - \mu_1(\sigma_2, -fu(0))K$, $L_1 = -K$, $r = -KF$, $\lambda = e$ and $\lambda_o = \mu_1(\sigma_2, -fu(0))$, then we can apply resutls concerning bifurcation from simple eigenvalue described in Theorem A2-2, to obtain properties concerning the local behavior of the bifurcating solutions. (Here, we have to use the well-known Fredholm alternative theorem, see e.g. [201], and use in Theorem A2-2

$N(L_o) = \{\text{span of } \phi_1 \text{ in } C^1(\bar{\Omega})\}$, where ϕ_1 is a positive principal eigenfunction for the simple eigenvalue $\mu_1(\sigma_2, -fu(0))$, and $R(L_o) = \{u \in C^1(\bar{\Omega}) : \int_\Omega u\phi_1 dx = 0\}$.) We can readily check that the conditions of Theorem A2-2 are all satisfied, and thus in a neighborhood of the bifurcation point $(\mu_1(\sigma_2, -fu(0), 0)$, all solutions (e,v) of (7.4-13) lie on a curve in $R\times C^1(\bar{\Omega})$ of the form $\{(\tilde{e}(\alpha), \phi(\alpha)) : -\delta \leq \alpha \leq \delta\}$ where $\tilde{e}(0) = \mu_1(\sigma_2, -fu(0))$ and $\phi(\alpha) = \alpha\phi_1$ + terms of higher "order" in α. From the fact that $\frac{\partial\phi_1}{\partial\eta} < 0$ on $\delta\Omega$ where η is the outward unit normal at the boundary, we thus conclude that for α sufficiently small and positive, the corresponding non-trivial solution v lies in the cone:

$$P = \{v \in C^1(\bar{\Omega}) : v(x) > 0 \text{ for } x \in \Omega, \frac{\partial v}{\partial\eta}(u) < 0 \text{ for } x \in \delta\Omega\}$$

Moreover, from Theorem A2-4, the closure of the set of non-trivial solutions (e,v) of (7.4-13) contains a component S (i.e. a maximal connected subset) such that either S joins $(\mu_1(\sigma_2, -fu(0)), 0)$ to ∞ in $R\times C^1(\bar{\Omega})$ or S joins $(\mu_1(\sigma_2, -fu(0)), 0)$ to $(\bar{\mu}, 0)$, where $\bar{\mu}$ is some other eigenvalue of L (i.e. $\bar{\mu}^{-1}$ is some other eigenvalue of K).

We will now show that we can actually deduce some more precise properties of the set S, in addition to those given by Theorem A2-4 explained above.

Theorem 7.4-2. *The component S contains a connected subset $S^+ \subset S - \{(\tilde{e}(\alpha), \phi(\alpha)) : -\delta \leq \alpha \leq 0\}$ with the following properties:*

(i) *S^+ is contained in* $R \times P$,

(ii) $\{\mu \in R : (\mu, v) \in S^+\} = (\mu_1(\sigma_2, -fu(0)), +\infty)$

Proof. (i) We include in the connected set S^+ solutions $(\tilde{e}(\alpha), \phi(\alpha))$, with $\alpha > 0$ sufficiently small. Such solutions are clearly in RxP. Suppose S^+ is not contained in RxP. Then there exists a sequence $\{(a_n, v_n)\}$ contained in $S^+ \cap (RxP)$ tending to a limit $(a_0, v_0) \in S^+ \cap (Rx\delta P)$ such that $(a_0, v_0) \neq (\mu_1(\sigma_2, -fu(0)), 0)$. Choose $M > 0$ such that $M - fu(0) > 0$ and $a_0 - gv_0 + f(u(v_0) - u(0)) + M > 0$ for all $x \in \Omega$. Thus v_0 satisfies

(7.4-14) $$[\Delta - (M - fu(0))]v_0 = -(a_0 - gv_0 + f(u(v_0) - u(0)))v_0 \leq 0$$

for all $x \in \Omega$. Since $v_0 \in \delta P$, either v_0 has an interior zero in Ω or $\delta v_0/\delta\eta$ has a zero on $\delta\Omega$. By the maximum principles Theorem 1.1-2 and Theorem 1.1-4, we conclude that $v_0 \equiv 0$ in $\bar{\Omega}$. Thus $(a_0, 0)$ is a bifurcation point of (7.4-13). We now show that $a_0 = \mu_1(\sigma_2, -fu(0))$. The functions $w_n \overset{(def)}{=} \frac{v_n}{\|v_n\|}$ satisfies

(7.4-15) $$w_n = \mu_n K w_n + KF(v_n)/\|v_n\|$$

Since K is compact, there exists a subsequence of $\{w_n\}$ (again denoted by $\{w_n\}$ for convenience) such that Kw_n converges in $C^1(\bar{\Omega})$. Since $\lim_{n\to\infty} KF(v_n)/\|v_n\| = 0$ in $C^1(\bar{\Omega})$, equation (7.4-15) implies that $\{w_n\}$ converges in $C^1(\bar{\Omega})$ to a function w_0 say, and

$$w_0 = a_0 K w_0$$

Moreover, we have $w_0 \geq 0$ since $v_n \geq 0$; and $w_0 \not\equiv 0$ since $\|w_n\| = 1$ for all n. Hence a_0 is an eigenvalue of K corresponding to a non-negative eigenfunction, and thus a_0 is exactly the first eigenvalue $\mu_1(\sigma_2,-fu(0))$. This is a contradiction and therefore we must have $S^+ \subset R \times P$.

(ii) For any $(\mu,v) \in S^+$, we have

$$-\sigma_2\Delta v - fu(0)v = \mu v - gv^2 + f[u(v) - u(0)]v \quad \text{in } \Omega \tag{7.4-16}$$

and $v=0$ on $\delta\Omega$. From Lemma 7.4-1 we have $u(v) \leq u(0)$, since $v \geq 0$. Thus, multiplying both sides of (7.4-16) by v and integrating over Ω, we obtain by means of the characterization of the first eigenvalue, as in Theorem 7.4-1, that $\mu \geq \mu_1(\sigma_2,-fu(0))$.

We now proceed to show that S^+ cannot approach ∞ for any finite value of μ by obtaining a priori bounds for the solution. Suppose (e,v) lies on S^+. Choose a constant $M(e)$ such that $e - gv + fu(0) < 0$ whenever $v > M(e)$. Let $U = \{x \in \Omega : v(x) > M(e)\}$. We have

$$-\sigma_2\Delta v = v(e - gv + fu(v)) \leq v(e - gv + fu(0)) \leq 0$$

for $x \in U$. Since $v(x) = M(e)$ for $x \in \delta U$, from the maximum principle, Theorem 1.1-1, we have $v(x) \leq M(e)$ for $x \in U$. To avoid contradiction, we thus have $v(x) \leq M(e)$ for $x \in \bar{\Omega}$. Consequently, the right hand side of (7.4-6) is bounded in $C(\bar{\Omega})$, the constant being dependent on e. Using $W^{2,p}(\Omega)$ bound by means of Theorem A3. and Sobolev's embedding theorem as before, we find that there exists constants $K(e) > 0$ such that $\|v\| < K(e)$ where $\| \;\|$ denotes the $C^1(\bar{\Omega})$ norm. From the bifurcation Theorem A2-4 and the proof of part (i), we may assume that S^+ is defined to be unbounded. Thus, if $\{\mu \in R : (\mu,v) \in S^+\}$ were bounded, the apriori bounds would imply that S^+ is bounded in $R \times C^1(\bar{\Omega})$, which leads to a contradiction. Consequently, $\{\mu \in R : (\mu, v) \in S^+\}$ is unbounded; and since this set is also connected, it must be equal to the interval $(\mu_1(\sigma_2, -fu(0)), +\infty)$. This completes the proof.

Theorem 7.4-2 shows that (7.4-6) has an unbounded continuum S^+ of positive solutions. When (e,v) is close to the bifurcation point $(\mu_1(\sigma_2, -fu(0)), 0)$, v is small and positive; thus by hypothesis (7.4-12), $u(v)$ is also positive. As e increases, v may increase and one may have $a \le \mu_1(\sigma_1, cv)$. This will cause $u(v) \equiv 0$ in $\bar{\Omega}$. That is, a nontrivial solution v of (7.4-6) may correspond to a solution of system (7.4-1) with $u \equiv 0$. The next theorem shows that this is precisely what will happen for large e.

Theorem 7.4-3. *Let* a, b, c, f *and* g *be fixed positive constants with* $a > \mu_1(\sigma_1, 0)$. *There exists a constant* $K > 0$ *such that if* $e > K$, *then every solution of* (7.4-1) *with each component non-negative in* $\bar{\Omega}$ *must have* $v \equiv 0$ *or* $u \equiv 0$ *in* $\bar{\Omega}$.

Proof. Let (u,v) be any solution of (7.4-1) with each component non-negative in $\bar{\Omega}$. Suppose that $v \not\equiv 0$, then v is the unique positive solution of the equation

$$-\sigma_2 \Delta v - fuv = v(e - gv) \quad \text{in } \Omega, \; v=0 \text{ on } \delta\Omega. \tag{7.4-17}$$

Let λ_1 denote the first (principal) eigenvalue of the problem $-\Delta u = \lambda u$ in Ω, $u=0$ on $\delta\Omega$; and let φ_1 denote the corresponding eigenfunction with $\max\{\varphi_1(x) : x \in \Omega\} = 1$. It is readily checked that if $e > \sigma_2\lambda_1$, then the function $g^{-1}(e - \sigma_2\lambda_1)\varphi_1$ is a lower solution for problem (7.4-17), and that any sufficiently large positive constant is an upper solution. Since v must be between the upper and lower solutions, we conclude that if $e > \sigma_2\lambda_1$ we have $v \geq g^{-1}(e - \sigma_2\lambda_1)\varphi_1 = k(e)\varphi_1$ in $\overline{\Omega}$, where $k(e) \overset{(def)}{=} g^{-1}(e - \sigma_2\lambda_1) \to \infty$ as $e \to \infty$.

Now, consider the eigenvalue problem

$$(7.4\text{-}18) \qquad -\sigma_1\Delta u + ck(e)\varphi_1 u = \lambda u \text{ in } \Omega, \; u=0 \text{ on } \delta\Omega.$$

The least eigenvalue $\lambda = \tilde{\lambda}_1(e)$ has the characterization:

$$\tilde{\lambda}_1(e) = \inf\{\int_\Omega \sigma_1|\nabla u|^2 + ck(e)\varphi_1 u^2 dx : u \in W_0^{1,2}(\Omega), \int_\Omega u^2 dx = 1\}$$

We now deduce that $\tilde{\lambda}_1(e) \to \infty$ as $e \to \infty$. Suppose not, then there exists a sequence $\{u_n\}$ in $W_0^{1,2}(\Omega)$ such that $\int_\Omega u_n{}^2 dx = 1$, $\int_\Omega |\nabla u_n|^2 dx$ is uniformly bounded and $\int_\Omega \varphi_1 u_n^2 dx \to 0$ as $n \to \infty$. Since $\{u_n\}$ is bounded in $W_0^{1,2}(\Omega)$, using Sobolev's embedding Theorem A3- we can select a subsequence which converges in $L^q(\Omega)$ for some $q > 2$. We can then further select a subsequence, again for convenience denoted by $\{u_n\}$, which converges pointwise almost everywhere to say u_0. Thus, using Fatou's lemma and $\int_\Omega \varphi_1 u_n^2 dx \to 0$, we conclude that $\int_\Omega \varphi_1 u_0^2 dx = 0$. Hence, u_0 must be the zero function. However, $\int_\Omega u_n^2 dx = 1$, which is contradictory, and consequently we must have $\tilde{\lambda}_1(e) \to \infty$ as $e \to \infty$.

Next, choose e large enough so that $\tilde{\lambda}_1(e) > a$. Then, from the characterization of first eigenvalue and comparing with (7.4-18), we find that the first eigenvalue of

$$-\sigma_1\Delta w + cvw = \lambda w \quad \text{in } \Omega, \; w=0 \text{ on } \delta\Omega$$

is greater than a. Hence the only non-negative solution of

$$-\sigma_1 \Delta u + cvu = u(a - bu) \quad \text{in } \Omega, \ u=0 \text{ on } \delta\Omega$$

is the zero function. We have proved that if e is large enough, then $v \not\equiv 0$ implies that $u \equiv 0$. This completes the proof of the theorem.

The last two theorems show that the only way the continuum of solutions S^+ can join the bifurcation point $(\mu_1(\sigma_2,-fu(0)),0)$ on the (e,v) plane to ∞ is by $u(v)$ becoming equal to zero for e sufficiently large. However, when $u(v) \equiv 0$, then clearly v is a solution of

$$-\sigma_2 \Delta v = v(e - gv) \quad \text{in } \Omega, \ v=0 \text{ on } \delta\Omega. \tag{7.4-19}$$

If we consider the bifurcation diagram on the $e-(u,v)$ plane, the continuum of solutions for (7.4-1) $\{(e,u(v),v) : (e,v) \in S^+\}$ must join up with the continuum of solutions $\{(e,0,v) : (e,v)$ is a solution of (7.4-9)$\}$. Solutions of (7.4-9) are discussed in Theorem 7.4-1 (ii). Within the range of e where there is solution of (7.4-1) with both u and v positive in Ω, it is not known whether such positive solution is unique. In other words, kinks may be possible in the bifurcation diagram.

We may rephrase some of the above results in the following form.

Theorem 7.4-4. Let a, b, c, f, g be fixed positive constants and $a > \mu_1(\sigma_1,0)$.

(i) There exists $\mu^* > \mu_1(\sigma_2,0)$ such that, for all value of the parameter e in the interval $(\mu_1(\sigma_2,-fu(0)),\mu^*)$, the boundary value problem (7.4-1) has at least one solution (u,v) with both components positive in Ω.

(ii) There exists $\hat{\mu} \geq \mu^*$ such that if $e > \hat{\mu}$, every non-negative solution (u,v) of (7.4-1) has at least one component identically equal to zero.

The only clarification needed for the above theorem is to observe that if $e \leq \mu_1(\sigma_2, 0)$, then the only non-negative solution of (7.4-19) is $v \equiv 0$. Thus if $(e,v) \in S^+$ and $e \leq \mu_1(\sigma_2, 0)$, then $u(v) \not\equiv 0$. Hence we can obtain part (i) of the theorem. The rest of the theorem follows from previous theorems.

We can express the results in ecological terms. Suppose that the intrinsic growth rate of the prey is larger than $\mu_1(\sigma_1, 0)$, then it will support a positive population in the absence of predator. When the intrinsic growth rate e of the predator is too low $(< \mu_1(\sigma_2, -fu(0)))$, then it is impossible for the predator to survive. (Note that $\mu_1(\sigma_2, -fu(0)) < \mu_1(\sigma_2, 0)$, and if $e > \mu_1(\sigma_2, 0)$ then the predator can survive without any prey). The prey and predator will co-exist in a range of value of e, with $e > \mu_1(\sigma_2, -fu(0))$. However, if the growth rate e of predator is too large $(> \hat{\mu})$, then the prey cannot survive in the presence of predator.

Other bifurcation results can be readily obtained by using the methods in this section, e.g. varying the parameter a rather than e. Other models can also be treated analogously (c.f. [25], [26], [151]).

Notes

The use of our particular type of Lyapunov function to study parabolic Volterra-Lotka system under Neumann boundary condition began with Rothe [197], Leung [134], William and Chow [229] and many others. Systematic use of graph theory to analyze such types of Lyapunov functions for large systems of ordinary differential equations was initiated by Redhaffer and Zhou [193], [194], and later generalized to parabolic partial differential systems by Redhaffer and Walter [191], [192]. Sections 7.2 and 7.3 are adapted from [191]. Related work using invariant, contracting rectangles and comparison method was done by McNabb [158], Chueh, Conley and Smoller [47], Conway, Hoff and Smoller [56], Rauch and Smoller [190], Lakshmikanthan [128], Chandra and Davis [44], Pao [178], [179], Fife and Tang [75], Zhou and Pao [231], Brown [33], [34], [35], Lazer and McKenna [132], and many others. Using different methods, Matano and Mimura [156] show that when the interacting coefficients do not satisfy the conditions in Theorem 7.3-6, spatially non-constant stable equilibrium can exist (cf. Section 4.6). Section 7.4 follows the work of Blat and Brown [25], using the bifurcation theorems of Crandall and Rabinowitz [59], and Rabinowitz [189]. Such bifurcations for ecological systems was motivated by the article of Leung and Clark [145], and Brown [32]. Further recent work along this line can be found in Blat and Brown [25], [26], and Li [151]. Further readings in bifurcation theory can be found in Chow and Hale [48], and Smoller [211].

CHAPTER VIII

Appendix

A1. <u>A-priori Bounds for Solutions, their Gradients and other Norms</u>

A-priori bounds for spatial and time derivatives of possible solutions of parabolic equations and systems are crucial for the existence proof for solutions of nonlinear problems. The following theorems in this section concerning intial value problem with Dirichlet boundary condition have been used in Chapter 2, e.g. Theorem 2.1-1. We consider the system:

$$\text{(A1-1)} \qquad \frac{\partial u_i}{\partial t} = \sum_{j,k=1}^{n} a_{jk}(x,t)\frac{\partial^2 u_i}{\partial x_j \partial x_k} + f_i(x,t,u_1,\dots,u_m) ,$$

for $(x,t) \in \Omega\times(0,T]$, $i=1,\dots,m$, where Ω is a bounded domain in R^n, with boundary condition

$$\text{(A1-2)} \qquad u_i(x,t) = g_i(x,t) \qquad \text{for } (x,t) \in \delta\Omega\times[0,T],$$

$i=1,\dots m$. We assume that each $g_i(x,t)$ has an extension $\hat{g}_i(x,t)$ which is in the class $C^{2,1}(\bar{\Omega}_T)$. Here, $C^{2,1}(\bar{\Omega}_T)$ is the set of all continuous functions in $\bar{\Omega}_T$, having $\frac{\partial}{\partial x_i}$, $\frac{\partial^2}{\partial x_i \partial x_j}$, $1\le i,j\le n$, $\frac{\partial}{\partial t}$ continuous in $\bar{\Omega}_T$.

<u>Theorem</u> A1-1. <u>Suppose that</u> $\delta\Omega\in H^{2+\alpha}$, $0<\alpha<1$, <u>the functions</u> $a_{j,k}(x,t)$ <u>and all their first partial derivatives with respect to</u> x <u>are continuous in</u> $\bar{\Omega}_T$, <u>and</u> $f_i(x,t,u_1,\dots,u_m)$ <u>are continous in</u> $\bar{\Omega}_T\times R^n$, $i=1,\dots,m$. <u>Further, assume that</u>

$$\nu\sum_{i=1}^{n}\xi_i^2 \le \sum_{j,k=1}^{n} a_{jk}(x,t)\xi_j\xi_k \le \mu\sum_{i=1}^{n}\xi_i^2 , \qquad \left|\frac{\partial a_{jk}}{\partial x_i}(x,t)\right| \le \mu,$$

for all $(x,t) \in \bar{\Omega}_T$, $(\xi_1,\xi_2,\ldots,\xi_n) \in R^n$, some positive constants ν and μ, $1\le j,k\le n$, $i=1,\ldots,n$; and for $(\sum_{i=1}^m u_i^2)^{1/2} \le M$, we have the bound

$$(\sum_{i=1}^m [f_i(x,t,u_1,\ldots,u_m)]^2)^{1/2} \le C.$$

(Here M, C are positive constants). Suppose that $u(x,t) = (u_1(x,t),\ldots, u_m(x,t))$, each $u_i \in C^{2,1}(\bar{\Omega}_T)$, $\max_{\bar{\Omega}_T} (\sum_{i=1}^m u_i^2(x,t))^{1/2} \le M$ and u is a solution of (A1-1) and (A1-2), then the $\max_{\bar{\Omega}_T} |\frac{\partial u_i}{\partial x_k}|$, $i=1,\ldots,m$, $k=1,\ldots n$, are bounded by a constant depending on ν,μ,M,C, the bounds on $a_{jk}(x,t)$ in $\bar{\Omega}_T$, the maximum of $|\hat{g}_i|$, $|\frac{\partial \hat{g}_i}{\partial x_k}|$, $|\frac{\partial^2 \hat{g}_i}{\partial x_j \partial x_k}|$ and $|\frac{\partial \hat{g}_i}{\partial t}|$ in $\bar{\Omega}_T$ $i=1,\ldots,m$, $1\le j,k \le m$, and $\{ \max(\sum_{i=1}^m \sum_{k=1}^n [\frac{\partial u_i}{\partial x_k}(x,t)]^2)^{1/2}: (x,t) \in (\delta\Omega\times[0,T])\cup(\bar{\Omega}\times\{0\}) \}$.

The above theorem is adapted from Theorem 6.1 in Chapter VII and Theorem 4.2 in Chapter V of [127]. It essentially estimates the gradients $\max_{\bar{\Omega}_T} |\frac{\partial u_i}{\partial x_k}|$ in terms of the maximum of the gradient at the lateral boundary surface $\delta\Omega\times[0,T]$ and at the initial time $\bar{\Omega}\times\{0\}$. (Note that in terms of the symbols of Theorem 6.1 in Chapter VII of [127], we may choose $\varepsilon(M)=0$, $P(|\vec{p}|,M)= C[1+|\vec{p}|]^{-1}$, where M and C are described above).

For an initial boundary value problem (A1-1) together with

(A1-3) $\quad u_i(x,t) = \psi_i(x,t)$, for $(x,t)\in (\delta\Omega\times[0,T])\cup(\bar{\Omega}\times\{0\})$, $i=1,\ldots,m$,

where $\psi_i(x,0)$ is continuously differentiable with respect to x in $\bar{\Omega}$ and each $\psi_i(x,t)$ can be extended in a way so that it is twice continuously differentiable with respect to x and once continuously differentiable with respect to t in $\bar{\Omega}\times[0,T]$, the following remark gives an estimate of the gradient $\frac{\partial u_i}{\partial x_k}$, $i=1\ldots,m$, $k=1,\ldots,n$ on the lateral boundary as well.

Remark A1-1. Assume that Ω, $\delta\Omega$, $a_{jk}(x,t)$, $f_i(x,t,u_1,\ldots,u_m)$ satisfy all the conditions in Theorem A1-1. Suppose that $u(x,t) = (u_1(x,t),\ldots, u_m(x,t))$, each $u_i \in C^{2,1}(\bar{\Omega}_T)$, $\max_{\bar{\Omega}_T} (\sum^m u_i^2(x,t))^{1/2} \le M$, and u is a solution of (A1-1), (A1-3), then the quantity $\{\max(\sum_{i=1}^{m}\sum_{k=1}^{n}[\frac{\partial u_i}{\partial x_k}(x,t)]^2)^{1/2}$: $(x,t) \in \delta\Omega\times[0,T]\}$ can be bounded by a constant depending on ν,μ,M,C, $\max_{\bar\Omega}|\frac{\partial\psi_i}{\partial x_k}(x,0)|$ for $i=1,..,m$, $k=1,..,n$, and the bounds of all $|\psi_i|$, $|\frac{\partial\psi_i}{\partial x_k}|$, $|\frac{\partial^2\psi_i}{\partial x_j\partial x_k}|$ and $|\frac{\partial\psi_i}{\partial t}|$ on $\delta\Omega\times[0,T]$.

The above remark is adapted from Lemma 6.1 in Chapter VII and Lemma 3.1 in Chapter VI. Together with Theorem A.1-1, it gives an estimate for $\max_{\bar\Omega_T}|\frac{\partial u_i}{\partial x_k}|$ for solutions of (A1-1) and (A1-3). With estimates for $|u_i|$ and $|\frac{\partial u_i}{\partial x}|$ in $\bar\Omega_T$, the next theorem gives further estimate for $|\frac{\partial u_i}{\partial t}|$ in $\bar\Omega_T$ and then for $|\frac{\partial u_i}{\partial x_k}|^{(\sigma)}_{\Omega_T}$ for some $0 < \sigma < 1$.

<u>Theorem</u> A1-2. <u>Let</u> $\delta\Omega \in H^{2+\alpha}$, $0<\alpha<1$, <u>and</u> $a_{jk}(x,t)$, $f_i(x,t,u_1,...,u_m)$ <u>satisfy the same properties as in Theorem</u> A1-1, <u>with</u> ν,μ,M <u>as described.</u> <u>Let</u> $|a_{jk}(x,t)|\le M_1$, $|\frac{\partial a_{jk}}{\partial x}(x,t)|\le M_1$, $|f_i(x,t,u_1,..,u_m)|\le M_1$ <u>for all</u> $1\le j,k\le n$, $i=1,..,n$, $i=1,..,m$, $(x,t)\in\bar\Omega_T$, $|\sum_{i=1}^{m}u_i^2|\le M$. <u>Futher</u> , <u>suppose that all</u>

$$\frac{\partial a_{jk}}{\partial t},\quad \frac{\partial^2 a_{jk}}{\partial x_i\partial t},\quad \frac{\partial f_i}{\partial u_s},\quad \frac{\partial f_i}{\partial t}$$

<u>are continuous in</u> $(x,t)\in\bar\Omega_T$, $|\sum_{i=1}^{m}u_i^2|\le M$; <u>and in there</u>, M_2 <u>is a bound for all of their absolute values.</u> <u>Suppose</u> $u(x,t) = (u_1(x,t),...,u_m(x,t))$, <u>each</u> $u_i\in C^{2,1}(\bar\Omega_T)$, <u>is a solution of</u> (A1-1) and (A1-3) <u>with</u>

$$\max_{\bar\Omega_T}(\sum_{i=1}^{m}u_i^2(x,t))^{1/2}\le M,\quad \max_{\bar\Omega_T}(\sum_{i=1}^{m}\sum_{k=1}^{n}[\frac{\partial u_i}{\partial x_k}(x,t)]^2)^{1/2}\le \hat M.$$

<u>Then</u>, <u>the</u> $\max_{\bar\Omega_T}|\frac{\partial u_i}{\partial t}|$ <u>and</u> $|\frac{\partial u_i}{\partial x_k}|^{(\sigma)}_{\Omega_T}$, <u>are bounded by a constant determined the quantities</u> $\nu,\mu,M,\hat M,M_1,M_2$ <u>and the bounds of all</u> $|\psi_i|$, $|\frac{\partial\psi_i}{\partial x_k}|$, $|\frac{\partial^2\psi_i}{\partial x_j\partial x_k}|$ <u>and</u> $|\frac{\partial u_i}{\partial t}|$ <u>on</u> $(\delta\Omega\times[0,T])\cup(\bar\Omega\times\{0\})$. (<u>Here,</u> $0<\sigma<1$ <u>is some constant determined</u>

by the same quantities).

The above theorem is adapted from Theorem 5.2, Chapter VII and Theorem 5.1 to Theorem 6.1, Chapter V of [127]. Theorem A1-1 and A1-2 are sufficient for applications in Chapters 2 and 3; more general version of these theroems are given in [127].

The remaining part of this section considers boundary conditions which are other than Dirichlet type. We will only be limited to scalar parabolic equations. However, the boundary conditions may even be nonlinear. A-priori bounds for the solutions, their gradients with respect to x, and their $|\ |_{\Omega_T}^{(1+\delta)}$ norms will be estimated. We consider the scalar equation:

$$\text{(A1-4)} \qquad \frac{\partial u}{\partial t} = \sum_{j,k=1}^{n} a_{jk}(x,t)\frac{\partial^2 u}{\partial x_j \partial x_k} + f(x,t,u) \qquad \text{for } (x,t) \in \Omega\times(0,T],$$

where Ω is a bounded domain in R^n with boundary condition

$$\text{(A1-5)} \qquad \sum_{j,k=1}^{n} a_{jk}(x,t)\frac{\partial u}{\partial x_k}\vec{\eta}.\vec{e}_j + \psi(x,t,u) = 0 \qquad \text{for } (x,t) \in \delta\Omega\times[0,T].$$

Here $\vec{\eta}$ is the outward unit normal at the boundary $\delta\Omega$ and $\vec{e}_j$ is the unit vector in the direction of x_k axis. (For example, if $a_{ii}=1$ for each i and $a_{ij}=0$ for $i\neq j$, then the boundary derivative term becomes $\frac{\partial u}{\partial \eta}$.)

Theorem A1-3. *Let* $\delta\Omega \in H^{2+\alpha}$, $0<\alpha<1$, *the functions* $a_{jk}(x,t)$ *satisfies*

$$\text{(A1-6)} \qquad \nu\sum_{i=1}^{n}\xi_i^2 \le \sum_{j,k=1}^{n} a_{jk}(x,t)\xi_j\xi_k \le \mu\sum_{i=1}^{n}\xi_i^2$$

for all $(x,t) \in \bar{\Omega}_T$, $(\xi_1,\ldots,\xi_n) \in R^n$, *some positive* ν *and* μ; *the derivatives* $\frac{\partial a}{\partial x_i}{}_{jk}$, $\frac{\partial a}{\partial t}{}_{jk}$, $\frac{\partial^2 a}{\partial t \partial x_i}{}_{jk}$ *all continuous in* $\bar{\Omega}_T$, $1\le j,k\le n$, $i=1,\ldots,n$, *and their absolute values all bounded by* μ *in* $\bar{\Omega}_T$. *Further, assume that*

ψ, $\frac{\partial\psi}{\partial x_i}$, $\frac{\partial\psi}{\partial u}$, $\frac{\partial^2\psi}{\partial u^2}$, $\frac{\partial^2\psi}{\partial x_i\partial u}$, $\frac{\partial^2\psi}{\partial t\partial u}$, $\frac{\partial\psi}{\partial t}$ are all continuous in the set $\Lambda = \{(x,t,u) \mid (x,t) \in \bar{\Omega}_T,\ |u| \le M\}$, $i=1,\ldots,n$, with their absolute values all bounded by μ in the set Λ; the functions $f(x,t,u)$, $\frac{\partial f}{\partial u}$, $\frac{\partial f}{\partial t}$ are all continuous in Λ, with their absolute values all bounded by μ in the set Λ. Suppose $u(x,t)$ is in $C^{2,1}(\bar{\Omega}_T)$, satisfying (A1-4) and (A1-5) with $\max_{\bar{\Omega}_T}|u| \le M$, then one has the estimates:

$$\max_{\bar{\Omega}_T}\Big[\sum_{i=1}^{n}\Big(\frac{\partial u}{\partial x_i}\Big)^2\Big]^{1/2} \le M_1, \qquad |u|^{(1+\delta)}_{\Omega_T} \le c,$$

where the constants M_1, c and $\delta > 0$ depend only on M, ν, μ described above, and the bounds for $|u(x,0)|$, $|\frac{\partial u}{\partial x_i}(x,0)|$, $|\frac{\partial^2 u}{\partial x_i \partial x_j}(x,0)|$ for all $x \in \bar{\Omega}$, $1 \le i, j \le n$.

The above theorem is used in section 3.2, and is adapted from Theorem 7.2 of Chapter V in [127]. The theorem is used for proving the existence of solution for initial-boundary value problems. The following two theorems are used in section 2.6.

Theorem A1-4. Let $\delta\Omega \in H^{2+\alpha}$, $0<\alpha<1$, and the functions $a_{jk}(x,t)$, $\psi(x,t,u)$ and $f(x,t,u)$ are continuous functions for $(x,t) \in \bar{\Omega}_T$, $-\infty < u < \infty$ satisfying the following conditions:

$$\text{(A1-7)} \qquad \nu\sum_{i=1}^{n}\xi_i^2 \le \sum_{j,k=1}^{n} a_{jk}(x,t)\xi_j\xi_k \le \mu\sum_{i=1}^{n}\xi_i^2$$

for all $(x,t) \in \bar{\Omega}_T$, $(\xi_1,\ldots,\xi_n) \in R^n$,

$$\text{(A1-8)} \qquad \begin{aligned} -u\psi(x,t,u) &\le c_1u^2 + c_2 && \text{for } (x,t) \in \delta\Omega\times[0,T],\ -\infty < u < \infty, \\ uf(x,t,u) &\le c_3u^2 + c_4 && \text{for } (x,t) \in \bar{\Omega}_T,\ -\infty < u < \infty, \end{aligned}$$

where ν, μ are positive constants, and c_i, $i=1,..,4$ are nonnegative constants. Let $u(x,t)$ be in $C^{2,1}(\bar{\Omega}_T)$ satisfying (A1-4),(A1-5), then the following estimate is valid for $(x,t) \in \bar{\Omega}_T$:

$$\text{(A1-9)} \qquad \max_{\bar{\Omega}_T} |u(x,t)| \le Ke^{\lambda t} \max\{ \sqrt{c_2}, \sqrt{c_4}, \max_{\bar{\Omega}} |u(x,0)| \}.$$

Here K, λ are constants determined only by ν, μ, c_1 and c_3.

The next theorem considers the question of existence of solution for initial-boundary value problem associated with (A1-4),(A1-5) together with an initial condition. It's proof essentially uses Leray-Schauder's Theorem 1.3-8, utilizing the estimates obtained from Theorem A1-4 and Theorem A1-3. The details are similar to that of the proof of Theorem 2.2-1. Let

$$Lu \overset{(def)}{\equiv} \frac{\partial u}{\partial t} - \sum_{j,k=1}^{n} a_{jk}(x,t)\frac{\partial^2 u}{\partial x_j \partial x_k} - f(x,t,u) \qquad , \text{ for } (x,t) \in \Omega\times(0,T],$$

$$L_0 u \overset{(def)}{\equiv} \frac{\partial u}{\partial t} - \mu\Delta u$$

$$\mathbb{B}u \overset{(def)}{\equiv} \sum_{j,k=1}^{n} a_{jk} \frac{\partial u}{\partial x_k} \vec{\eta}.\vec{e}_j + \psi(x,t,u) \qquad , \text{ for } (x,t) \in \delta\Omega\times[0,T].$$

$$\mathbb{B}_0 u \overset{(def)}{\equiv} \mu(\frac{\partial u}{\partial \eta} + u)$$

where $\mu > 0$ is described in (A1-6) and (A1-7).

Consider the family of problems:

$$\text{(A1-10)} \qquad \begin{aligned} &\tau Lu + (1-\tau)L_0 u = 0 && \text{in } \Omega\times(0,T], \\ &\tau\mathbb{B}u + (1-\tau)\mathbb{B}_0 u = 0 && \text{on } (x,t) \in \delta\Omega\times[0,T], \\ &u(x,0) = 0, \end{aligned}$$

for $\tau \in [0,1]$. (The first two equations for $\tau=1$ correspond to (A1-4) and (A1-5).)

Suppose that $\psi(x,0,0)|_{x\in\delta\Omega}=0$ is true, then the compatibility condition is satisfied for the initial and boundary data for $\tau=1$ (cf. section 1.3). Clearly, the compatibility condition is satisfied for all $\tau \in [0,1]$. Further, suppose that the coefficients of the operators L and $\mathbb{B}$ satisfy all the conditions (A1-7),(A1-8) in Theorem A1-4. Then the same set of μ, ν, c_i will satisfy the corresponding inequalities (A1-7),(A1-8) for the operators $L^{\tau} = \tau L + (1-\tau)L_0$ and $\mathbb{B}^{\tau} = \tau\mathbb{B} + (1-\tau)\mathbb{B}_0$ for all $\tau \in [0,1]$. Hence, one can use Theorem A1-4 to deduce a uniform estimate:

$$\text{(A1-11)} \qquad \max_{\bar{\Omega}_T} |u(x,t,\tau)| \le M, \qquad \tau \in [0,1],$$

for all possible solutions $u(x,t,\tau)$ of (A1-10) in the $C^{2,1}(\bar{\Omega}_T)$.

Next, raise μ if necessary, and assume additionally that a_{jk}, ψ and f satisfy all the conditions in the set $\{(x,t,u)|(x,t)\in \bar{\Omega}_T, |u| \le M\}$ as described in Theorem A1-3. It can then be readily verified that the functions $a^{\tau}_{ij} = \tau a_{ij} + (1-\tau)\delta^{j}_{i}u$, $\psi^{\tau} = \tau\psi + (1-\tau)\mu u$ and $f^{\tau} = \tau f$ satisfy the same corresponding conditions in Theorem A1-3 for all $\tau \in [0,1]$ with the same constants ν and μ as for $\tau = 1$. Therefore, from Theorem A1-3, we obtain the estimates:

$$\text{(A1-12)} \qquad \max_{\bar{\Omega}_T} \Big[\sum_{i=1}^{n} \Big(\frac{\partial u}{\partial x_i}(x,t,\tau)\Big)^2\Big]^{1/2} \le M_1 , \qquad |u(x,t,\tau)|^{(1+\delta)}_{\Omega_T} \le c$$

for all solutions $u(x,t,\tau)$ of (A1-10) in the class $C^{2,1}(\bar{\Omega}_T)$, where M_1 and c are the same for all $\tau \in [0,1]$. We are now ready to state the following theorem.

<u>Theorem</u> A1-5. <u>Suppose that all the following conditions are satisfied</u>:

(a) $\delta\Omega \in H^{1+\alpha}$, $0<\alpha<1$; $\psi(x,0,0) = 0$ for all $x \in \delta\Omega$.

(b) The functions $a_{jk}(x,t)$, $\psi(x,t,u)$, $f(x,t,u)$ satisfy all the hypotheses in Theorem A1-4.

(c) Let M be determined as described above leading to (A1-11). The functions $a_{jk}(x,t)$, $\psi(x,t,u)$ and $f(x,t,u)$ have all the further properties concerning their derivatives as described in Theorem A1-3, including their bounds in the set $\{(x,t,u) | (x,t) \in \bar{\Omega}_T, |u| \le M\}$, with μ increased if necessary.

(d) For $(x,t) \in \bar{\Omega}_T$, $|u| \le M$, the functions $\frac{\partial a}{\partial x_i}jk(x,t)$ are Hölder continuous in x with exponents α, $\frac{\partial \psi}{\partial x_i}$ are Hölder continuous in x, t with exponents α, $\alpha/2$ respectively and $f(x,t,u)$ is Hölder continuous in x with exponent α.

Then for each $\tau \in [0,1]$, the problem (A1-10) has a unique solution $u(x,t,\tau)$ in the class $H^{2+\alpha,1+\alpha/2}(\bar{\Omega}_T)$.

The last two theorems are adapted from Theorems 7.3 and 7.4 from Chapter V in [127].

A2. Some Bifurcation Theorems

In this section we state a few bifurcation theorems which are used in sections 7.4 in obtaining positive solutions in addition to the trivial solution. When the hypotheses of the implicit function theorem fails at a certain point, one might have more than one solution in its neighborhood. These theorems discuss sufficient conditions for such phenomenon to occur. To illustrate the situation, consider the function $F: R^2 \to R$ given by $F(\lambda,x) = \lambda - x^2$. At the point (0,0), we have $F(0,0) = F_x(0,0) = 0$. Thus one cannot apply the implicit function theorem to assert that the equation $F(\lambda,x)=0$ defines $\lambda = \lambda(x)$

uniquely in a neighborhood of $x = 0$. Indeed, we have two solutions for $F(\lambda, x) = 0$ for x as close to 0 as possible on the right.

To prepare for more general maps whose domain and range are not simply R^k, we let X, Y be Banach spaces and let $f \in C(O, Y)$ where O is an open subet of X. (Here, $C(O, Y)$ denotes the set of continuous function from O into Y). We say f is (Fréchet) differentiable at $a \in O$ if there exists a bounded linear transformation $df_a: X \to Y$ (denoted by $df_a \in L(X, Y)$) such that

$$\| f(a+\xi) - f(a) - df_a \xi \| = o(\|\xi\|) \qquad \text{as } \|\xi\| \to 0.$$

(Here $\| \ \|$ denotes both the norms in Y or X, whichever is appropriate). df_a is called the (Fréchet) derivative of f at a. We say $f \in C^1(O, Y)$ if the map $a \to df_a$ is continuous from O into $L(X, Y)$.

We define the second derivative $d^2 f_a$ of $f \in C(O, Y)$ at a point $a \in O$ to be the continuous bilinear form $d^2 f_a: X \times X \to Y$ which satisfies

$$\| f(a+\xi) - f(a) - df_a \xi - (1/2) d^2 f_a(\xi, \xi) \| = o(\|\xi\|^2) \qquad \text{as } \|\xi\| \to 0.$$

Here $d^2 f_a$ can also be interpreted as a bounded linear transformation from X into $L(X, Y)$, i.e. $d^2 f_a \in L(X, L(X, Y))$. We say $f \in C^2(O, Y)$ if the map $a \to d^2 f_a$ is continuous from $O \to L(X, L(X, Y))$.

Let B_1, B_2, B_3 be Banach spaces and U be open in $B_1 \times B_2$. Suppose $f: U \to B_3$ and $(u_1, u_2) \in U$, we let $U_1 = \{x \in B_1 | \ (x, u_2) \in U \}$. We say f is differentiable with respect to the first variable at (u_1, u_2) if the function $g(x) = f(x, u_2)$ is differentiable at $x = u_1$. We write $dg_{u_1} = D_1 f(u_1, u_2)$. Similarly we define $D_2 f(u_1, u_2)$ as the derivative with respect to the second variable. For simplicity, we first state a bifurcation theorem for mapping on a finite dimensional space, and assume $f: U \to R^n$ where U is an open subset of $R \times R^n$. Further, we assume

$$\text{(A2-1)} \qquad f(\lambda, 0) = 0 \qquad \text{for all } \lambda \in R.$$

Using Taylor's Theorem, we write

$$\text{(A2-2)} \qquad f(\lambda, u) = L_0 u + (\lambda - \lambda_0) L_1 u + r(\lambda, u),$$

where $L_0 = D_2 f(\lambda_0, 0)$, $L_1 = D_1 D_2 f(\lambda_0, 0)$, and $r \in C^2$ satisfies

$$\text{(A2-3)} \qquad r(\lambda, 0) \equiv 0, \; D_2 r(\lambda_0, 0) = D_1 D_2 r(\lambda_0, 0) = 0.$$

The following theorem describes "bifurcation from a simple eigenvalue".

Theorem A2.1. *Let U be an open subset of* $R \times R^n$, *and* $f \in C^2(U, R^n)$ *represented by* (A2-2) *with* r *satisfying* (A2-3). *Suppose that the null space of* L_0 *is spanned by* u_0, *and* $L_1 u_0$ *is not in the range of* L_0. *Then there is* $\delta > 0$ *and a* C^1*-curve* $(\lambda, \phi): (-\delta, \delta) \to R \times \{u_0\}^\perp$ *such that*
(i) $\lambda(0) = \lambda_0$, $\phi(0) = 0$, *and*
(ii) $f(\lambda(s), s(u_0 + \phi(s))) = 0$ *for* $|s| < \delta$.
Moreover, there is a neighborhood of $(\lambda_0, 0)$ *such that any solution of* $f(\lambda, u) = 0$ *is either on this curve or is of the form* $(\lambda, 0)$.

An outline of the proof goes as follows. Define

$$F(s, \lambda, z) = \begin{cases} s^{-1} f(\lambda, s(u_0 + z)), & \text{if } s \neq 0 \\ D_2 f(\lambda, 0)(u_0 + z), & \text{if } s = 0 \end{cases}$$

for (s,λ,z) in a subset of $R\times R\times\{u_0\}^{\perp}$. One verifies that $d_{(\lambda,z)}F(0,\lambda_0,0)(\xi,\eta) = \xi L_1u_0 + L_0\eta$, which is never 0 by assumption. Then one can proceed to use implicit function theorem to solve $F(s,\lambda,z) = 0$ uniquely and locally in the form $\lambda = \lambda(s)$, $z = \phi(s)$.

For an infinite dimensional version, we have the following theorem by Crandall and Rabinowitz [59], following earlier work by Sattinger.

Theorem A2-2. *Let* X *and* Y *be Banach spaces and* $f \in C^2(U,Y)$, *where* $U = S\times V$ *is an open subset of* $R\times X$ *containing* $(\lambda_0,0)$. *Let* $L_0 = D_2f(\lambda_0,0)$, $L_1 = D_1D_2f(\lambda_0,0)$; *and* $N(L_0)$, $R(L_0)$ *denotes the null space and range of* L_0 *respectively. Suppose that:*

(i) $f(\lambda,0) \equiv 0$ *for all* $\lambda \in S$,

(ii) $N(L_0)$ *is one-dimensional, spanned by* u_0,

(iii) *dim* $[Y/R(L_0)] = 1$,

(iv) $L_1u_0 \notin R(L_0)$.

Then the conclusions of Theorem A2-1 *are valid verbatim, except with* $\{u_0\}^{\perp}$ *replaced by* Z (*Here* Z *is any closed subspace of* X *with the property that any* $x \in X$ *is uniquely representable as* $x = \alpha u_0 + z$ *for some* $z \in R$, $z \in Z$).

The next two theorems lead to bifurcation results which are more general and also nonlocal.

Theorem A2-3. *Let* B *be a Banach space and* $f \in C(U,B)$, *where* U *is an open subset of* $R\times B$. *Assume that* f *is expressible as*

$$\text{(A2-4)} \qquad f(\lambda,u) = u - \lambda Lu + h(\lambda,u), \qquad \textit{where}$$

(a) $L: B \to B$ *is a compact, linear operator*,

(b) *the equation* $Lv = \rho v$ *has* $\rho = 1/\lambda_0$ *as an eigenvalue of odd* multiplicity.

(c) $h : U \to B$ is a compact operator.

(d) $h(\lambda, u) = o(\|u\|)$ as $u \to 0$, uniformly on bounded λ-intervals.

Then $(\lambda_0, 0)$ is a bifurcation point of $f(\lambda, u) = 0$.

To be specific, let $\Gamma : (\lambda, u(\lambda))$ be a curve of solutions of $f(\lambda, u) = 0$ with (λ_0, u_0) as an interior point on Γ. We say (λ_0, u_0) is a bifurcation point with respect to Γ if every neighborhood of (λ_0, u_0) in $R \times B$ contains solutions of $f(\lambda, u) = 0$ not on Γ.

Theorem A2-4. Suppose that all the hypotheses (a) to (d) in Theorem A2-3 are satisfied. Let G denotes the closure of the set of solutions of $f(\lambda, u) = 0$ with $u \neq 0$. Then G contains a component S which meets $(\lambda_0, 0)$, and either

(i) S is noncompact in U (Hence, if $U = R \times B$, the compactness of L and h together with formula (A2-4) imply that S is unbounded), or

(ii) S meets $u = 0$ in a point $(\bar{\lambda}, 0)$, where $\bar{\lambda} \neq \lambda_0$ and $1/\bar{\lambda}$ is an eigenvalue of L.

Theorem A2-3 is due to Krasnoselski and Theorem A2-4 is the global bifurcation theorem of Rabinowitz [189]. More detailed exposition can be found in Smoller [211].

A3. Sobolev Imbedding, Strong Solution, and $W^{2,p}(\Omega)$ Estimate

We now describe a few concepts and Theorems which had been mentioned in section 5.1 and section 7.4. Although these methods had not been emphasized, they are included here for convenient reference.

Let Ω be a bounded domain in R^n, $n \geq 2$. A function u defined almost everywhere on Ω is said to be locally integrable on Ω provided that $u \in L^1(\tilde{\Omega})$ for every measurable set $\tilde{\Omega}$ whose closure is contained in Ω. Let u be locally integrable in Ω and α be any multi-index; then a locally integrable function v in Ω is called the αth weak derivative of u if

$$\int_\Omega \varphi v dx = (-1)^{|\alpha|} \int_\Omega u D^\alpha \varphi \, dx \qquad \text{for all } \varphi \in C_0^{|\alpha|}(\Omega).$$

($C_0^{|\alpha|}(\Omega)$ is the subset of $C^{|\alpha|}(\Omega)$ consisting of all functions with compact support in Ω). We write $v = D^\alpha u$.

Definition. Let k be a nonnegative integer and let $1 \le p < \infty$. The space $W^{k,p}(\Omega)$ consists of all real functions $u \in L^p(\Omega)$ whose weak derivatives of order $\le k$ exist and belong to $L^p(\Omega)$. The space $W^{k,p}(\Omega)$ is normed by

$$\| u \|_{W^{k,p}(\Omega)} = \Big(\int_\Omega \sum_{|\alpha| \le k} |D^\alpha u|^p \Big)^{1/p} .$$

The space $W^{k,p}(\Omega)$ is a Banach space. If p = 2, it is a Hilbert space.

Theorem A3-1. Let Ω be a bounded domain in R^n, $n \ge 2$, with a locally Lipschitz boundary. Suppose that $(k-1)p < n < kp$, then we have an imbedding:

$$W^{j+k,p}(\Omega) \to H^{j+\alpha}(\bar{\Omega}), \qquad \text{where } 0 < \alpha \le k - n/p.$$

That is, each $u \in W^{j+k,p}(\Omega)$ can be redefined on a set of measure zero so that the modified function $\tilde{u}$ belongs to $H^{j+\alpha}(\bar{\Omega})$, and

$$|\tilde{u}|_\Omega^{(j+\alpha)} \le K \| u \|_{W^{j+k,p}(\Omega)} ,$$

with constant K *independent of* u. *Furthermore, the imbedding is compact.* The theorem is due to Sobolev, and the compactness and further extension are by Rellick and Kondrachov. For reference, see [89] and [1].

Let $Lu \equiv \sum_{i,j=1}^{n} a_{ij}(x)\frac{\partial^2 u}{\partial x_i \partial x_j} + \sum_{i=1}^{n} b_i(x)\frac{\partial u}{\partial x_i} + c(x)u$ with coefficients defined in a bounded domain $\Omega \subset R^n$, satisfying:

(i) $\sum_{i,j=1}^{n} a_{ij}(x)\xi_i\xi_j \geq \mu \sum_{i=1}^{n} \xi_i^2$ for all $x \in \Omega$, $\xi = (\xi_1,\ldots,\xi_n) \in R^n$, some $\mu > 0$, and $a_{ij}(x)$ are continuous in $\bar{\Omega}$;

(ii) $\sum_{i=1}^{n} |b_i(x)| + |c(x)| \leq K$ for all $x \in \Omega$, some $K > 0$, and $b_i(x)$, $c(x)$ are measurable functions.

We consider the elliptic Dirichlet problem:

(A3-1) $\quad Lu(x) = f(x) \quad$ in Ω,

(A3-2) $\quad u = 0 \quad$ on $\delta\Omega$,

where f is only assumed to be measurable. The smoothness of the coefficients of L and f here is not sufficient for us to apply the classical theory described in Chapter 1. We now describe a more general concept for solution.

Definition. *A function* u(x) *in* $W^{2,p}(\Omega)$ *is called a strong solution of* (A3-1) *if* (A3-1) *is satisfied almost everywhere in* Ω, *and the derivatives there are taken in the weak sense.* (Thus the classical solutions in Chapter 1 are also strong solutions).

Let $W_0^{1,p}(\Omega)$, $p > 1$, be the completion in the space $W^{1,p}(\Omega)$ of the subset of C^∞ functions with compact support in Ω. If the boundary $\delta\Omega$ is in C^1, then it can be shown that if $u \in C^1(\bar{\Omega})$ and $u = 0$ on $\delta\Omega$, then $u \in W_0^{1,p}(\Omega)$; conversely, if $u \in W_0^{1,p}(\Omega)$ and u is continuous in $\bar{\Omega}$, then $u = 0$ on $\delta\Omega$. (cf. [218] or [1] for proof). Therefore it is natural to introduce the following.

Definition. *A function $u(x)$ in $W^{1,p}(\Omega)$ is said to satisfy boundary condition (A3-2) in the generalized sense if $u \in W_0^{1,p}(\Omega)$.*

From the two definitions above, we thus say u is a strong solution of the Dirichlet problem (A3-1), (A3-2) if $u \in W^{2,p} \cap W_0^{1,p}(\Omega)$ and $Lu = f$ a.e. in Ω.

We state an existence theorem and an estimate of the $W^{2,p}(\Omega)$ norm of the solution.

Theorem A3-2. *Let Ω be a bounded domain in R^n, $n \geq 2$, with boundary $\delta\Omega$ in C^2. Let L be the operator defined above with coefficients satisfying conditions (i) and (ii) above, and $c(x) \leq 0$ in Ω. Then for any $f \in L^p(\Omega)$, $1<p<\infty$, there exists a unique strong solution of the Dirichlet problem (A3-1), (A3-2).*

Theorem A3-3. *Let Ω and L satisfy all the conditions of the above Theorem, and $1<p<\infty$. Then for all $u \in W^{2,p}(\Omega) \cap W_0^{1,p}(\Omega)$, we have*

$$\| u \|_{W^{2,p}(\Omega)} \leq \hat{K} \| Lu \|_{L^p(\Omega)}$$

for some constant $\hat{K}$ which depends on μ, K, the functions a_{ij} and the domain Ω.

For general p, the Theorems A3-2 and A3-3 are due to Agmon, Douglas and Nirenberg. The proof of the above theorems can be found in e.g. Agmon [3] and Friedman [84] for $p = 2$.

REFERENCES

[1] Adams, R., Sobolev Spaces, Academic Press, New York, 1975.

[2] Adams E. and Ames, W. F., 'On contracting interval iteration for nonlinear problems in R^n, Part I: Theory,' J. Nonlinear Analysis, 3 (1979), 773-794; 'Part II: Applications', Nonlinear Analysis, 5 (1981), 525-542.

[3] Agmon, S., Lectures on Elliptic Boundary Value Problems, D. Van Nostrand, Princeton, 1965.

[4] Alikakos, N. D. 'Remarks on invariance in reaction-diffusion equations,' J. Nonlinear Analysis, 5 (1981), 593-614.

[5] Ames, W. F., Nonlinear Partial Differential Equations in Engineering, Vol. I, 1965, Vol II, 1972, Academic Press, N. Y..

[6] Ames, W. F., Numerical Methods for Partial Differential Equations, second ed., Acadmic Press, N. Y., Thomas Nelson & Sons, London, 1977.

[7] Amman, H., 'On the existence of positive solutions of nonlinear elliptic boundary value problems,' Indiana Univ. Math. J., 21 (1971), 125-46.

[8] Amman, H., 'Fixed point equations and nonlinear eigenvalue problems in ordered Banach spaces,' SIAM Review, 18 (1976), 620-709.

[9] Amman, H., 'Existence and stability of solutions for semilinear parabolic systems, and applications to some diffusion-reaction equations,' Proc. R. Soc. Edinburgh, Sect. A: Math., 81 (1978), 35-47.

[10] Amman, H., 'Semigroups and nonlinear evolution equations,' Linear Algebra and its Appl., 84 (1986), 3-32.

[11] Amman, H. and Crandall, M. G., 'On some existence theorems for semilinear elliptic equations,' Indiana Univ. Math. J., 27 (1978), 779-90.

[12] Aris, R., The Mathematical Theory of Diffusion and Reaction in Permeable Catalysts, Vol. I and II, Oxford Univ. Press (Clarendon), London, 1975.

[13] Aris, R. and Zygourakis, K., 'Weakly coupled systems of nonlinear elliptic boundary value problems,' J. Nonlinear Analysis, 5 (1982), 1-15.

[14] Aronson, D. G., 'A comparison method for stability analysis of nonlinear parabolic problems,' SIAM review, 20 (1978), 245-264.

[15] Aronson, D. G. and Peletier, L. A., 'Global stability of symmetric and asymmetric concentration profiles in catalyst particles,' Arch. Rational Mech. Math., 54 (1974), 175- 204.
[16] Aronson, D. G. and Thames, H. D. Jr., 'Oscillation in a nonlinear parabolic model of separated cooperatively coupled enzymes,' Nonlinear Systems and Applications (Proc. Internat. Conf., Univ. Texas, Arlington, Texas,1976), 687-693, Acadmic Press, N. Y., 1977.
[17] Baiocchi, C. and Capelo, A., Variational and Quasivariational Inequalities, John Wiley & Sons, 1984.
[18] Bates, P. W. and Brown, K. J., 'Convergence to equilibrium in reaction-diffusion system,' J. Nonlinear Analysis, 8 (1984), 227-235.
[19] Beals, R., 'Partial-range completeness and existence of solutions to two-way diffusion equations,' J. Math. Phys., 22 (1981), 954-960.
[20] Bebernes, J. and Schmitt, K., 'Periodic boundary value problems for systems of second order differential equations,' J. Diff. Eqns.,13 (1973), 32-47.
[21] Belleni-Morante, A., 'Neutron transport with temperature feedback,' Nucl. Sci. Eng., 59 (1976), 56-58.
[22] Bellman, R., 'Functional equations in the theory of dynamic programming, V. , positivity and quasilinearity,' Proc. Nat. Acad. Sci. U.S.A., 41 (1955), 743-746.
[23] Birkhoff, G. and Varga, R., (eds), Numerical Solution of Field Problems in Continuum Physics, SIAM-AMS Proceedings II, 1980.
[24] Birkhoff, G. and Lynch, R. E., Numerical Solution of Elliptic Problems, SIAM Studies in Applied Math., 6, 1984.
[25] Blat, J. and Brown, K. J., 'Bifurcation of steady-state solutions in prey-predator and competition systems,' Proc. R. Soc. Edinburgh, Sect. A: Math., 97 (1984), 21-34.
[26] Blat. J. and Brown, K. J., 'Global bifurcation of positive solutions in some systems of elliptic equations,' SIAM J. Math. Anal., 17 (1986), 1339-1353.
[27] Bramble, J. H. and Hubbard, B. E., 'Approximation of derivatives by difference methods in elliptic boundary value problems,' Contributions to Differential Eqns., 3 (1964), 399-410.
[28] Bramble, J. H. and Hubbard, B. E., 'New monotone type approximations for elliptic problems,' Math. Comp.,18 (1964), 349-367.
[29] Bramble, J. H. and Hubbard, B. E., 'On a finite difference analogue of an elliptic boundary problem which is neither diagonally dominant nor non-negative type,' J. Math. and Phys., 43 (1964), 117-132.
[30] Brezis, H. and Lions, J. L., (ed.), Nonlinear Partial Differential Equations and Their Applications: College de France Seminar, Vol.I- VII, Research Notes in Mathematics, Pitman Advanced

Publishing Program, Boston, 1981-.
[31] Britton, N. F., Reaction-Diffusion Equations and Their Applications to Biology, Academic Press, New York, 1986.
[32] Brown, K. J., 'Spatially inhomogeneous steady state solutions for systems of equations describing interacting populations,' J. Math. Anal. Appl., 95 (1983), 251-64.
[33] Brown, P. N., 'Decay to uniform states in ecological interactions,' SIAM J. Appl. Math., 38 (1980), 22-37.
[34] Brown, P. N., 'Decay to uniform states in competitive systems,' SIAM J. Math. Anal., 14 (1983), 659-73.
[35] Brown, P. N., 'Decay to uniform states in predator-prey chains,' SIAM J. Appl. Math., 45 (1985), 465-78.
[36] Busoni, G., Capasso, V. and Belleni-Morante, A., 'Global solution for a problem of neutron transport with temperature feedback,' Nonlinear Systems and Applications, edited by V. Lakshmikantham, Academic Press, N. Y., 1977
[37] Cantrell, R., 'Multiparameter bifurcations problems and topological degree,' J. Diff. Eqns., 52 (1984), 39-51.
[38] Cantrell, R. and Cosner, C., 'Diffusive logistic equations with indefinite weights: population models in disrupted environments,' to appear.
[39] Cantrell, R. S. and Cosner, C., 'On the uniqueness and stablilty of positive solutions in Lotka-Volterra competition model with diffusion,' Houston J. of Math., 13 (1987), 337-352.
[40] Case, K. M. and Zweifel, P. F., 'Existence and uniqueness theorems for the neutron transport equation,' J. Math. and Phys., 4 (1963), 1376-1385.
[41] Case, K. M. and Zweifel, P. F., Linear Transport Theory , Addison-Wesley, Reading, Mass., 1967.
[42] Casten, R. G. and Holland, C. J., 'Stability proprerties of solutions to systems of reaction-diffusion equations,' SIAM J Appl. Math., 33 (1977), 353-364.
[43] Chan, C. Y., 'Existence, uniqueness, upper and lower bounds of solutions of nonlinear space-time nuclear reactor kinectics,' SIAM J. Appl. Math 27 (1974), 72-82.
[44] Chandra, J. and Davis, P. W., 'Comparison theorems for systems of reaction-diffusion equations.' Applied Nonlinear Analysis (V. Lakshmikantham, ed), Academic Press, N. Y., (1979).
[45] Chen, G. S. and Leung, A., 'Nonlinear multigroup neutron-flux systems: blow up, decay and steady states,' J. Math. Phys., 26 (1985), 1553-1559.
[46] Chen, G. S. and Leung, A.,'Positive solutions for reactor multigroup neutron transport systems: criticality problem,' to appear in SIAM J. Appl. Math.

[47] Cheuh, K. N., Conley, C. C. and Smoller, J. A., 'Positively invariant regions for systems of nonlinear diffusion equations,' Indiana Univ. Math. J., 26 (1977), 373-92.
[48] Chow, S. N. and Hale, J., Methods of Bifurcation Theory, Springer-Verlag, N. Y.,1982.
[49] Coddington, E. A. and Levinson, N., Theory of Ordinary Differential Equations, McGraw-Hill, New York, 1955.
[50] Cohen, D. S., 'Positive solution of nonlinear eigenvalue problem: application to nonlinear reactor dynamics,' Arch. Rat. Mech. Anal., 26 (1967), 305-315.
[51] Cohen, D. S., 'Multiple stable solutions of nonlinear boundary value problems arising in chemical reactor theory,' SIAM J. Appl. Math., 20 (1971), 1-13.
[52] Cohen, D. S. and Laetsch, T. W., 'Nonlinear boundary value problems suggested by chemical reactor theory,' J. Diff. Eqns., 7 (1970), 217-26.
[53] Collatz, L. Numerical Treatment of Differential Equations, 3rd ed., Springer, 1960.
[54] Conway, E., 'Diffusion and predator-prey interaction: pattern in closed systems,' Partial Differential Equations and Dynamical Systems' (W. E. Fitzgibbon, ed.), 85-133, Pitman, Boston, 1984.
[55] Conway, E., Gardner, R. and Smoller, J., 'Stability and bifurcation of steady-state solutions for prey-predator equations,' Advances in Appl. Math., 3 (1982), 288-334.
[56] Conway, E., Hoff, D. and Smoller, J., 'Large time behavior of solutions of systems of nonlinear reaction-diffusion equations,' SIAM J. Appl. Math., 35 (1978), 1-16.
[57] Cosner, C. and Lazer, A., 'Stable coexistence states in Volterra-Lotka competition model with diffusion,' SIAM J. Appl. Math., 44 (1984), 1112-32.
[58] Courant, R. and Hilbert, D., Methods of Mathematical Physics, Vols. I, II, Wiley-Interscience, New York, 1962.
[59] Crandall, M. and Rabinowitz, P., 'Bifurcation, perturbation of simple eigenvalues and linearized stability,' Arch. Rat. Mech. Anal., 52, (1973), 161-181.
[60] Cronin, J., Fixed Points and Topological Degree in Nonlinear Analysis, Math. Surveys, 11, Amer. Math. Soc., Providence, R.I., 1964.
[61] Dancer, E., 'On positive solutions of some partial differential equations,' Trans. Amer. Math. Soc., 284, (1984), 729-743.
[62] Dancer, E. and Hess, P., 'On stable solutions of quasilinear periodic-parabolic problems,' Pervenuto alla redazione 7 (1986),123-141.
[63] de Mottoni, P., 'Space structures of some migrating populations,' Mathematical Ecology (S. Levin and T. Hallam, eds), Lecture Notes in Biomathematics 54, 502-13, Springer, Berlin, 1984.

[64] de Mottoni, P., Schiaffino, A. and Tesei, A., 'On stable space-dependent stationary solutions of a competition system with diffusion,' private communications, (1984).
[65] Diaz, J. I., Nonlinear Partial Differential Equations and Free Boundaries, Research Notes in Math. 106, Pitman, Boton,106.
[66] Diekmann, O. and Temme, N. M., Nonlinear Diffusion Problems, Mathematisch Centrum, Amsterdam, 1976.
[67] Douglas, Jr. J., 'Alternating direction iteration for mildly nonlinear elliptic difference equations,' Numer. Math., 3 (1961), 92-98.
[68] Duderstadt, J. J. and Hamilton, L. J., Nuclear Reactor Analysis, John Wiley, N. Y., 1976.
[69] Duderstadt, J. J., and Martin, W. R., Transport Theory, John Wiley, N. Y., 1979.
[70] Esquinas, J. and Lopez-Gomez,'Optimal multiplicity in local bifurcation theory I: generalized generic eigenvalues,' J. Diff. Eqns., 71 (1988), 72-92.
[71] Ei, S. I., 'Two timing methods with applications to heterogeneous reaction-diffusion systems,' Hiroshima Math, J., 18 (1988), 127-160.
[72] Fife, P. C., 'Solutions of parabolic boundary problems existing for all time,' Arch. Rational Mech. Anal., 16 (1964), 155-86.
[73] Fife, P, C., 'Boundary and interior transition layer phenomena for pairs of second-order differential equations,' J. Math. Anal. Appl., 54 (1976), 597-521.
[74] Fife, P. C., Mathematical Aspects of Reacting and Diffusing Systems, Lecture Notes in Biomathematics, 28, Springer-Verlag, New York, 1979.
[75] Fife, P. C. and Tang, M. M., 'Comparison principles of reaction-diffusion systems: irregular comparison functions and applications to questions of stability and speed of propagation of disturbances,' J. Diff. Eqs., 340 (1981), 168-85.
[76] Fitzgibbon, W. E. and Walker, H. F., (eds.) , Nonlinear Diffusion, Reaearch Notes in Mathematics 14, Pitman, London, 1977.
[77] Forsythe, G. E. and Wasow, W. R., Finite Difference Methods for Partial Differential Equations, Wiley, N. Y., 1960.
[78] Freedman, H. I., Deterministic Mathematical Models in Population Ecology, Marcel Dekker, N. Y., 1980.
[79] Friedman, A., 'Interior estimates for parabolic systems of nonlinear elliptic and parabolic systems of partial differential equations,' J . Math. and Mech., 7 (1958), 43-60.
[80] Friedman, A., 'Boundary estimates for second order parabolic equations and their applications,' J. Math. and Mech., 7 (1958), 771-792.
[81] Friedman, A., 'Remarks on the maximum principle for parabolic

equations and its applications,' Pacific J. of Math., 8 (1958), 201-211.

[82] Friedman, A., 'On quasi-linear parabolic equations of the second order II,' J. Math. and Mech., 9 (1960), 539-56.

[83] Friedman, A., Partial Differential Equations of Parabolic Type, Prentice-Hall, Englewood Cliffs, New Jersey, 1964.

[84] Friedman, A., Partial Differential Equations, Academic Press, New York, 1969.

[85] Gantmacher, F. R., The Theory of Matrices, Chelsea Publ. Co., New York, 1959.

[86] Gardner, R., 'Asymptotic behavior of semilinear reaction-diffusion systems with Dirichlet boundary conditions', Ind. Univ. Math. J., 29 (1980), 161-90.

[87] Gardner, R., 'Comparison and stability theorems for reaction-diffusion systems,' SIAM J. Math Anal., 12 (1981), 603-616.

[88] Giaquinta, M., Multiple Integrals in the Calculus of Variations and Nonlinear Elliptic Systems, Annals of Math. Studies 105, Princeton Univ. Press, New Jersey, 1983.

[89] Gilbarg, D. and Trudinger, N., Elliptic Differential Equations of Second Order, second edition, Springer-Verlag, Berlin, 1983.

[90] Glasstone, S. and Sesonske, A., Nuclear Reactor Engineering, Van Nostrand, N. Y., 1981.

[91] Goh, B. S., 'Stability in models of mutualism,' Amer. Naturalists,113 (1979), 261-275.

[92] Hadeler, K. P., an der Heiden, U. and Rothe, F., 'Nonhomgeneous spatial distributions of populations', J. Math. Biol., (1974), 165-176.

[93] Hadeler, K. P., Rothe, F. and Vogt, H.,'Stationary solutions of reaction-diffusion equations', Math. Meth. Appl. Sci.. (1979), 418-431.

[94] Hale, J. , 'Bifurcation from simple eigenvalues for several parameter values', J. Nonlinear Analysis, 2 (1978), 491-497.

[95] Hale, J., 'Large diffusitivity and asymptotic behavior in parabolic systems,' J. Math. Anal. Appl., 118 (1986), 455-466.

[96] Hale, J. and Somolinos, A.,'Competition for fluctuating nutrient,' J. Math, Biol., 18, (1983), 255-280.

[97] Hastings, S. P., 'Single and multiple pulse waves for Fitzhugh-Nagumo equations,' SIAM J. Appl. Math., 42 (1982), 247-286.

[98] Hartman, P., Ordinary Differential Equations, Wiley, New York, 1964.

[99] Henry, D., Goemetric Theory of Semilinear Parabolic Equations, Lecture Notes in Math. 840, Springer-Verlag, Berlin, 1981.

[100] Hernandez, J., 'Some existence and stability results for solutions of nonlinear reaction-diffusion systems with nonlinear boundary conditions,' Nonlinear Differential Equations: Invariance, Stability,

and Bifurcations (P. deMottoni and L. Salvadori, eds), Academic Press, Toronto,1980.

[101] Hess, P.,'On positive solutions of semilinear-parabolic problems,' Lecture Notes in Math. , Springer, 1076 (1984), 101-114.

[102] Hess, P., 'Spatial homogeneity of stable solutions of some periodic-parabolic problems with Neumann boundary conditions,' J. Diff. Eqs. 68 (1987), 320-331.

[103] Hille, E. and Phillips, R., Functional Analysis and Semi-Groups, AMS Colloquium Publ., XXXI, 1957.

[104] Hirsch, M. W., 'Systems of differential equations which are competitive or cooroerative: I, Limit sets,' SIAM J. Math. Anal., 13 (1982), 167-179; 'II, Convergence almost everywhere,' SIAM J. Math. Anal.,16 (1985), 423-439.

[105] Hirsch, M. W., 'The dynamical systems approach to differential equations,' Bull. Amer. Math. Soc., 11 (1984), 1-64.

[106] Hoff, D., 'Stability and convergence of finite difference methods for systems of nonlinear reaction diffusion equations,' SIAM J. Numer. Anal.,15 (1978), 1161-1171.

[107] Hoff, E., 'Elementare Bemerkungen uber die Losungen partieller Differentialgleichungen zweiter Ordnung vom elliptischen Typus,' Sitzungsber. d. Preuss. Akad. d. Wiss., 19 (1927), 147-152.

[108] Hoff, E., 'A remark on linear elliptic differential equations of second order,' Proc. Amer. Math. Soc. 3 (1952), 791-793.

[109] Hoppensteadt, F., Mathematical Theories of Populations: Demographics, Genetics, and Epidemics, Reg. Conf. Ser. Appl. Math. 20, SIAM, Philadelphia, 1975.

[110] Howes, F. A., 'Some old and new results on singularly perturbed boundary value problems,' Singular Perturbations and Asymptotics, Edited by R. E. Meyer and S. V. Parter, Academic Press, N. Y., 1980.

[111] Howes, F. A., 'Boundary layer behavior in perturbed second-order systems,' J. Math. Anal. Appl., 104 (1984), 465-476.

[112] Huy, C. Y., McKenna, P. J. and Walter, W., 'Finite difference approximations to the Dirichlet problems for elliptic systems,' Numer. Math., 49 (1986), 227-237.

[113] John, F., Partial Differential Equations, Appl. Math. Sci. 1, 3rd ed., Springer-Verlag, N. Y.,1978.

[114] Jones, D. S. and Sleeman, B. D., Differential Equations and Mathematical Biology, George Allen and Unwin, London, 1983.

[115] Kalaba, R., 'On nonlinear differential equations, the maximum operation and monotone convergence,' J. Math. Mech., 8 (1959), 519-574.

[116] Kastenberg, W. E., 'A stability criterion for space-dependent nuclear-reactor systems with variable temperature feedback,' Nuc. Sci. Eng., 37 (1969),19-29.

[117] Kastenberg, W. E., 'Comparison theorems for nonlinear multicomponent diffusion systems,' J. Math. Anal. Appl., 29 (1970), 299-304.

[118] Kastenberg, W. E. and Chambre, P. L., 'On the stability of nonlinear space-dependent reactor kinectics,' Nucl. Sci. Eng., 31 (1968), 67-79.

[119] Keller, H. B., 'Elliptic boundary value problems suggested by nonlinear diffusion processes,' Arch. Rat. Mech. Anal., 35 (1969), 363-81.

[120] Korman, P. and Leung, A., 'A general monotone scheme for elliptic systems with applications to ecological models,' Proc. R. Soc. Edinburgh, 102A (1986), 315-25.

[121] Korman, P. and Leung, A., 'On the existence and uniqueness of positive steady states in the Volterra-Lotka ecological models with diffusion,' Applicable Analysis, 26 (1987), 145-159.

[122] Kuiper, H. J., 'Existence and comparison theorems for nonlinear diffusion systems,' J. Math. Anal. Appl. 60, (1977),166-181.

[123] Kuiper, H. J., 'Invariant sets for nonlinear elliptic and parabolic systems,' SIAM J. Math. Anal., 11 (1980), 1075-1103.

[124] Ladde, G., Lakshmikanthan, V. and Vatsala, A., 'Existence of coupled quasi-solutions of system of nonlinear elliptic boundary value problems,' J. Nonlinear Analysis, 8 (1984), 501-15.

[125] Ladde, G., Lakshmikanthan, V. and Vatsala, A., Monotone Iterative Techniques for Nonlinear Differential Equations, Pitman, Boston, London, 1985.

[126] Ladyzhenskaya, O. and Ural'ceva, N., Linear and Quasi-linear Elliptic Equations, Academic Press, New York, 1968.

[127] Ladyzhenskaya, O., Solonikov, V. and Ural'ceva, N., 'Linear and Quasilinear Equations of Parabolic Type,' American Math. Soc. Translation of Monograph, 23, Providence, R. I., 1968.

[128] Lakshmikantham, V., 'Some problems of reaction-diffusion equations,' Applied Nonlinear Analysis, Academic Press, New York, 1979.

[129] Lakshmikantham, V. and Leela, S., Differential and Integral Inequalities, Theory and Applications, Vol. II, Functional, Partial, Abstract and Complex Differential Equations, Academic Press, N. Y., 1969.

[130] Lazer, A., 'Some remarks on periodic solutions of parabolic differential equations', Dynamical Systems II, eds. Bednarek and Cesari, Academic Press, New York, 1982, 227-346.

[131] Lazer, A., Leung. A. and Murio, D., 'Monotone scheme for finite difference equations concerning steady-state prey-predator interactions, ' J. of Computational and Appl. Math., 8 (1982), 243-252.

[132] Lazer, A. and McKenna, J., 'On steady state solutions of a system of reaction-solution equations from biology,' J. Nonlinear Analysis, 6 (1982), 523-30.
[133] Leray, J. and Schauder, J., 'Topologie et equations fonctionelles,' Ann. Sci. l'Ecole Norm. Sup., 51 (1934), 45-78.
[134] Leung, A., 'Limiting behavior for a prey-predator model with crowding effects,' J. Math. Biol., 6 (1978), 87-93.
[135] Leung, A., 'Equilibria and stabilities for competing-species reaction-diffusion equations with Dirichlet boundary data,' J. Math. Anal. Appl., 73 (1980), 204-18.
[136] Leung, A., 'Stabilities for equilibria of competing-species reaction-diffusion equarions with homogeneous Dirichlet condition,' Funk. Ekv. (Ser. Interna,), 24 (1981), 201-10.
[137] Leung, A., 'Monotone schemes for semilinear elliptic systems related to ecology,' Math. Meth. Appl. Sci., 4 (1982), 272-85.
[138] Leung, A., 'A semilinear reaction-diffusion prey-predator system with nonlinear coupled boundary conditions: equilibrium and stability,' Indiana University Math J., 31 (1982), 223-241.
[139] Leung, A., 'Reaction-diffusion equations for competing populations, singularly perturbed by a small diffusion rate,' Rocky Mountain J. of Math., 13 (1983), 177-190.
[140] Leung, A., 'A study of 3-species prey-predator reaction-diffusions by monotone schemes,' J. Math. Anal. Appl., 100 (1984), 583-604.
[141] Leung, A., 'Nonlinear density-dependent diffusion for competing species interaction: large-time asymptotic behavior,' Proc. Edinburg Math. Soc., 27 (1984), 131-144.
[142] Leung, A. and Bendjilali B., 'N species competition for a spatially heterogeneous prey with Neumann boundary conditions: steady states and stability,' SIAM J. Appl. Math., 46 (1986), 81-98.
[143] Leung, A. and Chen, G. S., 'Positive solutions for temperature-dependent two-group neutron flux equations: equilibrium and stabilities,' SIAM J. Math. Anal. 15 (1984), 131-144.
[144] Leung, A. and Chen, G. S., 'Elliptic and parabolic systems for neutron fission and diffusion,' J. Math. Anal. Appl. 120 (1986) 655-669.
[145] Leung, A. and Clark, D., 'Bifurcations and large-time asymptotic behavior for prey-predator reaction-diffusion equations with Dirichlet boundary data,' J. Diff. Eqs., 35 (1980), 113-27.
[146] Leung, A. and Murio, D., ' Monotone scheme for finite difference equations concerning steady-state competing-species interactions,' Portugaliae Mathematica, 39 (1980), 497-510.
[147] Leung, A. and Murio, D., 'Accelerated monotone scheme for finite difference equations concerning steady-state prey-predator interactions,' J. Computational and Appl. Math., 16 (1986), 333-341.
[148] Leung, A. and Murio, D., 'L^2 convergence for positive finite

difference solutions of the diffusive logistic equation in two dimensional bounded domains,' International J. Comp. & Math. with Appl., 12A (1986), 991-1005.

[149] Levin, S., (ed.), Studies in Mathematical Biology, Part II: Populations and Communities, MAA Studies in Mathematics, Vol. 16, 1978.

[150] Levin, S., 'Models of population dispersal, differential equations and applications,' Ecology, Epidemics and Population Problems, edited by S. Busenberg and K. Cooke, Academic Press, 1981.

[151] Li, L., 'Coexistence theorems of steady states for predator-prey interacting systems,' Trans. Amer. Math. Soc., 305 (1988), 143-166.

[152] Lions, J. L. and Magenes, E., Non-Homogeneous Boundary Value Problems and Applications, Vol. I- III, Springer, Berlin, 1972-1973.

[153] Mann, W. R. and Wolf., F., 'Heat transfer between solids and gases under nonlinear boundary conditions,' Quart. J. Appl. Math., 9 (1951), 163-184.

[154] Matano, H., 'Asymptotic behavior and stability of solutions of semilinear diffusion equations,' Publ. RIMS Kyoto Univ., 15 (1979), 401-454.

[155] Matano, H.,'Existence of nontrivial unstable sets for equilibriums of strongly order-preserving systems,' J. Fac. Sci. Univ. Tokyo, 30 (1984), 605-673.

[156] Matano, H. and Mimura, M., 'Pattern formation in competition-diffusion systems in nonconvex domains,' Publ. RIMS, Kyoto Univ.,19 (1983),1049-1079.

[157] McKenna, P. and Walter, W., 'On the Dirichlet problem for elliptic systems,' Applicable Analysis, 21 (1986), 207-24.

[158] McNabb, A., 'Comparison and existence theorems for multicomponent diffusion systems,' J. Math. Anal. Appl., 3 (1961), 133-44.

[159] Mimura, M., 'Asymptotic behaviours of parabolic system related to planktonic prey and predator model,' SIAM J. Appl. Math., 37 (1979), 499-512.

[160] Mimura, M., Nishiura, Y. and Yamaguti, M.,'Some diffusive prey and predator systems and their bifurcation problems,' Annals of the N. Y. Academy of Sciences, 316 (1979), 490-510.

[161] Miranda, C., Equazioni Alle Derivate Parziali di Tipo Ellitico, Springer-Verlag, Berlin, 1955.

[162] Mitchell, A. R. and Griffith, D. G., The Finite Difference Methods in Partial Differential Equations, Wiley, Chichester, England, 1980.

[163] Mora, X., 'Semilinear parabolic problems define semiflows on C^k spaces,' Trans. Amer. Math. Soc., 278 (1983), 21-55.

[164] Murray, J. D., Lectures on Nonlinear Differential-Equation Models in Biology, Oxford Univ. Press (Clarendon), London.

[165] Nagumo, M. and Simoda, S., 'Note sur l'inegalite differentielle concernant les equations du type parabolique,' Proc. Japan Acad., 27

(1951), 536-539.
[166] Nelson, P. Jr., 'Subcriticality for transport of multiplying particles in a slab,' J. Math. Anal. Appl., 35 (1971), 90-104.
[167] Ni, W. M. and Serrin, J., 'Nonexistence theorems for singular solutions of quasilinear partial differential equations,' Comm. Pure and Appl. Math., 39 (1986), 379-399.
[168] Ni, W. M. and Takagi, I., 'On the Neumann problem for some semilinear elliptic equations and systems of activator-inhibitor type,' Trans. Amer. Math. Soc., 297 (1986), 351-368.
[169] Nicolis, G., 'Patterns of spatio-temporal organization in chemical and biological kinectics,' SIAM-AMS Proc., 8 (1974), 33-58.
[170] Nirenberg, L., 'A strong maximum principle for parabolic equations,' Comm. Pure and Appl. Math., 6 (1953), 167-177.
[171] Nishiura, Y., 'Global structure of bifurcating solutions of some reaction-diffusion systems,' SIAM J. Math. Anal., 13 (1982), 555-593.
[172] Noyes, R. and Jwo, J., 'Oscillations in chemical systems X: implications of cerium oxidation mechanism for the Belousov-Zhabotinskii reaction,' J. Amer. Chem. Soc., 97 (1975), 5431-3.
[173] Okubo, A., Diffusion and Ecological Problems: Mathematical Models, Springer-Verlag, Berlin, 1980.
[174] Oleink, O. A., 'On properties of some boundary problems for equations of elliptic type,' Math. Sbornik, N. S. 30 (1952), 695-702.
[175] Ortega, L., 'A Sturmian theorem for parabolic operators with periodic coefficients and applications,' Ph. D. thesis, Univ. Cincinnati, 1983.
[176] Othmer, H. G.,'Current problems in pattern formation,' Lectures on Mathematics in the Life Sciences, Vol. 9, Amer. Math. Soc., 1977.
[177] Pao, C. V., 'Neutron transport in a reactor system with temperature feedback,' Progress in Nuclear Energy, 8 (1981), 191-202.
[178] Pao, C. V., 'Co-existence and stability of a competition-diffusion system in population dynamics,' J. Math. Anal. Appl., 83 (1981), 54-76.
[179] Pao, C. V., 'On nonlinear reaction-diffusion systems,' J. Math. Anal. Appl., 87 (1982), 165-98.
[180] Pao, C. V., 'Comparison and stability of solutions for a neutron transport problem with temperature feedback,' SIAM J. Math. Anal., 14 (1983), 167-184.
[181] Pao, C. V., ' Monotone convergence of time dependent solutions for coupled reaction-diffusion systems,' Proc. Conf. Trends in theory and practice of nonlinear differential equations (V. Lakshmikantham, ed.), North-Holland, (1985), 455-65.
[182] Payne, L. E. and Philippin, G. A., 'Comparison theorems for a class of nonlinear elliptic boundary value problems,' J. Nonlinear Analysis, 9

(1985), 787-797.

[183] Parter, S. V., 'Mildly nonlinear elliptic partial differential equations and their numerical solutions,' Numerische Math., 7 (1965), 113-128.

[184] Parter, S. V., Numerical Methods for Partial Differential Equations, ed., Academic Press, N. Y., 1979.

[185] Pazy, A. and Rabinowitz, P., 'A nonlinear integral equation with applications to neutron transport theory,' Arch. Rat. Mech. Anal. 32 (1969), 226-246.

[186] Petrovesky, I. G., Lectures on Partial Differential Equations, Wiley-Interscience, N. Y., 1954.

[187] Polya, G., 'Sur une intepretation de la methode des differences finies qui peut fournir des bornes superieures ou inferieures,' C. R. Acad. Sci., Paris, 235 (1952), 995-997.

[188] Protter, M. H. and Weinberger, H., Maximum Principles in Differential Equations, Prentice Hall, Englewood Cliffs, N. J., 1967.

[189] Rabinowitz, P. 'Some global results for nonlinear eigenvalue problems,' J. Funct. Anal., 1 (1971), 487-513.

[190] Rauch., J. and Smoller, J., 'Qualitative theory of the FitzHugh-Nagumo equations,' Adv. in Math., 27 (1978), 12-44.

[191] Redheffer, R. and Walter, W., 'On parabolic systems of Volterra predator-prey type,' J. Nonlinear Analysis, 7 (1983), 333-47.

[192] Redheffer, R. and Walter, W., 'Solution of the stability problem for a class of generalized Volterra prey-predator systems,' J. Diff. Eqs., 52 (1984), 245-263.

[193] Redheffer, R. and Zhou, Z. M., 'Global asymptotic stability for a class of many-variable Volterra prey-predator systems,' J. Nonlinear Anal., 5 (1981), 1309-1329.

[194] Redheffer, R. and Zhou, Z. M., 'A class of matrices connected with Volterra prey-predator equations,' SIAM J. Alg. Discr. Meth., 3 (1982), 122-34.

[195] Redlinger, R., 'Uber die C^2-Kompaktheit der Bahn von Losungen semilinear parabolischer Systeme,' Proc. R. Soc. Edingurgh, A93 (1983), 99-103.

[196] Richtmyer, R. D. and Morton K. W., Difference Methods for Initial-Value Problems, Interscience, N. Y., 1967.

[197] Rothe, F., 'Global Solutions of Reaction-Diffusion Equations,' Lecture Notes in Mathematics, vol.1072, Springer-Verlag, Berlin, 1984.

[198] Rothe F. and de Mottoni, P., 'A simple system of reaction-diffusion equations describing morphogenesis: asymptotic behaviour,' Ann. Mat. Pure Appl., 122 (1979), 141-57.

[199] Saaty, T. L., Modern Nonlinear Equations, McGraw-HIll, N. Y., 1967.

[200] Sattinger, D., 'Monotone methods in nonlinear elliptic and parabolic

equations,' Ind. Univ. Math. J., 21 (1972), 979-1000.
[201] Sattinger, D., 'Topics in Stability and Bifurcation Theory,' Lecture Notes in Mathematics, vol. 309, Springer-Verlag, Berlin, 1973.
[202] Schauder, J., 'Der Fixpunktsatz in Funktionalraumen,' Studia Math., 2 (1930), 171-180.
[203] Schauder, J., 'Uber lineare elliptische Differentialgleichungen zweiter Ordnung ,' Math. Zeit., 38 (1934), 257-282.
[204] Schauder, J., 'Numerische Abschatzungen in elliptischen linearen Differentialgleichungen,' Studia Math., 5 (1934), 34-42.
[205] Schiaffino, A. and Tesei, A., 'Competition systems with Dirichlet boundary conditions,' J. Math. Biol., 15 (1982), 92-105.
[206] Schmitt, K., 'Boundary value problems for quasilinear second order elliptic equations,' J. Nonlinear Analysis, 2 (1978), 263-309.
[207] Sleeman, B. D., 'Analysis of diffusion equations in biology,' Bull. IMA 17 (1981), 7-13.
[208] Sperb., R., Maximum Principles and Their Applications, Academic Press, New York, 1981.
[209] Smith, H. L., 'On the asymptotic behavior of a class of deterministic models of cooperating species,' SIAM J. Appl. Math., 46 (1986), 368-375.
[210] Smith, H. L., 'Systems of ordinary differential equations which generate an order preserving flow, a survey of results,' SIAM Review, 30 (1988), 87-113.
[211] Smoller, J., Shock Waves and Reaction-Diffusion Equations, Springer-Verlag, New York, 1983.
[212] Smoller, J., Tromba A. and Wasserman, A., 'Nondegenerate solutions of boundary-value problems,' J. Nonlinear Anal., 4 (1980), 207-215.
[213] Stakgold, I. and Payne, L. E., 'Nonlinear problems in nuclear reactor analysis,' Proc. Conf. on Nonlinear Problems in Physical Sciences and Biology, Lecture Notes in Math. 322, Springer, N. Y., 1977, 298-307.
[214] Szarski, J., Differential Inequalities, PWN, Polish Sci. Publ., Wasaw, 1965.
[215] Terman, D., 'Threshold phenomena for reaction-diffusion system', J. Diff. Eqn. 47 (1983), 406-443.
[216] Thames, H. D., Jr. and Elster, A., 'Equilibrium states and oscillations for localized two enzyme kinetics: model for circadian rhythms,' J. Theor. Biol., 59 (1976), 415-427.
[217] Travis, C. C. and Post, W. M., 'Dynamics and comparative statics of mutualistic communities,' J. Theor. Biol.,78 (1979), 553-571.
[218] Treves, F., Basic Linear Partial Differential Equations, Academic Press, New York, 1975.
[219] Tsai, L. Y., 'Periodic solutions of nonlinear parabolic differential equations,' Bulletin Inst. Math. Academia Sinica, 5 (1977), 219-47.

[220] Tsai, L. Y., 'Existence of solutions of nonlinear elliptic systems,' Bulletin Inst. Math. Academia Sinica, 8 (1980), 111-27.

[221] Turner, V. L. and Ames, W., F., 'Two sided bounds for linked unknown nonlinear boundary conditions of reaction-diffusion,' J. Math. Anal. Appl., 71 (1979), 366-378.

[222] Walter, W., Differential and Integral Inequalities, Springer-Verlag New York, 1970. (German Edition: Differential und Integral Ungleichungen, 1964)

[223] Waltman, P., Competition Models in Population Biology, CBMS-NSF Regional Conf. Series in Appl. Math. 45, 1984.

[224] Wake, G., 'Non-linear heat generation with reactant consumption' Quart. J. Math. Oxford, 22 (1971), 583-95.

[225] Wasow, W., 'The capriciousness of singular perturbations,' Nieuw Arch. Wisk. 18 (1970), 190-210.

[226] Weinberger, H., 'Lower bounds for higher eigenvalues by finite difference methods,' Pacific J. Math. 8 (1958) , 339-368.

[227] Weinberger, H., Variational Methods for Eigenvalue Approximation, Regional Conference Series in Applied Mathematics, SIAM, 15, Philadelphia, Pa., 1974.

[228] Westphal, H., 'Zur Abschatzung der Losungen nichtlinearer parabolischer Differentialgleichungen,' Math. Z., 51(1949), 690-695.

[229] Williams, S. and Chow, P. L., 'Nonlinear reaction-diffusion models for interacting populations,' J. Math. Anal. Appl., 62 (1978), 157-69.

[230] Yosida, K., Functional Analysis, Springer, Berlin (3d ed.), 1971.

[231] Zhou, L. and Pao, C. V., 'Asymptotic behaviour of a competition-diffusion system in population dynamics,' J. Nonlinear Analysis, 6 (1982), 1163-84.

[232] Zweifel, P. F., 'A generalized transport equation,' Transport Theory and Statistical Physics, 11 (1982), 183-198.

Index

9 780792 301387